National Electrical Safety Code Handbook

Sixth Edition

Allen L. Clapp, P.E., P.L.S.
Editor

A Discussion of the National Electrical Safety Code Grounding Rules, General Rules, and Parts 1, 2, 3, and 4 of the 3rd (1920) through 2007 Editions of the National Electrical Safety Code, American National Standard C2

Published by
Standards Information Network
IEEE Press

Trademarks and Disclaimers

IEEE believes the information in this publication is accurate as of its publication date; such information is subject to change without notice. IEEE is not responsible for any inadvertent errors.

Library of Congress Cataloging-in-Publication Data

National electrical safety code handbook: a discussion of the grounding rules, general rules, and parts 1, 2, 3, and 4 of the 3rd (1920) through 2007 editions of the National electrical safety code, American national standard C2 / Allen L. Clapp, editor. — 6th ed.

   p. cm.

   ISBN 0-7381-4930-6
   1. Electric engineering—Safety measures—Standards—United States.
   I. Clapp, Allen L. II. American National Standards Institute. National electrical safety code.
   TK152.N345 2006

   621.319'240218--dc22

                                                                                    2006045816

IEEE
3 Park Avenue, New York, NY 10016-5997, USA

Copyright © 2006 by The Institute of Electrical and Electronics Engineers, Inc.
All rights reserved. Published August 2006. Printed in the United States of America.

Michelle D. Turner, Program Manager, Document Development
Jennifer A. McClain, Managing Editor

IEEE Press/Standards Information Network publications are not consensus documents. Information contained in this and other works has been obtained from sources believed to be reliable, and reviewed by credible members of IEEE Technical Societies, Standards Committees, and/or Working Groups, and/or relevant technical organizations. Neither the IEEE nor its authors guarantee the accuracy or completeness of any information published herein, and neither the IEEE nor its authors shall be responsible for any errors, omissions, or damages arising out of the use of this information.

Likewise, while the author and publisher believe that the information and guidance given in this work serve as an enhancement to users, all parties must rely upon their own skill and judgement when making use of it. Neither the author nor the publisher assumes any liability to anyone for any loss or damage caused by any error or omission in the work, whether such error or omission is the result of negligence or any other cause. Any and all such liability is disclaimed.

This work is published with the understanding that the IEEE and its authors are supplying information through this publication, not attempting to render engineering or other professional services. If such services are required, the assistance of an appropriate professional should be sought. The IEEE is not responsible for the statements and opinions advanced in the publication.

Review Policy

The information contained in IEEE Press/Standards Information Network publications is reviewed and evaluated by peer reviewers of relevant IEEE Technical Societies, Standards Committees and/or Working Groups, and/or relevant technical organizations. The authors addressed all of the reviewers' comments to the satisfaction of both the IEEE Standards Information Network and those who served as peer reviewers for this document.

The quality of the presentation of information contained in this publication reflects not only the obvious efforts of the authors, but also the work of these peer reviewers. The IEEE Press acknowledges with appreciation their dedication and contribution of time and effort on behalf of the IEEE.

To order IEEE Press Publications, call 1-800-678-IEEE.

Print: ISBN 0-7381-4930-6   SP1149

See other IEEE standards and standards-related product listings at:
http://standards.ieee.org/

Dedications

To **Oscar C. "Chuck" Amrhyn**, currently Chair of the NESC Main Committee and the Executive Subcommittee and a member of the subcommittees on Overhead Clearances (SC4); Underground (SC6); and Scope, Applications, Definitions and Coordination (SC1). Chuck replaced Slim Glancy as representative of the telephone group for the 1984 Edition, first serving on the Overhead Clearances Subcommittee and the Overhead General Subcommittee (Secretary 1987 Ed.; Chair 1990 Ed.). He joined SC1 for the 1990 Edition. A longtime member of the Interpretations Subcommittee and a student of the English language, Chuck has been instrumental in helping us refine the code language to better inform users of the intended applications. Chuck took the lead on the difficult metrification of the code requirements and assisted greatly in the process of refining the numbers used in the code to assure that required calculations would have the desired level of accuracy. He was also responsible for some of the experiments on cable movement under load that helped us understand the dynamic reactions and appropriately refine code requirements.

To **Donald E. Hooper,** currently Chair of the Interpretations Subcommittee and a member of the subcommittees on Overhead Clearances (SC4) and Scope, Applications, Definitions and Coordination (SC1). Don joined the Overhead Clearances Subcommittee for the 1981 Edition and quickly became one of the *go-to* people for resolving difficulties in developing or refining complex requirements. When the subcommittee reached an impasse, it was not unusual for the subcommittee to lock Don and me in a separate room to see if we could not find a way to satisfy the variety of concerns and interests involved. Don's wordsmithing capability, his knowledge of the industries involved, his ability to think outside the box and determine the real underlying causes of difficulties leading to expressions of concern by others, and his ability to look ahead for future pitfalls were extremely helpful during the process of changing the specification system for clearances that occurred in the 1990 Edition. He followed me as Chair of the Interpretations Subcommittee and has continued to provide yeoman service to the Code and subcommittee members by his careful and thoughtful review of member responses and preparation of draft answers to Interpretations.

Editor and Reviewers

Editor: Allen L. Clapp, P.E., P.L.S., *Raleigh, NC*
Member and Past Chair, NESC Committee
Chair and Past Secretary, NESC Subcommittee 1
Member and Past Acting Secretary, NESC Subcommittee 4
Member and Secretary, NESC Subcommittee 5
Member and Past Chair, NESC Interpretations Subcommittee

Reviewers:

Charles C. Bleakley, *Conyers GA*
Secretary, NESC Subcommittee 1
Chair, NESC Subcommittee 7
Member, NESC Interpretations Subcommittee

Johnny B. Dagenhart, P.E., *Raleigh, NC*
Chair, NESC Subcommittee 2
Member, NESC Subcommittee 1
Member, NESC Interpretations Subcommittee

Jack Christofersen, P.E., *Plymouth, MN*
Chair, NESC Subcommittee 3
Member, NESC Subcommittee 1
Member, NESC Interpretations Subcommittee

Eric K. Engdahl, P.E., *Columbus, OH*
Secretary, NESC Subcommittee 4
Alt Member, NESC Subcommittee 1

Frank A. Denbrock, P.E., *Jackson, MI*
Chair, NESC Subcommittee 5
Member, NESC Subcommittee 1
Member, NESC Interpretations Subcommittee

James R. Tomaseski, *Washington, D.C.*
Secretary, NESC Subcommittee 8
Member, NESC Subcommittee 1
Member, NESC Interpretations Subcommittee

Acknowledgments

An effort of the magnitude of this handbook cannot be accomplished without the counsel and assistance of a number of interested, knowledgeable parties. I have been especially fortunate to have the guidance and suggestions of both the NESC subcommittee chairs and secretaries who have served as peer reviewers and the NESC Secretaries Conrad Muller (1st Edition), Vincent Condello (2nd and 3rd Editions), Sue Vogel (4th and 5th Editions) and Bill Ash (6th Edition), as well as the NESC subcommittee members who have shared comments and suggestions for improvement of the Handbook.

Not enough can be said for the patience of my wife, Anne; the excellent professional support of David Castranio and Torrie Wilson; and the research assistance, general review, and counsel of John B. Dagenhart, P.E. I appreciate the assistance of Madeleine Reardon Dimond in assuring consistent use of terms, doing the final layout, and making other helpful suggestions to improve the readability of this work.

I owe an extreme debt of gratitude to these individuals, as well as to the more than 20 000 students of our seminars on NESC requirements, for their timely suggestions for improving the usefulness and accuracy of this document.

If you have suggestions for improvement of this handbook, please send them to me (marked Attention: NESC Handbook Editor) at the address below:

Allen L. Clapp, P.E., P.L.S.,
Editor NESC Handbook
Power & Communication Utility Training Center
6112 Saint Giles Street
Raleigh, NC 27612

Abstract

The development and application of the requirements of the Grounding Rules, General Rules and Parts 1, 2, 3, and 4 of ANSI C2-2002, the National Electrical Safety Code (NESC) are discussed and illustrated. Where the requirements of the 2007 Edition of the NESC differ from those of the 3rd, 4th, 5th, 6th, 1973, 1977, 1981, 1984, 1987, 1990, 1993, 1997, or 2002 Editions, the changes are clearly indicated. Sections of the text are identified by the NESC rule to which they refer; rule numbers that differ from those in an earlier edition are cross-indexed. Requirements of earlier editions for which no similar requirement exists in the 1973 or later editions, such as requirements for radio installations, either are not discussed or are discussed in less detail. In many cases, the evolution of rules from inception to the present is provided.

The discussions and illustrations in this document are developed from the texts of all prior editions of the NESC, the published official Discussions of the 5th and prior editions, the unpublished Discussion of the 6th Edition, all official Interpretations of the Code, the Rationales issued with public drafts, Change Proposals and Subcommittee Recommendations (including Comments and final Subcommittee decisions thereon), and the editor's and reviewers' knowledge of items considered during revision of the 1973 and later editions.

This document is intended specifically to aid users in understanding and correctly applying the requirements of the Grounding Rules, General Rules, and Parts 1, 2, 3, and 4 of the 2007 Edition of the NESC. It also is intended to aid those users in jurisdictions where earlier editions of the Code have been adopted or otherwise used by the administrative authority, or when considering facilities constructed under earlier editions of the NESC. It is especially useful to users of new or nonstandard designs, or construction, operation, and maintenance methods, for which specific requirements have not been detailed in the Code, as an aid in assuring that such installations and activities are consistent with the intent of the Code.

Introduction

Early electric supply and communications systems were isolated systems serving a specific town or area. They were constructed without standardization of clearances, strengths of materials, construction methods, or operation, thus causing problems for vehicles and electrical workers traveling from one area to another. These problems were further compounded as consumer use increased and smaller systems were linked together to take advantage of economies of scale; an action that would be safe in one area might not be in another. In addition, some installations were found by a 1919 joint survey of the National Bureau of Standards (NBS) and the National Electric Light Association to be constructed in a less than desirable manner.

In response to these problems, the National Bureau of Standards had started in 1913 to develop the National Electrical Safety Code in order to bring consistency and safety to the design, construction, operation, and use of electric supply and communications installations throughout the United States. The requirements of the original Code were based upon engineering theory and generally accepted good practice. They were codified after extensive research and public review, a practice that continues today. By the 3rd Edition (October 21, 1920), the text and application of the requirements were well defined. With the exception of several significant changes in the late 1930s and early 1940s, the requirements of the 3rd Edition continued with only minor changes until the early 1970s.

By the late 1960s, it was apparent that many areas of the Code needed significant revision to reflect recent advances in materials, designs, uses, and construction and operation techniques. Because of changes in the operations of the National Bureau of Standards, the NBS asked in 1972 to be relieved of its Secretariat duties. The Institute of Electrical and Electronics Engineers, Inc. (IEEE) was chosen as the new Secretariat.

Part 1 was extensively revised in 1971 and reprinted in the 1973 and 1977 Editions. Parts 3 and 4 were completely revised in the 1973 Edition and reprinted in the 1977 Edition. Although work was begun on revising Part 2 at the same time, the necessary revisions were extensive enough to require public drafts in 1973, 1975, and 1976, before the 1977 Edition was approved. Since the extensive revisions of the 1977 Edition was published, the NESC has been revised on a frequent, scheduled basis. The 1981 and later Editions include revisions to each of these parts.

The 1981 Edition marked the first time that all parts of the NESC were revised on the same schedule. The new three-year revision cycle allowed similar provisions in each of the individual parts to be consolidated into new Section 1—Introduction, containing the general rules applying throughout the Code, and Section 3—References. The existing Definitions section became Section 2.

A new round of intensive review of existing requirements by numerous working groups began with the 1984 Edition, especially in the area of design clearances. While several significant changes were made in the 1984 and the 1987 Editions, the 1990 Edition (1) completely revised the method of specifying clearances above ground and to buildings and other installations, (2) completely revised the work rules in Part 4 for clarity and ease of revision, and (3) partially revised Section 1 to aid utilities and others in understanding their responsibilities under the Code. The 1993 Edition continued that work, as did the 1997. The 1997 Edition also revised the strengths and loadings requirements in Section 24–26. The 2002 Edition substituted new 3-second gust wind data for the older fastest-mile wind data.

During the original preparation of this Handbook, every document known to exist concerning the codification of the NESC through the 1984 Edition was reviewed, including all past editions of the NESC, the Official Interpretations, the Official Discussions issued by the National Bureau of Standards (the first Secretariat of the NESC), previous drafts of various editions, and subcommittee minutes from 1984 and earlier discussions. Extensive discussions were held with living subcommittee members from the 1960 6th Edition and later editions, some of whom also provided access to personal notes from meetings, including a draft of an Official Discussion of the 6th Edition that was never published and early, unpublished drafts of published Discussions.

During preparation of the 2nd and later Editions of the NESC Handbook, all Official Interpretations, Change Proposals, Preprints, Comments, and meeting minutes have been reviewed by the Editor and Subcommittee Reviewers to provide guidance to code users. In addition, many of the discussions in this Handbook came directly from subcommittee requests to provide information that has been considered by the subcommittees during the review process to aid code users in understanding why the code requirements exist in their present forms and help them to determine when and how rules apply to specific local situations.

The assistance of NESC subcommittee officers and members during the intensive process of developing and updating this Handbook over the decades has been instrumental in helping to both assure accuracy and make this Handbook be a practical, useful historical text.

This document does not include the exact and complete text of NESC requirements; it is intended to be used as a companion to the Code as an aid in understanding the intended application of the text of the NESC rules. No statement herein should be considered to be an official requirement or an official interpretation of the NESC. The requirements of the Code are solely contained in the document published as **American National Standard C2, National Electrical Safety Code** by the Secretariat of the Code, the Institute of Electrical and Electronics Engineers, Inc. Bound copies of past Interpretations are available from the IEEE http://standards.ieee.org/nesc/nescproducts.html and authorized resellers, including the Utility Bookstore of the Power & Communication Utility Training Center http://www.pcutraining.com/. Recent Interpretations are available for download at the NESC Zone on the web site of the IEEE Standards Department http://standards.ieee.org/nesc/. The NESC Archives containing the initial formation documents, code books and discussions issued from 1913 through 1971 are now available on a Compact Disc from IEEE and from authorized resellers.

The code cycle was lengthened to four years for the 1997 Edition to allow more time for subcommittee review. The cycle was further lengthened to 5 years in 1996.

A Tentative Interim Amendment (TIA) process allows interim changes if they are deemed sufficiently critical. Copies of TIAs are available from the NESC Zone of the IEEE web site: http://standards.ieee.org/nesc/.

Contents

xxvii

Section 1. Introduction to the National Electrical Safety Code

(This section was created in the 1981 Edition, Rules 010–015 were generally contained previously in the introductory rules of each separate part of the Code (Rules 100, 102, 103, 200, 201, 202, 210, 211, 300, 301, 302, 303, 310, 311, and 400). When all parts of the Code were simultaneously revised for the first time in the 1981 Edition, these rules were collected in one place and revised for uniformity to eliminate redundant language, and to increase the clarity and specificity of requirements so as to increase the understandability of the NESC.)

010. Purpose

(This rule was formed in 1981 from previous Rules 100, 200, 202, 210, 211, 300, 310, 311, and 400.)

In the 1977 and later editions of the National Electrical Safety Code (NESC), it is made clear by choice of wording that the purpose of these rules is the practical safeguarding of persons during the installation, operation, or maintenance of overhead and underground supply and communication lines and their associated equipment. The NESC Subcommittees made every effort to emphasize that it is not merely enough that an installation be *possible*—it must be *practical* as well—to qualify as a requirement of the Code. It is unfortunate that earlier editions sometimes used the word "practicable" and that some individuals instigating legal actions have tried to infer that the word was intended to convey the meaning "possible." It is clear from the official Discussion of the very earliest codified edition, the 2nd Edition, that general practicality of installation was intended. This emphasis on "practicality," as opposed to the extreme requirement of "possibility," is especially noted in Rule 202—*Design and Construction* of the 2nd Edition and its Discussion. The language of that rule is as follows: "*202—Design and Construction. All electrical supply lines and equipment shall be of* suitable design and construction for the service and conditions under which they are to be operated, and all lines shall be so installed and maintained as to reduce the life hazard as far as practicable."

The language of the 2nd Edition Discussion is as follows: "*This rule...strikes the keynote of the code. There is no intention of requiring or even recommending more expensive construction than good practice requires and good business justifies. But it must be remembered that the public in the end pays whatever extra cost is caused by requiring safer and better construction, and hence the public may rightly require a good degree of safety in the construction...*"

Rules 101, 201, and 301 of the 3rd and later editions included either exactly or substantially the following language: "*The rules shall apply to all installations except as modified or waived by the proper administrative authority. They are intended to be so modified or waived whenever they involve expense not justified by the protection secured, or for any other reasons are impracticable; or whenever equivalent or safer construction can be more readily provided in other ways.*"

It is clear that the original codifiers intended to achieve a reasoned balance between the public's needs for both safe and economical utility service, reflecting both the expected degree of a problem and the degree of difficulty in solving the problem. That balance has been continued in the intervening years, as operating conditions have changed and new equipment and installation types have become available. Although these words no longer appear in the NESC, their effect does. The practical experience of the intervening years has led to the inclusion of more stringent requirements in some areas and more relaxed requirements in others. As a result, the NESC is itself the compilation of design, installation, operation, and maintenance requirements that have been shown over the entire history of utility construction to be appropriate to "reduce the life hazard as far as practicable."

The NESC comprises specific actions required in recognition of specific conditions. These actions are based upon the potentially conflicting activity that is normally encountered or reasonably anticipated. For example, in all areas except those limited to pedestrians or restricted-height vehicles, the clearances above grade plan for a 4 m (14 ft) high truck (see NESC Appendix A). Vertical clearances are based upon the reference distance based on potentially conflicting activity plus the clearance building block that includes appropriate mechanical and electrical components based upon the part, conductor, or cable above the area. Where the conditions encountered in a given local situation are those specified within the NESC, the required actions constitute good practice for the specific conditions.

Where the local conditions differ in some particular way from those specified in the NESC, it is the responsibility of the appropriate party to recognize the differences in conditions with actions that constitute good practice under such differing conditions. Such practice may be reflected in the design of the installation, the construction practices, the maintenance practices, the operating practices, or some combination of the above, as applicable for the given local conditions. An example of such an area is a lumber yard, where fork lifts are normally encountered or reasonably anticipated with vertical extensions exceeding a 4.0 m (14 ft) truck. In such a case, the expected height of the forklift can be added to the appropriate mechanical and electrical component from Table A-1 of NESC Appendix A to produce the appropriate clearance. However, the better way to perform the same task would be to add the difference between the expected conflicting activity and the applicable reference dimension from NESC Appendix Table A-1 (i.e., a 4.0 m [14 ft] truck in this case) to the clearance in the applicable table, thus recognizing the difference in conditions. The result is the same, but it avoids any problem with pulling the wrong mechanical and electrical component from NESC Appendix Table A-2, which is a more complicated table than Table A-1.

It is important to note that the NESC recognizes the limitation on expected activities around electrical facilities required under federal and state OSHA regulations and high-voltage line safety acts. Those performing acts around power lines have a personal responsibility to plan and control their actions so as to avoid contact with power lines.

The rules for lines differ from those for stations. In stations, the apparatus, equipment, and wires are confined to limited areas where access is restricted to trained personnel. In these latter cases, the safeguarding of persons by (1) actual enclosure of the current-carrying parts, (2) use of barriers, or (3) elevation of such parts beyond reach is not only desirable but generally feasible.

With overhead lines, on the other hand, the wires and equipment are not confined to limited areas and, with few exceptions, are not under constant observation by trained personnel. Safeguarding by enclosure is feasible with underground lines and, in fact, is in most cases essential to operation. For overhead lines, however, isolation by elevation generally must be depended upon for the safety of persons in the vicinity. The elevations required for effective isolation of overhead lines must be greater than ordinarily would be required inside buildings; the voltages are usually higher, and the height of expected traffic is usually greater.

Practice and experience have determined reasonable limits for elevation of lines and equipment and for the necessary strength of their construction. These rules are intended to include the more important requirements from the standpoint of safety, both to the public and to utility workers. Clearance requirements are determined relative to the degree of hazard involved, and strength requirements necessary to meet the required clearances are determined by (1) the degree of safety problem presented by the installation and (2) the mechanical loads to which it is assumed the lines may be subjected.

The NESC is a performance code, not a set of design specifications. The NESC construction rules specify *what* is to be performed, not *how* it is to be accomplished. For example, to meet the vertical clearance required above a corn field, either (1) taller structures spaced farther apart or (2) shorter structures spaced closer together may be used. The NESC is indifferent to what type of structures or materials are used, as long as applicable clearances and strength requirements are met.

The NESC addresses the matters required to effect reasonable and adequate safety in the construction, operation, and maintenance of electric supply and communications facilities. It is not intended to provide, and the rules do not provide, such detailed requirements as are needed for construction specifications. In many particulars, the rules do not require as substantial or as expensive construction as many companies have found it expedient or desirable to provide for service reliability or reasons other than safety.

In essence, the rules of the NESC give the basic requirements of construction that are necessary for safety. If the responsible party wishes to exceed these requirements for any reason, he may do so for his own purpose, but need not do so for safety purposes. For example, if the combination of required pole placement and overhead clearance requirements indicated that a 11.4 m (37.5 ft) pole would be needed, a 12.2 m (40 ft) pole could be used. Since poles are inventoried in 1.50 m (5 ft) increments for economy purposes, the additional 0.8 m (2.5 ft) of conductor attachment height would be for economy purposes; it is not required for safety. Thus, even though older editions of the Code sometimes used the word "minimum" for clearance or other requirements, the wording generally used in later editions is "not less than" to indicate the basic amount that is required for *safety* purposes.

The 1990 Edition of the NESC was specifically editorially revised to delete the use of the word "minimum" because of intentional or inadvertent misuse of the term by some to imply that the NESC values were some kind of minimum number that should be exceeded in practice; such is not the case. The NESC is the best information that we have available about what needs to be done and what must not be done in various circumstances; it is based on the experiences of hundreds of thousands of installations located in and serving areas with a variety of conditions in a variety of ways. The NESC is *the* national standard for safety in the installation, maintenance, and operation of electric supply and communication system facilities.

Rule 010 is a general statement of the purpose of the Code; the bulk of the rules are concerned with applying this principle in detail to the various construction situations. Where a specific rule provides detailed requirements for particular conditions, the general "purpose" rule is considered to be superseded by the specific requirements.

NOTE: Where an individual rule or subrule consists of an overarching paragraph and several distinctive subparts, both the overall requirements and the applicable subrequirements must be met.

While it is not entirely possible to eliminate the possibility of hazard to life or equipment resulting from the negligence of persons in the vicinity of electrical or mechanical objects or devices, it is possible to reduce the exposure of personnel and equipment to such hazards by using appropriate construction methods and work practices.

According to the National Safety Council publication "Accident Facts—1990," the number of deaths of utility workers *and* the public in contact with energized electric supply utility facilities of generating plants and transmission/distribution lines has been reduced to two-thirds of the annual total number of deaths due to excessive heat and one-fifth of the annual total number of deaths due to falling objects.

The rules of the NESC detail the requirements that are practical and necessary to reduce exposure to known or expected hazards to personnel or equipment. To that end, the NESC Subcommittees have been diligent in the development and analysis of data concerning (1) the construction, operation, and maintenance of lines and equipment and (2) the problems and benefits of each method.

The Code is prepared by a diversified group of active participants; they represent a wide variety of public and industry viewpoints and bring to the codification process a great depth of experience covering the entire field of utility system construction, operation, maintenance, and use. The process is public, and proposed changes are widely distributed, so that interested parties may comment and provide additional data. These rules, therefore, reflect the considered judgment of a wide body of expertise. The rules are reviewed on a regular basis; they are revised, as necessary, to reflect changes in materials or methods and, as experience indicates, to recognize changes in the nature and degree of problems presented.

011. Scope

(This rule was formed in 1981 from previous Rules 101, 201, and 301.)

This rule details the coverage of the NESC. The Code covers supply and communication lines, equipment, and associated work practices employed by a *public* or *private* electric supply, communications, railway, or similar utility in the exercise of its function as a utility. The NESC no longer covers electric fences, radio installations, or utilization equipment (see the National Electrical Code® [NEC®])[1] except as covered in Part 1 or Part 3. It does not cover mines, ships (see U.S. Coast Guard requirements and IEEE Std 45™ *IEEE Recommended Practice for Electric Installations on Shipboard* [ANSI]), aircraft, automotive equipment, or railway rolling stock.

The difference between the facilities involved in the *utility* function (covered by the NESC) and those involved in the *utilization* function (covered by the NEC) was amplified in the 1990 Edition. This language was again revised in the 1993 Edition to clearly state that these requirements apply to *public* and *private* utility systems.

In the 1980s and early 1990s, electricians started a controversy over whether area lights installed by an electric utility and fed off the distribution system could only meet the NESC or had to meet the NEC. Such installations have always been covered by the NESC and exempted from the NEC. The 1996 NEC revised its Article 90-2(b)(5) to exclude lighting associated with an electric distribution system that is under the exclusive control of an electric utility and is located on or along public highways, streets, roads, etc., or outdoors on private property by established rights such as easements. As a practical matter, customers generally grant either specific or "blanket" easements to utilities when

---

1. Refer to the edition of the NEC called out in the applicable edition of the NESC. The NEC is published by the National Fire Protection Association, Batterymarch Park, Quincy, MA 02269, USA (http://www.nfpa.org). Copies are also available from the Institute of Electrical and Electronics Engineers, Inc., 445 Hoes Lane, Piscataway, NJ 08854, USA (http://standards.ieee.org/).

applying for area lighting. If the electrical system feeding the lighting comes directly off the utility distribution system, it is clear that the NESC applies to such installations. However, if the lights are fed off the customer service entrance equipment, or if the customer has access to a switch to control the lightning, the NEC will govern. This was clarified in rule 011C of the 2002 NESC.

Both the NESC and the NEC cover some equivalent facilities, such as service drops, because they could be maintained by the customer or the utility. Depending upon local ordinances, if the installation is under qualified control (such as in some large industrial and large commercial complexes), the utility delivery system portion of such installations would be entirely under the NESC until such point as they connected to the utilization wiring system (such as at a building weatherhead on an aerial service), at which point the NEC would take over.

In 2002, the NESC added an explanatory note under Rule 011B referencing the *service point* as the point where the NEC picks up from the NESC. The service point (point of delivery) between the NESC- and the NEC-covered facilities is easy to determine for overhead service. The connectors form the service point between the NESC-covered utility service drop conductors and the NEC-covered premises-wiring service entrance conductors located at the weatherhead. The NEC allows the NESC-covered utility meter to be located in the NEC-covered service entrance conductor run and, in a fine print note (FPN), exempts the metering from NEC application.

In an underground service, the underground service cable can be under either code, depending upon ownership and control. In a typical installation where the utility installs the service drop cable underground from the transformer (or underground secondary bus cable) to the building and brings it up to the meter base, the service drop is covered by the NESC. If the customer ran the cable from the building out to the utility transformer *and maintained ownership and control over the service drop*, the NEC would govern. In some situations, the customer's electrical workers will initially install the underground service cable out to a utility transformer pad and the customer will transfer ownership to the utility which will own, control, and maintain the service drop from then on. In such cases, the NESC applies.

Where practical, the NESC incorporates other codes and standards by reference to avoid duplication and promote consistency among standards; likewise, certain NESC requirements are incorporated by reference within other codes, such as the NEC. Section 3 includes a list of the standards and codes referenced in the NESC.

012. General Rules

(This rule was formed in 1981 from previous rules 102, 200, 201, 202, 210, 211, 300, 303, 310, and 311.)

Rules 012A and 012C were in one paragraph until Rule 012B was added in 1993. The required construction is intended to be in accordance with good practice and, indeed, to set a standard of good practice in many respects: see Rule 012A. Safety is promoted by uniformity in practice; this, in turn, tends to avoid confusion and misunderstanding, both in construction and operation.

It is not sufficient to provide only against possible hazards in new construction. Deterioration in materials of construction makes it essential that adequate safety be preserved by inspection and maintenance. Certain rules in Section 26 specify quantitatively the amount of deterioration permissible before replacement but, in general, this must depend upon the good judgment of those in charge. This subject is further considered in Rule 214.

When Rule 012 was created in the 1981 Edition from prior similar rules located in the different parts of the Code, it was specifically reworded to the current language to remove references to "conditions under which the line is to be operated." The previous language had been misinterpreted by some to mean that utilities, as agents of the ratepayers, were required to provide clearances for any activity that could possibly occur. It must be recognized that it is not only impractical but absolutely impossible to provide special clearances or other construction for every location where it is *possible* for a negligent or impaired human to contact a utility installation with a vehicle or with a crane, antenna, metal ladder, extended paint-roller handle, irrigation pipe, portable conveyor, or other special apparatus. See additional discussion under Rule 010.

The 1997 Edition further clarified this issue in Rule 012C by requiring good practice for the *conditions known at the time* by those responsible for the construction or maintenance of the communications or supply lines and equipment. In essence, if the utility has knowledge that a condition not specified in the Code will be normally encountered or is reasonably anticipated, the utility should use good practice to reflect the differences (if any) in those conditions and those specified in the Code. On the other hand, the utility cannot be expected to be clairvoyant.

The operators or erectors of apparatus having a capability of contacting power lines, or other utility lines, have a responsibility to take special care to avoid damaging, or otherwise interrupting the service of, utility installations or other facilities in the vicinity of their work or operations. Such operators or erectors are strongly advised or, in some cases, mandated to consult with representatives of affected utilities *prior* to the use of such apparatus. Some states have "high-voltage acts" or "crane laws" that prohibit the use of tools, equipment, or conductive objects within stated distances of electric supply lines without first notifying the operator of the lines and receiving a clearance to work near the lines. This requirement is federally mandated by the Occupational Health and Safety Administration (OSHA) regulations. Under OSHA, employees are required to inspect job sites for power lines and determine by observation or otherwise if there are power lines (exposed or concealed) in the work area before beginning any construction work (29 CFR 1926,416(a)) and, if power lines are found, employees must be told where the lines are located (exposed or concealed), how to avoid the lines, and the consequences of not avoiding the lines. In addition, OSHA requires the employer to put up appropriate warning signs. If the work will require employees to bring a conductive object within 3 m

(10 ft) of a power line, the employer must notify the utility and take appropriate action to assure the safety of its employees.

The requirements of the NESC apply to the entity performing the work. Rule 012B recognizes that many public and private utilities contract with another party to perform some or all work, often including supervision and inspections, relating to a particular job. Although recognized by the codifiers for decades, the responsibilities of contractors (rather than owners or operators in many cases) to meet NESC requirements was explicitly stated for the first time in the 1993 Edition.

In many contracts, an owner utility retains a right to stop a job, or otherwise alter the course of work, if it finds unsuitable work being performed or identifies a need for different work to be performed. Such a contract does not relieve a contractor from responsibility for ensuring that appropriate training, tools, and supervision are provided to employees to ensure safe work and compliance with NESC requirements. Likewise, a utility that serves another public or private utility with bulk power or other service has no duty for ensuring compliance of the other utility with NESC requirements.

013. Application

(This rule was formed in 1981 from previous Rules 102, 202, and 302).

Rules are written to cover *general* cases and, for the described circumstances, are the governing requirements. *EXCEPTIONS* provide for *specific* conditions under which the rule is not or may not be applicable; no preference is intended, only the differentiation between the general and special cases.

The rules are intended to be observed completely in new work under usual conditions. In order for the rules to provide for special cases without undue burden on ratepayers, alternatives or exemptions are sometimes provided. Since the requirements of the Code may not be practical during emergency or temporary conditions, and since these requirements reflect considered judgment of the appropriate uses of current materials and construction methods, the NESC includes provisions that (1) allow the rules to be modified or waived by the proper authority for temporary or emergency installations (see Rule 014) and (2) allow experimentation with new materials or construction methods (see Rule 013A2).

013A. New Installations and Extensions

Rule 013A directly recognizes that *if* there is a controlling authority such as a state public utility commission, such authority *may* have the right to waive or modify NESC rules in their jurisdiction. The provisions now found in Rule 013A have changed over the years as more specificity has been added in the NESC as to expected actions under various conditions. For example, the limits imposed on clearances and strengths of emergency and temporary installations are now specified in Rule 14 and Rule 230A. The requirements of the NESC have been well planned to consider the full effects of these actions under the specified circumstances; Rule 013A1 thus requires equivalent safety to be achieved using other methods, systems, work methods, etc., when an NESC rule is modified by an administrative authority. The EXAMPLE was added in the 1993 Edition.

Rule 013A2 is not intended to allow a utility system operator to disregard these rules. It recognizes the need for serious experimentation with new methods, systems, etc. It requires qualified supervision, usually by a registered professional engineer who is competent in the area of work being performed. Appropriate record keeping that will allow careful and complete analysis of the results is intended. The 2002 Edition required equivalent safety and agreement between all parties involved for experimentation to occur in or on a facility.

013B. Existing Installations

It is not appropriate to add facilities to existing installations without ensuring that the new facilities meet applicable Code rules. However, it is also not appropriate to require that existing facilities be modified to meet current rules just because an addition is being made. The Code recognizes the relative necessity and practicality of the conversion of existing systems. In general, the new Code applies if it's new; if it's existing, or being added to an existing installation, the Code in effect at the time of the original construction (or a subsequent edition with which the installation has been brought into compliance) applies.

Replacement of existing construction to secure compliance of the entire installation with changes in subsequent editions of the Code would, in most cases, involve unwarranted expense; such replacement is not required by the NESC. When, however, an extension or reconstruction is being carried out that is of relatively large proportion, it may be advisable to reconstruct certain other portions of the installation to comply with the current rules and suitably safeguard the altered installation. In some cases it will be feasible and proper to reconstruct, as far as necessary, the entire installation to comply with the rules. The safety of existing installations that do not conform to current requirements can, in some instances, be improved by the proper placing of guards and signs. This method of safeguarding may be attended to with small expense and is often effective, especially for rarely seen installations with which many workers may no longer be familiar. Such treatment was required by early editions of the Code unless the administrative authority determined that the increase in safety was not worth the expense.

In considering the application of new rules to existing installations, it is evident that some rules can be made effective at once without unwarranted expense. Frequently, this further assistance in safeguarding workers and the public will significantly improve service reliability. Such reconstruction can usually be accomplished most economically at a time when important extensions or reconstructions are being undertaken for reasons *other* than accident prevention, as noted above. It is recognized that during most utility maintenance activities, only one set of specifications will be available, and existing facilities will be "upgraded" over time as a routine matter.

On the other hand, when extensions or reconstructions are undertaken, it may sometimes be impractical to comply fully with revised rules. For example, the arrangement of the crossarms on a single new pole so as to have the supply wires above communication wires, when the other poles of the existing line still continue with the arms in the reverse relation, might add to the danger instead of reducing it. Alternatives that would not be considered appropriate for new installations may often be reasonable and appropriate for existing ones.

As increased experience with supply and communication installations has matured the Code over the years, and as formerly nonconforming installations have been retired or replaced, the Code requirements relating to reconstruction of facilities have reflected these changes. For example, Rule 201B—*Realization of Intent* of the 6th and prior editions indicated that the new rules should be applied "in full to all new installation, reconstructions, and extensions, except where for special reasons any rule is shown to be impracticable or where the advantage of uniformity with existing construction is greater than the advantage of construction in conformity with" the new rules. The obvious intention was to discontinue outdated construction practices and to apply the new Code when adding or altering conductors or equipment, except in special cases. As in previous editions, the 6th Edition continued the use of Rule 201A—*Intent, Modification*, which stated the intention that the rules should be "modified or waived whenever they involve expense not justified by the protection secured or for any other reasons are impracticable; or whenever equivalent or safer construction can be more readily provided in other ways." See the discussion of Rule 010.

One of the reasons for the particular wording of old Rules 201A and B (and the similar rules in the other parts) was that lines that dated from the pre-Code era still existed in many areas of the country. As a result, essentially all overhead facilities built prior to the 1977 Edition should be expected to be in conformance with the requirements of the 6th Edition.

The revision of Part 1 *Installation and Maintenance of Electric Supply Stations and Equipment* in the 1971 Edition required application of the rules "in full to all new installations, alterations, reconstructions, and extensions." In short, the new edition was intended to apply to any installation that was not limited to maintenance replacement, except that this was the first revision that allowed the so-called *grandfather clause* to be applied to existing electric supply station installations when the code edition changed (see Appendix E *NESC Grandfather Clause*).

When the 1973 Edition created Part 3 *Installation and Maintenance of Underground Electric Supply and Communication Lines*, it also added a grandfather clause for underground lines. Similarly, the 1977 Edition added the grandfather clause for overhead lines (see Appendix E *NESC Grandfather Clause*).

The revision of Part 2 in the 1977 Edition recognized the maturing character of the utility industries. Although Rule 202B of the 1977 Edition continued to use language similar to that of old Rules 201A and B, the 1977 Edition restricted the use of waivers with Rule 202C— *Waiver*, which only allowed waiver "in cases of emergency, temporary installations, or installations which are soon to be discarded or reconstructed…"

In the 1981 Edition, the applicability rules of the various parts of the NESC were consolidated into a new Section 1—*Introduction to the National Electrical Safety Code* and several word changes were made. In Rule 013A, the successor to old Rule 202B1, the word "reconstruction" was dropped; this word had only caused confusion between maintenance replacements (which are not intended to be required to be subject to a new code provision) and new installations and extensions (which are subject to new code requirements). In Rule 013B, the successor to old Rule 202B2, a new paragraph 013B2 was added.

Rule 013B3 (Rule 013B2 of the 1981–1987 Editions) was intended to state the intention of the Code with respect to *other* facilities when conductors or equipment of the Code are added, altered, or replaced on an existing structure. This entire area of the Code was editorially revised in the 1990 Edition to clearly indicate the requirements that have been intended since 1977. Rule 013B1 now reflects that the latest edition contains the best knowledge of appropriate requirements. If an installation meets the present requirements, it is acceptable—regardless of what provisions may have been in effect at the time of its construction. Thus, when work on an existing structure is completed, it may meet the current edition requirements or those of a previous applicable edition.

The addition of new facilities does not require changing the existing line facilities, as long as the existing facilities (including the structure itself), *after the new addition*, still meet the strength, clearance, and other requirements of an earlier edition of the Code that is applicable *even if the existing facilities do not meet present code requirements*. However, if for example, the structure or the supply conductors would not, after the addition of communication cable, meet the grade of construction and strength requirements of the edition of the Code that was in effect at the time of their installation, the addition would not be allowed. If a problem exists in obtaining required clearances from existing facilities, nonconforming existing facilities may be moved on the existing structure. If the latter is the case, the modification is required to meet the Code requirements of the applicable edition unless the structure is replaced with a larger, stronger, or taller unit (see below).

The language of Rule 013B3 was carefully chosen to require that the *resulting installation* meet the applicable edition whenever conductors or equipment are added, altered (rearranged), or replaced on an *existing structure*. The two key issues are:

(1)   The rule only applies whenever an existing structure is being modified by the addition or replacement of conductors or equipment or facilities on an existing structure are being moved around (such as moving a neutral up or communication cable down to accommodate another communication cable. If a new structure is involved, Rule 013A applies and the current edition must be used, unless an existing structure is being replaced as a maintenance replacement under Rule 013B2.

(2)   When the work is complete, the entire *resulting installation* will be inspected to assure that all of the installation meets the applicable edition of the NESC.

This language was added to address two issues. First, one code edition must be used for the whole structure, including all of its supported facilities; installers cannot *cherry-pick* code provisions from different editions. Second, the whole installation needs to be inspected for potential problems, such as conductors with excess sag/pulled too tight or out of place, broken insulators, loose guys, etc., that might affect code compliance and present a potential safety hazard to public personnel around the installation or the next workers on the installation. This requirement is a complement of Rules 121A, 214A2 and 313A2, as well as Rule 230I, and it helps to limit the opportunity for changes or damage that occur after initial installation to cause a later problem.

There are several reasons that the language in Rule 013B3 was added to require the entire structure and supported facilities to be inspected when workers are working on the installation. There were a number of instances of installations with energized jumpers out of place that might endanger a line worker during storm restoration at night. In other instances, guys had been damaged, leaning poles had dropped cables, or conductors below required clearances. Similarly, improper guying or tensioning of conductors or cables had caused either lower cables to sag below the required ground clearance or upper conductors to sag down too close to lower ones at midspan (or had pulled a lower cable so tight that it approached an upper conductor too close under design conditions). If the lowest conductor or cable in an overhead span is caught by a truck or other vehicle and lines or structures are broken, all occupiers of a joint-use structure can be adversely affected. Even though the contact may be with a relatively benign communication cable, the potential hazard can be electrical (if power lines are severed or otherwise brought down low enough for contact) or physical (if structures fall or other cables or conductors are brought down low enough to be contacted by a subsequent vehicle).

As a result, it is important that workers of any utility working on a structure also take the time to check clearances between wires and between wires and the ground in the adjacent spans for obvious problems. However, nothing in the language of Rule 013B3 is intended to require one utility to inspect another utility's facilities under the normal inspection rules (unless by agreement another utility assumes that responsibility). In other words, while at the site, workers should inspect the facilities on that structure for obvious problems that may adversely affect the safety of their own installation. There is no intention of requiring detailed inspections of hardware, equipment, etc., or a pole wood integrity inspection, for example, each time work is performed on the structure. If problems are found, they should be repaired or reported and scheduled for repair by the appropriate personnel. There is also no requirement under Rule 013B3 to inspect neighboring structures. For example, if the facilities in an electric supply station are owned and maintained by multiple parties, such as the transmission side and power transformer (high side bay) by a generation or transmission utility and the distribution protection system and outgoing facilities (low side bay) by a distribution utility, there is no duty for the generation or transmission utility to inspect the distribution facilities or vice versa (see IR 405 issued 28 April 1987).

If the existing structure is replaced in kind, regardless of the reason, it is generally considered maintenance; it may be replaced without affecting other existing facilities, if the resulting installation would conform to the applicable edition of the Code. Existing transformers may be replaced with larger transformers if the strength requirements of the applicable Code edition are met. The fact that several structures or other installations within an existing line or section are replaced at one time does not negate this allowance, except that Rule 202 requires a replaced structure to meet the current edition of Rule 238C. However, replacement with stronger, larger, or taller units to meet strength or clearance requirements related to an addition(s) to the structure(s) is not considered maintenance and, therefore, requires conformance to the current Code requirements. Existing facilities may be rearranged on an existing or maintenance replacement structure as long as they meet the requirements of the applicable edition.

A frequent question concerns the appropriate edition for facilities that are temporarily relocated for highway reconstruction work and are returned to their original placement after the work. If trenches are dug beside the poles and the pole butts are kicked over to the new location, the previously applicable edition applies; if new structures are used, the current edition applies. In any case, different clearances may be required in the temporary location if the conditions differ, or some new structures may have to be used if the terrain features are different.

This rule plainly states the intentions of the Code with respect to application of new or revised rules to existing installations. It should be stressed that, in general, the edition of the NESC that is applicable to a given installation is the edition that was in effect at the time of construction (see Rule 016). In later years, a subsequent edition will be applicable if the facilities are "upgraded." A change of voltage of an existing line does not affect the applicable edition unless the change cannot be made under the applicable older edition; if the structures need to be changed out to accommodate the voltage change, the current edition would then apply. The utility always has the option of meeting current requirements, whether current requirements are the same, greater, or lesser than those in effect at the time of original construction or a subsequent applicable edition.

In the 1993 Edition, the word *currently* was added to Rule 013B2 to eliminate confusion with the so-called "grandfathering" of existing installations. See the discussion of Rule 230I and Appendix E *NESC Grandfather Clause* for more detail.

The intention of the Code is that good practice be met for the conditions in place. For example, the designer of a line crossing a roadway can either install the line with enough extra clearance to allow for future road resurfacing or can plan to raise the line when the resurfacing occurs. A line over water must meet the requirements for sailboat clearances, if sailing is expected. If such a line originally was over pasture land that is now flooded, and the original structures were tall enough to allow the installation to meet good practice for sailboating areas, the previously applicable edition can remain applicable if so desired. Otherwise the structure would need to be changed out to meet the present edition.

013C. Inspection and Work Rules

The 2007 Edition added this rule to clarify which inspection rules and work rules apply to both existing and new facilities. Rule 013A requires the use of the current edition for physical construction attributes of new facilities. Rule 013B1 *allows* the current edition to be used for the physical construction attributes of existing facilities built prior to the current edition, if so desired, but it does not mandate using the current edition. Rule 013B3 allows a choice of the current edition or the previously applicable edition for the physical construction attributes when adding, altering, or replacing facilities on existing installations. However, regardless of which edition of the NESC is being used for the physical construction requirements (such as location, clearances, grounding, loadings, strengths, etc.) against which existing facilities are inspected, the inspection requirements contained in the current edition shall be used to determine the responsibilities of the utility(ies) involved. Similarly, the work rules in the current edition must always be used, regardless of whether the work is on an existing or new installation. The grandfather clause of Rule 013B2 applies only to the physical attributes of the installations, not to the inspection or work methods.

014. Waiver

(This rule was formed in 1981 from previous Rules 102, 202, and 302.)

Good judgment must be exercised in constructing temporary installations to meet the requirements of these rules. Safety to employees and others must not be overlooked; yet in some cases the strength and arrangement (not clearances) of construction may appropriately be very different from that required for permanent installations because the expense of complete compliance would often be prohibitive, unnecessary, and inappropriate. For example, temporary installations may not encounter the worst weather conditions. One of the considerations is the required reliability of a temporary or emergency installation.

In many cases, it will be necessary for the person in charge to decide which rules should be waived, as a decision must often be made quickly. Such decision is, of course, subject to review by the proper authority, and the person making it must assume responsibility for the consequences. Where the construction involves other utilities, as at crossings and with joint use of poles, it is intended that the appropriate officials or other representatives of such utilities should be notified before action is taken. However, the Code carefully specifies the few conditions where clearances may be reduced by agreement; others cannot be waived.

Although earlier editions of the Code allowed waivers to be given for the use of different construction requirements than those in the Code, the 1977 and later editions have specifically limited waivers to emergency or temporary installations. Prior to the 1977 Edition, the Code was updated on a sporadic basis and an expanded waiver allowance was appropriate. However, with the maturation of the Code and its scheduled, frequent revisions, such waiver is no longer appropriate. Rule 013A2 allows experimentation; if the results are favorable and convincing, it is expected that such methods or conditions would be recognized by the Code as part of the frequent revision process.

The result of these requirements is increased uniformity of construction without undue penalty to installations that meet the general conditions. It is felt that, if such nonstandard construction cannot, after careful experimentation and documentation, survive the rigorous examination of the codification process, such construction should be discontinued and, therefore, it is not appropriate that waivers for such construction exist. Thus, in the 1990 Edition, the limits of the waiver authority given under both emergency and temporary conditions were completely specified.

In both emergency and temporary conditions, strengths are required to meet Grade N. This recognizes that these installations are not expected to be in place long enough for significant deterioration to occur. In some cases, seasonal design loadings may not be expected.

Grade N requires consideration of the loadings that are expected to occur during the life of the installation.

In *emergency* installations only, certain clearances are allowed to be reduced during the term of the emergency. The reduced clearances are specified for cables and for open supply conductors of 0–750 V. For *temporary* installations, no decrease in clearances is allowed.

Specifications for the reduced clearances allowed during emergencies were added in the 1990 Edition; they were moved to Rule 230A in the 1993 Edition. For open conductors above 750 V, the utility is allowed some flexibility; Rules 014A1b and 014A1c allow unspecified reductions, but *appropriate recognition to the difference in voltage* is required. This recognizes that the safety afforded by traffic signals and highway lighting during emergency times is often so great that it is worth the short-lived clearance changes to decrease the time required to reinstate these services. A good example of this problem occurred at the time of the revisions to the Second Edition of this Handbook. In the cleanup of Hurricane Hugo in the Charlotte, North Carolina area, members of the National Guard and the police were struck by vehicles while directing traffic in intersections without power for the signals.

015. Intent

(This rule was formed in 1981 from previous Rules 102, 202, and 302.)

This rule clarifies the intent of the use of "shall," "should," "RECOMMENDATIONS," "NOTES," and "EXAMPLES."

The difference between a "shall" requirement and a "should" requirement is, in essence, the difference between "possible" and "practical." For a "shall" requirement, the requirement is expected to be met in all conditions. A "should" requirement recognizes that the requirement may not be practical in all cases; it is intended to be mandatory where practical. Where a "should" requirement is not practical, the installation should be designed, installed, and maintained in a manner that is consistent with the prevailing conditions and in accordance with Rule 012. The 2002 Edition clarified that EXCEP-

TIONS have always been intended to have the same force and effect as the main rule. Similarly, footnotes to tables are an integral part of the table. It is only NOTES to rules and EXAMPLES to rules or tables that are purely informative and not considered to be part of the code.

016. Effective Date

(This rule was new in 1981.)

This rule recognizes that the design and approval processes may be so lengthy for major facilities that it is impractical to make a change in design or construction in response to a revision of the NESC. Obviously, where responsive changes can be made before construction without undue burden, they should be made, but it must be recognized that they are not required if either design or approval was started before the effective date of the revision. This rule was initiated in the 1981 Edition. The 180-day period before the effective date recognizes that a time lag is often required to obtain copies of the new editions, review standards, train workers, and implement the change. Previously no effective date was specified.

The new editions of the Code have always been intended to be able to be used when they are issued. The rule was revised in the 1990 Edition to clarify that a new edition *may* be used on and after its publication date, but is not required to be used until the 180-day grace period has elapsed. On occasion, a greater lag time is given to allow for full implementation; such was the case with the cable-marking requirements of Rule 350G of the 1993 Edition, which were initially delayed until 1 January 1994, and subsequently delayed until 1 January 1996.

See Table H16-1 for the publication dates and effective dates of the various editions of the NESC. Many early editions were revised piecemeal; that is, individual Parts were issued when revised, rather than waiting until all were revised. This table omits the dates of the several NBS Handbooks that reprinted the various Parts of editions in groups.

Table H16-1 Effective Dates

Parts	Edition	Publication Date	Effective Date
Work Rules	1	1 Aug 1914	NS*
1–4	2	15 Nov 1916	NS
1–4	3	31 Oct 1920	NS
1	4	5 Feb 1926	NS
2	4	15 Apr 1927	NS
3	4	12 Mar 1926	NS
4	4	15 July 1926	NS
5	4	15 July 1926	NS
1	5	8 May 1940	NS
2	5	23 Sep 1941	NS
3	5	23 Jan 1940	NS
4	5	13 Oct 1938	NS
5	5	1 Dec 1939	NS
6	5	17 Apr 1940	NS
2 only	6	1 Nov 1961	NS
2 (Supp. 1)	6	15 Dec 1965	NS
2 (Supp. 2)	6	Mar 1968	NS
3, 5, 6	Deleted	1970	NS
1	(6) 1971	Jun 1972	NS
3,[†] 4	1973	20 Jul 1973	NS
2	1977	28 Feb 1977	NS
1–4	1981	5 Sep 1980	180 days
1–4	1984	26 Sep 1983	180 days
1–4	1987	1 Aug 1986	180 days
1–4	1990	1 Aug 1989	180 days
1–4	1993	1 Aug 1992	180 days
1-4	1997	1 Aug 1996	180 days

* This is the year that underground rules moved from Section 29 to Part 3. The previous Part 3 had been Utilization Wiring now contained in the National Electrical Code (NEC).

† Not specified.

017. Units of Measure

(This rule was new in 1984.)

Metric values were introduced in the 1984 Edition for information only; the customary inch-foot-pound values governed with respect to rule requirements until 1990, when either system was allowed to be used. The metric values are not identical equivalents to the customary values; the metric values have been rounded to provide convenient working numbers.

In the 1993 Edition, the intention of Rule 017B was originally clarified to indicate that the required dimensions of items such as ground rods be considered to be nominal dimensions, and that the tolerances allowed by applicable standards are acceptable by the Code. During that revision cycle, existing standards for ground rods were considered and the dimensions were found to be appropriate for utility grounding. However, at a later date, NEMA GR-1 was revised to allow lesser dimensions of ground rods. As a result, the grounding rules were revised in the 2007 Edition to specifically state dimensional requirements and the language of this rule was also revised to delete the reference to ground rods and apply *nominal values* only to dimensions not specified in the NESC.

The 1997 Edition reversed the order of the values to put metric first, but either may still be used (see Handbook Appendix C—*Metric Conversions used in the NESC and NESC Handbook*).

018. Method of Calculation

(This rule was new in 2007.)

In 2007, a coordinated effort was made to use values with appropriate decimal places in rules, tables and calculations. Each of the areas of the NESC was reviewed and appropriate requirements were placed therein. For example, the results of calculations required by the overhead clearances rules in Section 23 must be rounded up to specified digits. If there is no specific requirement for rounding the results of calculations required by the NESC in a particular rule or section, the result is now required by Rule 018 to be rounded off to the nearest significant digit.

Section 2. Definitions of Special Terms

This section contains definitions that are special or are otherwise necessary to the understanding and use of the NESC. For terms not specifically defined, see IEEE 100, *The Authoritative Dictionary of IEEE Standards Terms, Seventh Edition* (ANSI). For all other definitions, the standard dictionary definition is intended.

Because these definitions are generally self-explanatory, further discussion is not included here, except for the following items:

(1) Prior to 1970, the Code included requirements relating to radio installations and included a definition of *antenna conflict*. This definition was carried forward in later editions as a result of an editing error. The Code was not specific as to clearances or other construction requirements relating to antennas. Since the definition served no purpose, it was removed.

(2) This section includes several definitions relating to voltage. Unless otherwise indicated, the term *voltage* as used in the Code refers to root-mean-square (rms) voltage. Where crest voltage is specified, such as in the calculation of alternate clearances under Rule 232D, a voltage value of 1.414 times the rms value is intended to be used. Most of the tables in the Code use phase-to-ground voltages. Where circuits are not effectively grounded, the highest nominal voltage available between any two conductors is to be used. For example, if a 19.9/34.5 kV three-phase circuit is effectively grounded, 19.9 kV would be used in the tables; otherwise, 34.5 kV is to be used. The rules use nominal voltages through 50 kV to ground; above 50 kV, the maximum operating voltage is to be used.

(3) Line conductors and cables, as well as equipment, are classified as either *supply* or *communication* and are intended to be located accordingly in their respective spaces. Items not meeting the definition of a communication line are considered as supply. However, in some special cases, certain supply cables may be allowed

to be located in the communication space on a structure. Neither supply nor communication line conductors or cables are allowed in the communication worker safety zone required between the supply space and communication space (see 2002 Rule 238E) either at the pole structure or in the midspan. Vertical runs of supply cable meeting applicable rules are, however, allowed to traverse the communication worker safety zone on a pole. In addition, certain safety-related facilities (luminaires and traffic signals) are allowed in the communication worker safety zone, when necessary to accommodate mounting height restriction for these items.

(4) When fiber-optic cables began to be commonly used, they were not identified specifically as to their intended treatment. The difficulty in identifying appropriate treatment lay in the fact that, while the fibers themselves were of dielectric material and were not a safety concern from the voltage perspective, they were frequently accompanied by a metallic messenger or sheath or both, which obviously could form a path for the flow of current. Today, some fiber-optic cables have metallic "talk" pairs of ordinary, telephone type for use in trouble shooting. The wording of the definition and specific rules have been revised several times in an attempt to clarify the intended use of such systems. Although the definition of a fiber-optic supply cable would not appear to allow placement in the communication area of the pole, other rules allow treatment of fiber-optic supply as fiber-optic communication under specified conditions. In such cases, the fiber-optic supply cable cannot be placed between the supply and communication spaces on an overhead structure; any transition must occur on one structure and meet the requirements for a vertical conductor of its type. In the 1990 Edition, the definitions of fiber-optic cables were revised and, in 1993, Rule 224 and Table 235-5 were revised to explicitly limit such placement. (see also Rule 230F).

(5) Definitions for **in service** and **out of service** were added in the 1993 Edition to limit the opportunity for misinterpretation of Rule 214 and similar rules regarding inspection requirements.

(6) The requirements for "effectively grounded" are sometimes confused with the requirements for "multiple grounding" as used in Section 9. The multiple grounding requirements of Section 9 require not less than 4 or 8 ground connections in each 1.6 km (mile) line segment. However, the definition of effectively grounded does not depend upon a particular number of grounds but rather on the adequacy of the connected grounds and their ability to take surge current away fast enough to limit voltage buildup to required levels. Depending upon the type of electrodes used and the soil resistance, the number of ground connections required to meet multiple grounding requirements of Section 9 may be sufficient or insufficient to meet the effectively grounded definition.

(7) The 1997 Edition revised the definition of **vault** to further differentiate a vault from an electric supply station. The 2002 Edition further clarified the distinction between the enclosure systems. A key requirement is limitation of access to vaults to *qualified* personnel, whereas the access to electric supply stations may also include *authorized* personnel. A main difference in the rules specifying conditions in vaults (Part 3) versus supply stations (Part 1) is that vertical clearances are *specified* for supply stations and are not specified for vaults. The requirements for guarding are similar, but less detailed in the vault rules of Part 3.

(8) Although no definition for *surge-protection wire* (or *overhead static wire)* was presented here until **shield wire/conductor** was added in 2002, there was and are definitions of supply lines and communication lines. Surge-protection wires are grounded conductors, but they are not considered to be line conductors. Many are not continuous or do not have a direct connection to a circuit or form a part thereof. However, an effectively grounded

neutral line conductor can be used as an overhead shield wire. The rules specifying clearances or other requirements make this differentiation. Shield wires may, or may not, meet the multigrounding and effective grounding that would allow them to be connected to co -function as a distribution neutral. In 2007, overhead ground wire, static wire, surge protection wire, and shield wire were all defined and related to each other.

(9) New definition for *neutral conductor* and *multiple grounded/multiple grounded system* were added in 2002.

(10) The 2002 Edition redefined **de-energized** as **disconnected** and added an information note. This change was coordinated with changes in Part 4 to refer to **de-energized and grounded** to more specifically detail requirements. Merely disconnecting does not necessarily make it safe to touch.

(11) The definition of **readily climbable** was completely revised in 2002 to specify in detail what is or is not considered to be a readily climbable supporting structure. It was further revised in 2007 by defining both *readily climbable* and *not readily climbable* under *supporting structure*.

(12) The definition of **qualified** was expanded in 2002 to require training and demonstration of knowledge.

(13) Single grounded, unigrounded, and ungrounded systems were defined in 2002.

Section 3. References

(This section was added in the 1984 Edition.)

From 1981 through 1993, Section 3 included in one place the standards that are referenced within various other sections of the Code. They form a convenient reference for checking library copies for up-to-date versions of other standards that are specified in the current edition of the Code.

In 1997, the references were split into two parts. Section 3 now includes only those standards that form a part of the NESC to the extent called out in the rules. Other standards that are cited for information or documentation purposes are now shown in a new bibliography designated as NESC Appendix B.

Section 9. Grounding Methods for Electric Supply and Communications Facilities

090. Purpose

The purpose of Section 9 is to provide practical *methods* of grounding for use where grounding is required as a means of safeguarding employees and the public from injury that may be caused by electrical potential on electric supply or communications facilities. The *requirements* to ground items are found in Parts 1–4.

The object of protective grounds on electric circuits or equipment, as required by the rules of the NESC, is to keep some point in the electric circuit or equipment at, or as near as practical to, the potential of the earth in the vicinity. Grounding helps to prevent harm to persons or damage to property in the event of accidental contact by persons with conductive equipment casings or enclosures, guys, conduit, etc.; direct or near hits by lightning; accidental contact of high-voltage conductors with low-voltage conductors; breakdown between primary and secondary windings of transformers; etc.

In order of descending effectiveness, ground systems serve to (1) enhance prompt operation of system fault-protective devices and (2) minimize the exposure of personnel to electrical potential.

The ideal condition would be to have a grounding electrode with a resistance to ground so small that the voltage to ground would be held to a small value under any condition. In many situations, however, this is not practical due to either high soil resistivity or very low circuit impedance. In such cases, a high degree of protection is obtained if the grounding electrode has a low enough resistance to ground to ensure the current flow required to promptly operate protective devices and remove the source of the potential (see Rule 096—*Ground Resistance*).)

Under high-capacity ground fault or lightning conditions, substantial voltages may develop between locations on the earth's surface only a few feet apart, due principally to the very appreciable resistance of the earth itself. Good grounding alone will not remove this hazard; additional means are required. Where there is a high probability that personnel may be exposed to large step potentials resulting from the operation of fault-current or other protective devices, such as in a supply substation, the effective potential may be minimized by the use of properly spaced buried grid conductors and by covering the earth with coarse crushed rock in the critical areas.

091. Scope

Section 9 specifies the proper methods to be used in the grounding of electrical circuits and electrical equipment (neutrals, transformer cases, switchboard frames, motor frames, conduit, etc.) when such grounding is required. The circuits and equipment that are required to be grounded are specified in other sections of the Code.

NOTE: Not all circuits and equipment are required to be grounded.

092. Point of Connection of Grounding Conductor

092A. Direct-Current Systems That Are to Be Grounded

It may appear that the restricted number of ground connections permitted on direct-current (dc) circuits does not provide quite the same assurance against loss of protection as is provided by the multiple grounds recommended for alternating-current (ac) distribution circuits. There are, however, a few factors that offset in large measure the apparently less adequate protection on dc circuits. One of these is the fact that such circuits are largely underground or confined to private premises and, hence, are not so much exposed to high voltages as are ac circuits. In addition, large ground wires are usually installed; their location at stations under controlled access and expert supervision reduces the chance of breakage. The benefits from reduction of the possibility of electrolytic damage, which might occur if multiple grounds were required or permitted, are sufficient to warrant the restriction of the number of ground connections.

The Third and Fourth Editions of the NESC specify that a ground connection on three-wire dc *distribution* systems is to be made to the system neutral at one or more supply stations. The ground connection for a two-wire dc *distribution* circuit is only to be made at one station; otherwise, the grounded side of the circuit is to be insulated from the ground. Ground connections at individual services or within a building are prohibited. The Fifth and Sixth Editions generalize the rule by eliminating the reference to two-wire dc systems. No mention of nondistribution dc systems is made in these editions.

Beginning with the 1977 Edition, the original provisions were reworded and retained for dc systems of 750 V or less. For higher-voltage dc systems that are to be grounded, a grounding connection to the neutral is required at *both* the supply and load stations. One, but not both, of these connections may be through surge arresters; the other station neutral must be effectively grounded.

Beginning with the 1981 Edition, the ground or grounding electrode is allowed to be located external to or remote from each of the stations. This provision is useful where needed to reduce electrolytic damage to electric supply or other facilities from the flow of direct current through the ground. An EXCEPTION was added in the 1993 Edition to allow one ground connection to serve both a supply station and load station where the stations are not geographically separated, as in back-to-back converter stations.

092B. Alternating Current Systems That Are to Be Grounded

Ground connections at all building entrances served by any particular secondary circuit are desirable, since they (1) permit ready means for inspection and testing and (2) because of their number, they provide good insurance against the entire loss of the ground connection. Since the resistance of multiple grounds varies very nearly inversely as their number, a larger number will more readily open automatic protective devices in case of accident and provide a greater degree of safety.

The requirements of this rule were degeneralized in the 1977 Edition into (1) those affecting low-voltage circuits of 750 V and below and (2) those affecting higher-voltage circuits. The earlier requirements essentially remain for the low-voltage circuits. Where grounding is required, a ground connection is to be made at the transformer (source) and at the line side of all service equipment.

The wording of the requirements regarding the point of grounding of a two- or three-phase circuit was revised for clarity in the 1977 Edition. At the same time, the requirement that a secondary distribution system to be grounded shall have at least one additional ground connection, other than at a service, was removed. The last paragraph of Rule 092B1 now requires grounding connections at the source and at the line side of all service equipment (see Figure H092B1). Requirements for additional grounds are included in Rule 097C and Rule 097D. Requirements for ground electrode resistance values are in Rule 096.

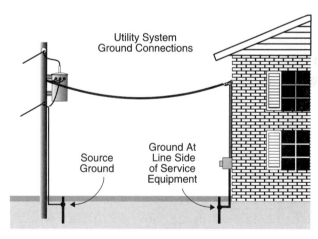

Figure H092B1
Utility system ground connections

The requirements for grounding three-phase, three-wire systems were clarified in the 1981 Edition. Regardless of whether the circuit is derived from a delta-connected or an ungrounded-wye-connected transformer installation not used for lighting, the grounding connection may be to any of the circuit conductors or to a separately derived neutral. If a *phase* conductor is grounded, it is *not a neutral* and does not meet the requirements of Rule 230E1.

For *nonshielded* conductors or cables of over 750 V, the 1977 Edition specifies that the grounding connection, where required, *shall* be made at the source. Beginning in the 1987 Edition, the system neutral may be used as a connecting link between a grounding conductor and the source transformer. Connections *may* also be made along the length of the neutral if the neutral is a system conductor. This rule allows multigrounded neutrals but does not require neutrals to be multigrounded.

For the first time, in the 1977 Edition, grounding connections for various kinds of shielded cables and cable installations were specified. Where underground cables are connected to overhead lines, any required shield grounds must be bonded to any available surge arrester grounds; otherwise, a separate ground may be used. Detailed requirements for both limiting the exposure of personnel to hazardous potentials and protecting the system integrity are specified.

Rule 092B2b(3) does not require grounding of a splice in a manhole when the splice is effectively insulated for the voltage that may appear on the surface of the splice.

Where a separate grounding conductor is used, it is required to be run along with circuit conductors in order to minimize inductive reactance and limit hazardous conditions arising from accidental violation of the cable insulation or faults down the line. If a conduit made of magnetic materials is used, Rule 092B3 requires the auxilliary grounding conductor to be placed in the same duct (hole) with the energized conductors; in the alternative, the grounding conductor must be bonded to each end of the magnetic conduit enclosing the grounding conductor. If the conduit is nonmagnetic, any duct of a multiduct conduit may be used.

Rule 092B3 allows an auxiliary grounding conductor to be installed along a cable route (for such purposes as limiting corrosion on the cable concentric neutrals, etc.), but it does not allow the use of the adjunct grounding conductor as a substitute for the concentric neutral required by Rule 350B. Special care must be taken when auxiliary grounding conductors are used with conduit systems. If a conduit surrounding the energized conductors is of magnetic material (steel), the auxiliary neutral must be run *inside* the conduit (see Figure H092B3-1), unless it is bonded on each end (see Figure H092B3-2).

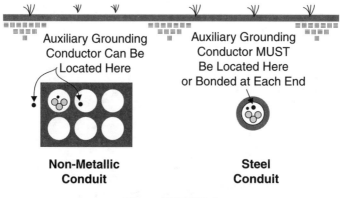

Figure H092B3-1
Location of auxiliary grounding conductor

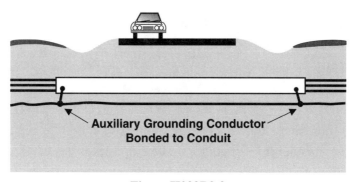

Figure H092B3-2
EXCEPTION to Rule 92B3

092C. Messenger Wires and Guys

(With the introduction of these requirements for messenger wires and guys in the 1977 Edition, the previous Rule 92C—Current in Grounding Conductor was renumbered to Rule 92D.)

Conductors of different impedances may have different voltage drops over the same length of conductor (see Figure H092C1). If the impedance difference is great (such as between a power neutral and a communication messenger or between a large and a small messenger, or if the distance is great, the potential difference can create a significant safety hazard. When such problems were observed on utility systems, new specifications were added in the 1977 Edition for the grounding of messenger wires and guys.

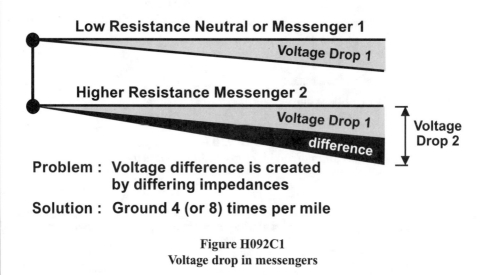

Figure H092C1
Voltage drop in messengers

Where messenger wires are large enough to meet the requirements for system grounding conductors, they are likewise required to be grounded at a minimum of four connections in each and every 1.6 km (1 mi) in order to be considered as "effectively grounded" for clearance purposes. Where messengers are smaller or otherwise inadequate, at least eight connections to ground per 1.6 km (1 mi), exclusive of service grounds, are required to limit exposure of equipment and personnel to hazardous conditions arising from line faults, lightning, or other surge conditions.

These considerations are found in several places in the Code. For example, the 1990 rules for grounding the concentric neutral of direct-buried supply cables with fully insulating jackets placed with random separations from communication cables (see Rule 354) require eight ground connections in each and every 1.6 km (1 mi) containing such cables in random separation; each individual phase conductor must have a copper concentric neutral conductor with a conductance not less than half that of the phase conductor. On portions of the line where the random separation does not exist, the normal four ground connections in each mile requirement applies.

Rule 215C2 requires guys on structures carrying open supply conductors above 300 V (277/480) or if subject to accidental energization from movement of these conductors or by movement of the guy be grounded or insulated. Where guys are required to be grounded, they must be grounded at the structure and must be connected to an effectively grounded conductor or structure. These requirements recognize that, if a guy strand is accidentally severed at the ground, the guy may be energized through contact with a live supply conductor. Additionally, if a conductor or its attachment breaks, the guy may be energized by a falling conductor. In either case, by grounding the conductor at the structure, the circuit-protective devices can be operated by such contact. This rule was revised in the 1987 Edition to specify allowed connections.

A new Rule 092C3 was also added in the 1987 Edition specifying how common grounding of messengers and guys on the same structure is to be accomplished if both are to be grounded (see Rule 215C3). These requirements were editorially revised in the 1990 Edition to state more clearly *how* this common grounding is to be accomplished. These requirements recognize the real problems that can occur from having multiple messengers at different potentials or messengers at different voltage potentials from nearby guys (see Figure H092C3).

This problem can be especially acute at crossing poles, which is why *bonding* is required on crossing poles. Where the route of one line crosses the route of another line and the facilities are supported at the crossing on a common structure, a significant voltage potential can exist between messengers of the different lines unless they are bonded together. The actual bonding and grounding connections may be at other locations, but the messenger must be *bonded* at the crossing structure (see Figure H092C3).

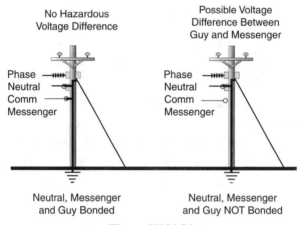

Figure H092C3
Bonding of guys, messengers and neutrals

092D. Current in Grounding Conductor

(This rule was numbered Rule 092C prior to the 1977 Edition. The Fifth Edition requirements of Rule 092E—Service Conduit were included within Rule 092D—Equipment and Wire Raceways; these requirements were placed into Rule 093C in the 1977 Edition.)

Rule 092D refers to actions required in the case of objectionable flows of current over a grounding conductor. Before the 2007 Edition, the word *objectionable* was undefined in the NESC; it was left to the designer's discretion, utilizing good design and operating practice, to appropriately identify and remedy the situation. In 2007, *objectionable current* was defined under normal conditions to be a level set by either the owner/operator of the electrical or communication utility system or by the authority having jurisdiction.

Current in a grounding conductor resulting from operation of a protective device is specifically "not objectionable;" it is integral to the operation of such devices and is intended to occur for short periods. A new NOTE was added in 2007 to reinforce the knowledge that some amount of current will always be present on grounding conductors of an operating AC electrical system.

Where multiple grounding is used, there generally will be some circulating current between the different ground connections. These currents may arise from unbalanced loads, improper connection or loss of ground wires, or other reasons. A fraction of an ampere, or even several amperes on circuits of large capacity, may not be a serious matter. In other cases, however, such flow may be disturbing to the service, as is sometimes the case (or, more frequently, alleged without foundation to be the case) in computer rooms or around dairy barns in which cows are connected to milking systems. In essence, the mere presence of an electrical current on a grounding conductor is not objectionable; in fact, it may be expected to occur in some measurable degree in many places and circumstances. The current must be at such a level that can be demonstrated to be the cause of a problem. It is recognized that interrupting the circulating current between the primary neutral and the secondary neutral may not solve the problems at dairy barns and may actually cause other problems.

Voltage/current-related problems at dairy barns, industrial plants, or commercial installation are most often related to National Electric Code (NEC) violations, unbonded building construction, and other building-related problems that produce voltage gradients at entrances or in building floors. While it is generally both infeasible and unnecessary to ascertain the circulating current flow at every ground location, installations near areas that are often known to present specific problems (such as milking barns without adequate voltage gradient control, pipelines, electric railways, conduits, etc.) may need special attention to limit damage to equipment or uncomfortable conditions for personnel or animals. In the case of alleged problems at dairy barns, sanitation conditions and issues related to feed quality are often found to be the real cause of milking problems.

In the case of computer rooms, operational problems are often found to be caused by airborne conductive fibers from floor systems, shorting out circuit boards etc. NEC Article 250.6 (formerly 250-6) contains similar requirements to NESC Rule 092D. NEC Article 250.6D (formerly 250-6(d)) specifies that "currents that introduce noise or data errors in electronic equipment shall not be considered the objectionable currents addressed in this section." Indeed, early electronic micro processes used in some industrial equipment did not include adequate filters and algorithms to ensure satisfactory operation with normal power frequency and harmonics on power lines. Today, electronic equipment that needs to be sensitive as part of its operation is usually installed with power quality control equipment appropriate to meet its needs.

The advantage in permanency and reliability, which results from the use of a number of grounds on a given circuit feeding a considerable area, will generally warrant the use of multiple grounds on alternating-current secondaries, notwithstanding the possible existence of a slight interchange of alternating current over these connections. Heating or electrolysis from such small alternating currents is generally negligible. A value of interchange current that would not be harmful with alternating current, however, might be sufficient to cause damage if it were direct current, often due to corrosion problems.

If the protective ground connection normally carries current, it is part of a closed circuit. As a result, this *can* be an undesirable type of ground for a number of reasons under certain circumstances. Direct current, in particular, may cause electrolytic damage if it is not confined wholly to the metallic circuit and the utilization devices designed for use with the direct current. Multiple grounds from a neutral wire of a dc, three-wire circuit may, if the dc circuit is unbalanced, cause earth currents and produce electrolytic damage by reason of such earth currents. Even alternating current, if in large amounts or continued for long periods, may unnecessarily deteriorate the ground connection. However, such a current could only result from a fault or from excessive unbalancing of three-wire, ac circuits with multiple ground connections, and such unbalancing would be expected to soon be detected and corrected. With made electrodes, the surrounding soil may be dried under such conditions. This condition can be serious and, with dc neutrals, might result in corrosive destruction of the grounding wire and loss of the protection afforded by the made electrode.

An objectionable flow of current over a grounding conductor may be due to any one of several reasons. For example, if electric railway returns are located in close proximity to water pipes or other grounds, part of the railway current may be carried through the supply conductors themselves from one ground connection to another. The result may be the deterioration and ultimate failure of such ground connections from electrolysis or drying of the ground.

In this respect, it might be well to consider cases in which the high-voltage side of a distribution or station transformer is grounded. Where transformer banks consisting of three single transformers connected in wye on the high-voltage side have the neutral point grounded, a certain amount of current will flow in this ground connection because of the third-harmonic voltage present. This current may be of considerable magnitude unless proper methods are employed to control it. Methods of control are left to the designer.

Station transformer banks may also have their secondary windings connected in wye and the neutral point grounded. In some older systems the neutral wire was not carried out of the station as the fourth wire of a three-phase system, as when the load supplied was almost exclusively a power load. In some systems, where lighting was supplied, it used to be an occasional practice to install a single-phase transformer so that one side of its primary winding was connected to one of the phase wires and the other side to the ground. This resulted in a continual flow of earth current at all times, varying from the small excitation current under no-load conditions to a maximum at full load or under fault conditions. If a made electrode was used, this flow of current could result in enough drying of the soil, in dry seasons, to cause the soil immediately adjacent to the artificial ground to become nonconducting. As a result, the potential of the ground connection could be raised much above ground and even approach that of the line. It is evident that a very serious condition of hazard could be produced if the high-voltage potential is brought down to the ground line. Should a rain occur at such a time, there is danger of the pole burning because of current flow across the surface of the pole. *Such a flow of current would be considered objectionable.*

As a result of such problems, Rule 215C in the Fourth Edition (1941) introduced the prohibition against ground returns in urban areas; it recommended against them in rural areas. They were prohibited in any location in the Sixth Edition (1961); that prohibition has been retained in subsequent editions. Further, beginning in 1977, Rule 096 and Rule 097 required the neutral of a multigrounded *system* to be carried throughout the system. This allows transformer cases, cable sheaths, etc., to be connected directly to the neutral and enhance the operation of the system protection devices in the event of conductor failure, transformer winding failure, or cable failure. The NESC appropriately recognizes and allows the earth to become a parallel part of a return path for distribution systems, but prohibits the earth from being the sole return path.

Objectionable direct current can generally be eliminated by following one of the procedures recommended in the rules by either omitting or changing ground connections. The prohibition of removing the system ground from the source transformer was added in the 1977 Edition.

092E. Fences

Fences may be subject to imposition of undesirable voltage potentials from various sources, such as falling conductors, operation or failure of equipment in supply stations, circulating currents, and lightning-induced step potentials. Such exposure is site-specific. In order to limit the potential on fences in certain situations, the NESC requires these fences to be grounded; limited specifications for the *methods* of fence grounding were included in Rule 092E—*Fences* beginning with the 1977 Edition. The 1997 Edition revised the rule to delete specific requirements as to locating ground connections at line crossings and not more than 45 m (150 ft) on each side; instead, the user is referred to IEEE Std 80™ *Guide for Safety in AC Substation Grounding* and industry practices for guidance in grounding electric supply station fences. The fence grounding design is specifically required to limit touch, step, and transferred voltages. The previous language was generally adequate for many of the older, smaller stations, but it may not provide required safety around today's high-fault-current stations.

The rules are intentionally restrictive concerning requirements for bonding jumpers at gates and other fence openings, bonding of separate barbed strands, and bonding to station grounds or other effective grounds. Past experiences have shown that gate-hinging mechanisms and gate-roller mechanisms are not effective conductors.

These rules are not complete specifications for many situations and Rule 012 (good practice) may require that additional measures be taken. For example, in the 1977–1993 Editions, the fence was required to be grounded only in the area where conductors cross over the fence; this takes care of a conductor falling and induction at a crossing location, and it also takes care of touch potentials resulting from ground currents in that area. However, it did not address touch potentials in other sections of the fence that result from ground currents or the operation of enclosed equipment; thus additional ground connections may be appropriate.

The requirements for bonding fence gates, bonding across fence openings, and grounding at each side of gates or other openings in electric supply station fences apply *regardless* of whether a line crossing is nearby.

Rule 093C6 of the 1977 Edition added a requirement to ground *each* conductive post. The 1997 Edition rearranged and revised Rule 092E, bringing in that portion of Rule 093C6 to clarify the intended connections, depending upon the conductivity of the post.

Typical fence construction uses metallic posts with the fence mesh and barbed wire strands (if used) attached with metallic stretcher clamps at corner posts and gate posts. At intermediate posts, mesh is tied to the posts and the barbed wire rides in a slotted support bar. If the posts are conductive, the posts are the preferred grounding connections, but all posts may not be required by IEEE Std 80 or industry practices to be grounded for any particular station design. Where the fence posts are conductive and the fence mesh is electrically connected to the fence with tight tension support systems, the grounding connection must be made to as many posts as required to meet appropriate standards and practices. If the tension support system is loose, additional connections to the mesh are appropriate

Where the posts are nonconductive, the fence mesh *and* any barbed wire must be directly connected to each grounding conductor.

Where the corner and gate posts are conductive and grounded, and mechanical clamps for barbed wire used at these posts maintain the barbed wire taut in a manner as to provide a solid electrical contact, the clamps typically also serve to electrically bond the barbed wire to the grounded posts and no further bonding is required. Where barbed wire is not mechanically clamped to grounded, conductive posts but, instead, rides in a loose slot or is loosely wrapped around the post, an electrical bonding connection between the barbed wire and the grounded post or grounding conductor is required.

093. Grounding Conductor and Means of Connection

093A. Composition of Grounding Conductors

Copper is the usual material for grounding conductors. Aluminum might be used in some rare instances, such as where aluminum conductors are used on outdoor lines and are not in contact with earth or concrete. However, the use of aluminum underground is not appropriate (see Rule 093E5). Copper-covered steel is suitable. The corrodibility of iron and steel makes them generally unsuitable for grounding conductors, especially where installed in damp or moist locations where corrosion is likely to occur, but they are occasionally used, especially with galvanized steel fences and in areas where soil characteristics reduce the life of underground copper faster than galvanized steel.

Fuses, circuit breakers, and switches are not permitted in the grounding conductor except under the conditions mentioned in the Code. The loss of the ground connection through operation of a fuse, circuit breaker, or switch would often defeat the purpose of the ground.

In the 1977 Edition, the specification that the structural metal frame of a building or structure may serve as the grounding conductor to an acceptable grounding electrode was added. The EXCEPTION to allow removal of the grounding conductors for test purposes under competent supervision was also added. In the 2002 Edition, metallic electrical equipment cases were allowed to be part of the grounding conductor. Obviously, where anything other than a continuous conductor is used, the electrical continuity of the joints must be maintained. In the 1987 Edition, the use of a surge-arrester disconnector to disconnect the grounding conductor from a surge arrester was recognized, but it was also cautioned that the base of the arrester may still remain energized at line potential.

Because of the importance and special problems associated with surge arresters and ground detectors, the NESC requires conductors used to ground these items be as short, straight, and free from sharp bends as practical. During a typical lightning discharge the high current and short duration (i.e., high frequency) combined with additional ground lead length may significantly affect the protective capability of lightning arrestors. Prior to the 1977 Edition, the requirement of freedom from sharp bends was freestanding. The wording of the 1977 Edition joined that requirement with the requirements of shortness and straightness under the aegis of "practicality." It should be noted that the change in wording in the 1977 Edition was not a change in requirement but was an improvement in wording consistency; it is almost always practical to keep sharp bends out of such conductors.

An EXCEPTION was added in the 1993 Edition to allow opening devices in the grounding conductor of high-voltage direct-current (HVDC) systems for the purpose of changing from a remote electrode to a local electrode.

093B. Connection of Grounding Conductors

(The requirements under this rule number were added in the 1977 Edition. The previous requirements relating to size and capacity were moved to Rule 093C—Ampacity and Strength.)

Rule 093B encourages the use of continuous grounding conductors without joint or splice in order to prevent discontinuity in the conductor. As a result of past problems, the 1977 Edition specified the means by which grounding conductors are to be connected, if such connection is necessary. Special attention to material characteristics and installation conditions is required, especially where two different materials are being joined. When joining dissimilar materials, connectors that are rated for joining both materials should be used. Additional care may need to be taken with placement of dissimilar materials. For example, an aluminum grounding conductor segment can be placed above a copper segment, but to the reverse; copper salts can deteriorate aluminum.

093C. Ampacity and Strength

(These requirements were included in Rule 093B prior to the 1977 Edition.)

The appropriate size of grounding conductors is determined principally by mechanical considerations and by short-time ampacity requirements. Grounding conductors are more or less liable to mechanical injury and must therefore be strong enough to resist any strain or abrasion to which they likely are to be subjected. This is especially true where the grounding conductors are located in parking lots or exposed to abrasion from other sources, such as riding lawn mowers, forklift trucks, or similar vehicles. Generally accepted and satisfactory practice in electrical construction has been to place the *minimum* size at AWG No. 8 copper (or conductor with a tensile strength not less than No. 8

AWG copper) for general service or system grounding. The use of larger sizes is left to the discretion of the designer (see Rule 093D). Note that at least AWG No. 6 copper or No. 4 aluminum is required for grounding surge arresters.

Where grounding conductors are protected from mechanical injury or potential fault currents are large, the size of the grounding wire is determined more by the amount of current it may be required to carry than by mechanical considerations. This current, in turn, is determined by the available short-circuit current and the time-current characteristics of the overcurrent protection equipment in the circuit. Rule 093C1 requires (where determinable) the use of short-time ampacity for single-grounded systems, and Rule 093C2 requires the use of continuous ampacity for multigrounded systems. For a single-grounded system (i.e., one ground), the grounding conductor must be able to withstand the maximum anticipated current. However, for a multigrounded system, it is recognized that the multiplicity of grounds results in a sharing of current between electrodes. Therefore the maximum ampacity of any single-grounding conductor on a multigrounded system need not be greater than one-fifth of the ampacity of the neutral conductor.

It should be noted that service requirements differ between types of installations; this rule specifies these particularities, including exceptions.

Accessibility of the ground connection, where it is attached both to the equipment and to the ground, is an important matter; this enables the connections to be inspected after the equipment has been installed. On the other hand, if the connections are concealed, corrosion and deterioration could not be detected and remedied. Corroded connections sometimes render the ground ineffective, and thus its purpose is defeated. In some instances, plumbers and others have had occasion to disconnect grounding wires and clamps for the purpose of repairing piping. They frequently have been left disconnected but, in some cases, the fact that the point of connection was accessible disclosed this neglect and resulted in prompt remedy. However, it is recognized that it is appropriate to locate some grounding connections underground, such as ground-rod connections (see Rule 094B2c). Locating ground rods below grade removes the danger if a line worker falls.

It is not intended that each length of conduit and each piece of equipment separately be grounded by independent grounding wires. Where a metal conduit or raceway system is employed, it is sufficient to properly bond the different sections together, either by separate bonds or through the junction boxes by scraping off paint or other coverings and screwing the bushings and locknuts tight. Galvanized conduit and fittings may provide proper electrical continuity between the separate sections. However, because of vibration or workmanship problems, it must be recognized that locknuts and bushings are not always reliable for making electrical connections; dependence should not be placed upon them where the potential involved exceeds 150 V (see Rule 093B).

Surge-protection devices serve to conduct surge currents into the earth to minimize both equipment damage and personnel exposure. Due to the high frequencies involved in lightning surges, the inductance of the grounding conductor may be of greater importance than the resistance of the system, including that of the electrode. Levels of grounding resistance developed by application of these rules may not be sufficient to meet some technical performance requirements, such as line lightning protection for sensitive equipment, and additional measures may be required.

The 1977 Edition added to this rule a requirement to ground all conductive fence posts in electric supply stations. This requirement was relaxed in the 1997 Edition to require connection only to the number of posts required to achieve the intended grounding level under industry standards; the fence post grounding requirement was moved to Rule 092E in 1997.

093D. Guarding and Protection

(This rule was renumbered from Rule 093C in the 1977 Edition.)

Where there is only a single grounding connection on a circuit, the path of the grounding conductor should be as far as possible out of reach of persons, and as much care should be taken to prevent contact of persons with it as ordinarily would be taken with a low-voltage circuit conductor. For these reasons, such single grounding conductors are required to be guarded. It should be noted that this guarding requirement applies to all types of single-grounded systems, but not to multi-grounded systems.

For example, if a lamp post is grounded effectively, and the return side of the lamp is connected only to the post so that the post becomes a single grounding conductor (a practice that existed decades ago but is discouraged because the post itself becomes a live conductor if the ground connection is lost), the lamp post itself must be guarded. Note that Rule 215C and Rule 314C prohibit the use of the earth as the sole return and this type of installation has not been allowed in urban areas since 1941 (Fifth Edition) and anywhere since 1961 (Sixth Edition). If, however, there is a separate return conductor for the lamp, which happens to be interconnected with the grounded lamp post, the post would not be considered a single grounding conductor and would not be required to be guarded.

Where there are several grounding connections to a circuit, there is less likelihood of having a substantial potential on a grounding conductor. Where the grounding conductor is part of a multiple grounding system meeting the requirements of Rule 096A3, guarding is not required; where it is exposed to mechanical damage, it is to be attached on the surface of the structure, preferably on a protected side. By the same token, the location of a ground rod and its connection to the grounding conductor should be protected (see Figure H093D).

When a lightning surge travels down a grounding conductor toward an electrode and the grounding conductor passes through, but does not attach to, a steel or iron conduit, the steel conduit effectively produces an electrical choke coil; the resulting back pressure emf (electromotive force) limits the amount of current that can travel to the electrode. Thus, nonmagnetic guards are required by this rule when guarding grounds of lightning-protection equipment *unless the guard (made of magnetic material) is bonded at both ends to the grounding conductor.* The 1997 revision of Rule 239D superseded part of this rule, because

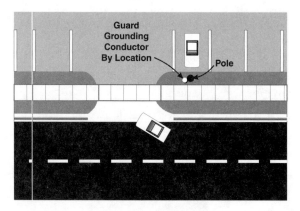

Figure H093D
Protection of grounding conductor by location

Rule 239D requires *all* metal *conduits* containing lightning protection wires to be bonded at both ends. Neither rules requires metal U-guards to be bonded at *both* ends, but the intention of Rule 215 is that at least one bond be made. When the NESC started, only wood molding and iron pipe were available for this use; in modern times, nonmagnetic materials such as PVC are easily available in various sizes and types for this application, thus eliminating worry about the continuity of the end connections required for magnetic metal. The separate statement of requirements for protection of grounds in indoor installations was dropped in the 1977 Edition.

The use of polyethylene-covered wire as a grounding conductor does *not* meet the guarding requirements of Rule 093D1 & Rule 093D2. The definition of *guarded* indicates that some form of adequate mechanical protection is to be provided where guarding is required. Further, the NOTE under the definition of *guarded* in Section 02 states that wires that are insulated, but not otherwise protected, are not considered to be guarded unless there are specific exemptions in the applicable rules. Rule 093D does not contain any exemptions that would allow use of a covered conductor as the only means of protection where guarding is required.

093E. Underground

(This rule was renumbered from Rule 093D in the 1977 Edition; the previous Rule 093E was renumbered to Rule 093F.)

The special problems of underground installations, including strain caused by settling of earth and corrosion, are addressed in this rule. Particular attention is paid to the problems caused by circulating currents. As electric supply systems get larger and the diversity of underground installations increases, these problems may be expected to increase in frequency and severity unless appropriate prior planning is employed. The specific requirements of the NESC were expanded in the 1977 Edition and notes concerning materials usage were added. The 1997 Edition deleted the previous specified requirement for welded, brazed, or compression connectors in favor of more general "nondesign-specific" language requiring appropriate corrosion resistance, permanence, mechanical characteristics, and ampacity. This is not so much a specification of new requirements as it is a specification of the capabilities of the old systems, if they were appropriately sized. The original language was placed in the Code because of failures of bolted splices underground. However, there are some relatively new bolted systems that have proven to have permanence underground. It is the responsibility of the designer to choose the correct splice to match the requirements of the conditions involved. The new language matches that already existing in Rule 095A. This rule prohibits having "looped magnetic elements" separating the grounding conductor and the phase conductors. The concern is that IR losses in the metal from either fault current or current induced from normal load current may generate significant heat in those elements. This heat may compromise the mechanical integrity of such material. In the case of concrete-encased steel, large levels of current in the steel can vaporize moisture in the concrete. The force of this vaporization may cause the failure of the concrete.

094. Grounding Electrodes

Rule 094 contains requirements for the grounding electrodes to which grounding conductors are to be attached. The paragraphs were rearranged in the 1977 Edition to include *existing electrodes* and *made electrodes*, former requirements were respecified, and significant NOTES and EXCEPTIONS were added. Rule 094D—*Grounds to Railway Returns* was deleted in the 1977 Edition.

094A. Existing Electrodes

For many years metallic water-piping systems with solidly bonded connections were considered to constitute a superior means of grounding electrical circuits and equipment. The resistance of such water-pipe grounds ordinarily was less than 0.25 Ω; most measure 0.1 Ω or less. With the development of new materials, however, the use of metallic conductive pipe by the water industry has become virtually extinct. Even old metallic systems are often repaired by inserting lengths of nonconducting pipe. For this reason, the reliance on water piping as a grounding electrode was substantially de-emphasized in the 1977 revision of the Code. Until 2007, it was recognized that, where dependability can be ensured, an extensive water system is still the best available ground, but it was also recognized that such dependability was becoming increasingly elusive. For these reasons, there has been a substantial industry trend toward multigrounded neutral systems, with the electric utility assuming the responsibility for providing an extensive grounding system over a large area, with adequate neutral conductivity. As of 2007, metallic water piping systems are no longer shown as being preferred.

The use of gas-piping systems as grounding electrodes where water-piping systems were not available was allowed, but discouraged, in the Fifth Edition. Allowance of the use of gas-piping systems as grounding electrodes was discontinued with the Sixth Edition (see Rule 095B).

Where extensive metallic public piping systems are not available, the grounding connection should be made in a manner that secures the most effective ground. Frequently, there are buried structures such as local piping systems, building frames and steel-reinforced foundations, well casings, and the like that would be more effective than separately driven or buried artificial grounds. Care also should be taken to ascertain that such electrodes will be effective during all seasons of the year; dry seasons may tend to render some of these electrodes ineffective, especially in sandy areas. (See Rule 094B for the requirements that foundations must meet to qualify as an acceptable electrode.) In some situations, two or more of such structures can be bonded together. This will not only provide a lower-resistance ground, but will also lessen the change of difference of potential within the premises.

Particular attention is called to the need for ensuring bonding between reinforcing bars in concrete foundations and steel structural members, if the steel structure is to be considered as a grounding conductor. Rule 094A does not specify the required conductor size for bonding, but Rule 095A specifies the type of bonding required and includes size requirements. Where foundations are separated, as in the case of a four-legged steel tower, the structure is considered adequate electrical connection; separate bonding of the foundations is neither required nor prohibited.

The metal covering of metal-clad buildings exposed to accidental contact with circuits should be grounded to limit personnel exposure, especially where the covering is insulated from ground by a wooden or masonry foundation.

094B. Made Electrodes

A "made electrode" is an electrode of any form buried in the ground for the special purpose of attaching a grounding conductor to it. Access to the grounding connection for assurance of its integrity may require connection to a made electrode as well as, or instead of, the water-piping system. It often may be appropriate for one or more made electrodes to be utilized in addition to a water-piping system.

The type and number of made electrodes required depends upon the type of soil conditions encountered. The sizes, shapes, and materials specified for the various conditions result from long experience with various types of installations. Many of the types are not equivalent to others and require multiple installations to achieve the same effect.

Made electrodes may be constructed of driven rods or pipes; buried conductors, plates or strips of metal; or combinations thereof (such as an acceptable foundation reinforcing rod cage). Driven rods or pipes are most generally used. They are required to be long enough to be driven to a depth of at least 2.45 m (8 ft). A layer of dry soil on the surface, of course, necessitates a greater length of pipe to achieve a satisfactory resistance, and may require additional rod sections. However, after 2.45–3.6 m (8–12 ft) of conducting soil has been penetrated, increased length may not give proportionate decrease of resistance unless penetration of a new soil strata of lower resistance is achieved. If a further decrease in resistance is needed, it is usually more economical to use several grounds in parallel because, if they are separated by an adequate distance, the total resistance varies approximately inversely with the number.

Rule 094B2 requires driven rods to extend to the 2.45 m (8 ft) depth level unless rock is hit or, beginning in 1977, to the 2.3 m (7.5 ft) depth inside the doughnut ring pad under pad-mounted equipment (leaving 150 mm [6 in] above ground for attachments).

Due to a controversy that developed in 2001 when NEMA standard GR-1 allowed minimum ground rod dimensions below those previously found by the NESC to be appropriate, the 2007 NESC added specific decimal dimensions for the minimum cross-sectional diameters of ground rods of different constructions to assure continuation of past good practices for grounding utility systems. NEMA Std GR-1 was changed in 2005 to match the impending 2007 NESC specifications.

The 2002 Edition allowed other dimensions or configurations of grounding electrodes to be used if supported by a qualified engineering study. This is especially useful when surface or subsurface constraints make meeting normal rule requirements impractical. For example, Rule 094B2b requires a minimum 1.8 (6 ft) spacing between multiple ground rods, and implies a vertical orientation of the ground rods. In some cases, other electrode arrangements, which provide equivalent or superior performance, are desirable due to their ease of installation, reduced impact on lawns and sidewalks, or other advantages. An actual example of such an alternate ground rod configuration, including a supporting analysis, is given below to illustrate the desirability of broadening this Rule.

Example: Inverted-V Ground Rod Geometry

Based on the following engineering analysis, pairs of 3.0 m (10 ft) long, 19 mm (3/4 in) diameter ground rods, driven at an angle of 45° ± 15° with the vertical (included angle between 60° and 120°), may be used where multiple ground rods are required. With this geometry, the rods are close to each other at the top, and are a minimum of 3.0 m (10 ft) apart at the bottom, forming an inverted-V. This arrangement has the advantage of not requiring a long trench to interconnect the two grounds rods. This arrangement is also, in some cases, less intrusive to lawns, sidewalks, etc.

Estimates of relative ground resistances show that the inverted-V arrangement using 3.0 m (10 ft) long, 19 mm (3/4 in) diameter rods has a lower resistance to ground than the vertical configuration using 2.45 m (8 ft) long, 16 mm (5/8 in) diameter rods with 1.8 m (6 ft) spacing. Thus, the inverted-V configuration is at least equivalent to (or better) than that currently required by Rule 094B2.

The resistance to ground of the inverted-V ground rod configuration and that of the "NESC configuration" were estimated using the Integrated Ground System Design (IGS) computer program that was developed by Professor Sakis Meliopoulos at the Georgia Institute of Technology. Professor Meliopoulos ran the IGS program for NU on 23 January 1997 to calculate the resistance to ground of several configurations, including an inverted-V arrangement (two 3.0 m [10 ft] long, 19 mm [3/4 in] diameter rods inclined 30° with the vertical) and the NESC configuration (two 8 ft long, 16 mm [5/8 in] diameter vertical rods, 1.8 m [6 ft] apart). The results indicated that the NESC arrangement had a resistance of 73.66 Ω, while the inverted-V arrangement had a resistance of 72.08 Ω; both cases arbitrarily assumed a 30 000 Ω-cm

---

soil resistivity. (A soil resistivity of 30 000 Ω-cm is assumed throughout this discussion. This parameter is simply a linear scaling factor that affects all of the resistance values by the same relative amount. Therefore, the particular choice of soil resistivity is immaterial with regard to comparisons of relative grounding effectiveness.) The IGS computer analysis also determined that the inverted-V configuration is approximately equivalent to two, 3.0 m (10 ft) long, 19 mm (3/4 in) diameter vertical rods 300 mm (1 ft) apart.

A second, more detailed analysis of the inverted-V configuration (3.0 m [10 ft] long, 19 mm [3/4 in] diameter rods inclined 30° with the vertical), was provided by Dr. Meliopoulos on 21 April 1998. This analysis examined the effects of varying the angle of inclination, as well as the distance between the tops of the rods. The results indicate that any angle between approximately 20° and 60° with the vertical, yielded an acceptable resistance to earth, i.e., a resistance that was less than or equal to that of the vertical configuration using 2.45 m (8 ft) long, 16 mm (5/8 in) diameter rods, 1.8 m (6 ft) apart. The results also showed that increasing the separation at the top of the inclined rods from 0 to 300 mm (1 ft) provided slight reductions in resistance—in the order of 2%. Resistance values of less than 60 Ω were obtained for the inverted-V configuration. The minimum resistance appeared to occur at an inclination of 45° with the vertical.

As a check on the results of the IGS computer model, the resistances of the "NESC configuration" (two 2.45 m [8 ft] long, 16 mm [5/8 in] diameter rods 1.8 m [6 ft] apart) and that of two vertical 3.0 m (10 ft) long, 19 mm (3/4 in) rods, 300 mm (1 ft) apart (conservatively approximating the inverted-V configuration) were calculated using the Dwight formulas as found in ANSI/IEEE Std 142™ *IEEE Recommended Practice for Grounding of Industrial and Commercial Power Systems* ("IEEE Green Book"). These formulas allow the calculation of resistance to ground for various arrangements of electrodes, given the electrode lengths, diameters, and spacing and the soil resistivity.

The Dwight formula for calculating the resistance to ground of two vertical ground rods separated by a distance less than their length ($s < L$) is approximated by the following:

$$R = \frac{\rho}{(4\pi L)}\left(In\frac{4L}{a} + In\frac{4L}{s} - 2 + \frac{s}{2L} - \frac{s^2}{16L^2} + \frac{s^4}{512L^4}\right)$$

where

R = resistance to ground, in Ω,

ρ = soil resistivity, in Ω-cm

L = ground rod length, in cm,

a = ground rod radius, in cm, and

s = spacing between ground rods, in cm

Using the previous Dwight formula, two vertical rods, 3.0 mm (10 ft) long, 19 mm (3/4 in) diameter, spaced 1 ft apart (conservatively approximating the inverted-V configuration) have a calculated resistance of 69.65 Ω. In comparison, two 2.45 mm (8 ft) long, 16 mm (5/8 in) diameter vertical rods spaced 1.8 m (6 ft) apart ("NESC configuration") are found to have a resistance to ground of 69.78 Ω. These calculations are performed in an EXCEL spreadsheet as summarized in Table H094-1 (Configuration Nos. 1 and 2).

Table H094-1. Inputs for Dwight Calculations

Configuration No.	Ground Rod Configuration	ρ, Ω-cm	a, cm	L, cm	s, cm	R, Ω
1	two 3.0 m (10 ft) L, 19 mm (3/4 in) dia., vert, s = 3.05 m (1 ft)	30 000	0.953	304.8	30.5	69.65
2	two 2.45 m (8 ft) L, 16 mm (5/8 in) dia., vert., s = 1.83 m (6 ft)	30 000	0.794	243.8	183	69.78

In the example given previously, a comparison of the resistance to earth of an inverted-V ground rod arrangement with that of a pair of vertical rods meeting the current NESC minimum requirements indicates that the inverted-V is as effective or superior to the vertical arrangement. The inverted-V is a preferred configuration for some applications due to its practicality and relative ease of installation.

It is clear from Rule 094B4a that pole-butt plates and wire wraps normally are not considered to provide effective grounding electrode functions except in *some* areas of very low soil resistivity. The three biggest reasons for this are that a driven ground has approximately twice the exposed surface area as a butt ground, acts as a lineal source instead of a point source, and has a greater driven depth. Additionally, pole movement may reduce both surface contact and pressure. Also, the wicking action of the pole will sometimes dry the soil around the pole enough to increase the ground resistance intolerably. In homogeneous soil, a butt plate has a resistance to ground several times higher than that of a driven rod. In those limited cases, as determined by Rule 096, Rule 094B4a allows *two* such electrodes meeting the requirements of Rule 094B4b or 094B4c to count as one made electrode for certain requirements (*not* including transformer locations) *only* in areas of low soil resistivity. Soil resistivity of 3000 Ω-cm or less meets this requirement for low soil resistivity; this is discussed in the Preface to the 15 August 1973 Unapproved Draft (Preprint). Transformer locations require a grounding electrode of more substance than a butt plate or wire wrap.

NOTE: Designers should be cautioned that *pole-butt plates and other systems exist in the market place that do not meet the specifications required for such apparatus* by these rules. They sometimes are called "pole protection assemblies" or "pole grounding plates" and are often employed in areas of high lightning occurrence as part of a pole lightning-protection system or are used as an economical means of enhancing the system's overall grounding.

It requires nearly a 1.8 m (6 ft) length of 25 mm (1 in) diameter pipe to provide 0.185 m^2 (2 ft^2) of superficial area, or a 3.6 m (12 ft) length for 0.4 m^2 (4 ft^2). For 31 mm (1-1/4 in) diameter pipe, the respective lengths are 1.4 m (4.5 ft) and 2.7 m (9 ft) respectively; for 38 mm (1-1/2 in) diameter pipe, the respective lengths are 1.2 m (4 ft) and 2.5 m (8 ft).

The size of plates need hardly be greater than 0.9 m^2 (10 ft^2). Larger sizes may provide for a greater rate of dissipation of energy in case of current flow, but added area after the first 3.0 m (10 ft^2) does not result in anything like a proportionate decrease of resistance. If it is necessary to attain a resistance much less than that provided by a plate of medium size, say 0.6 – 0.9 m^2 (6 – 10 ft^2), it would be better to use several plates in parallel, placing them well apart.

The resistance of grounds made with buried strips varies almost inversely to the length of the strip. This type of ground is best suited to rocky locations where the top soil is shallow, because the strips can be laid in trenches to almost any length and give the least resistance for the amount of metal used of any of the different types.

Note that there are tradeoffs between exposed area and depth of burial. A 2.5 m (8 ft) driven rod of 12 mm (1/2 in) diameter has 975 cm^2 (151 in^2) of exposed surface averaging over 1.2 m (4 ft) in depth and extending to 2.5 m (8 ft) or more in depth (Rule 094B2). In contrast, a buried wire (Rule 094B3a) or buried strip (Rule 094B3b) has 3940 cm^2 (611 in^2) or 4650 cm^2 (720 in^2) of exposed surface, respectively, at a depth of 450 mm (18 in). A buried plate (Rule 094B3c) of 1860 cm^2 (288 in^2) of exposed surface has a depth of 1.5 m (5 ft) or more. These relative exposed areas/depths contrast sharply with the smaller pole-butt plate (Rule 094B4a, Rule 094B4b) requirements of 465 cm^2 (72 in^2) at approximately 1.5 m (5 ft) depth depending upon pole depth. Since the wire wrap (Rule 094B4a, Rule 094B4c) is only exposed to the ground on one side, it has even less exposure area and is at a lesser average depth (see Figure H094-1).

Where rock does not permit driving an 2.5 m (8 ft) rod to its full depth, the rod is frequently driven at an angle to fully expose the surface to the soil. However, the shallower the soil layer, the shallower is both the average depth of burial and the maximum burial depth. At some point, the rod begins to look more like a buried strip and additional square inches will be needed to achieve the desired effect.

Materials most commonly used as electrodes for artificial grounds are galvanized rods and pipes, copper-covered steel rods, and copper

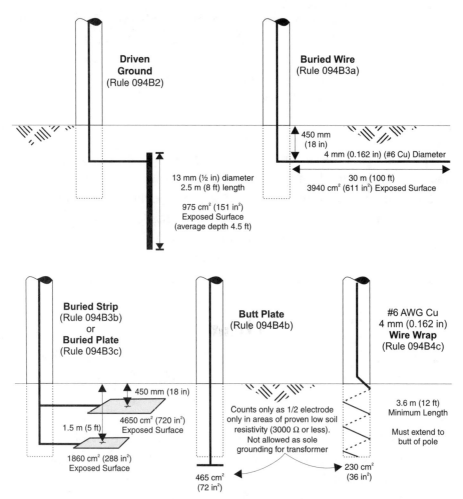

Figure H094-1
Relationships of made electrodes showing electrode contact area versus depth

plates and strips. Galvanized-iron or cast-iron plates may be used, but this is less advisable because of the possibility of corrosion of the galvanized iron, which, in the case of a plate, is difficult to detect without digging it up. Corrosion of driven pipes can, on the other hand, readily be detected near the surface with very little labor.

In the 1977 Edition, the requirement was added that all outer surfaces of made electrodes must be conductive, prohibiting coatings of paint, enamel, or other insulative materials.

The requirements for concrete-encased electrodes, whether a part of a structural foundation or separately installed, which are to be used as a grounding electrode, were detailed in the 1977 Edition.

094B7. Directly embedded metal poles

For the first time, the 2007 Edition specified the conditions under which a directly embedded metal pole would be considered to be an acceptable grounding electrode. During the deliberations on this subject, concerns were raised about the use of nonconductive coatings, increased ground contact resistance due to pole movement during storms, and corrosion. As a result, the backfill must be native earth or conductive grout, not less than 1.5 m (5 ft) of embedded length must be exposed to earth without nonconductive coating, and both pole diameter and metal thickness requirements are given. Since nonconductive coatings used to protect against corrosion are typically used only near the ground line (where the combination of moisture and oxygen is more conducive to corrosion) and since poles tend to pivot around a point approximately one-third of the distance from the butt to the ground level, the bottoms of directly embedded metal poles are generally well connected to earth, so long as the backfill is reasonably conductive.

Although the rule uses the term *metal*, aluminum is not considered acceptable and weathering steel is usually not acceptable. As a result, with present technologies, this rule generally will only apply to galvanized steel poles. NOTE 2 reminds the user to consider the structural and corrosion concerns prior to using metal poles as a grounding electrode.

Different lengths or configurations than those specified are allowed with a qualified engineering study.

095. Method of Connection to Electrode

(This rule was rearranged and respecified in the 1977 Edition.)

095A. Ground Connections

Before the 1977 Edition, the NESC specifications were generally applicable to grounding connections made to water piping. In the 1977 Edition, the specifications were expanded significantly to reflect current equipment availability and necessary practice for connection of grounding conductors to various electrodes. During recent years, there has been a notable development in the equipment available for making ground connections, and there are now a number of suitable devices on the market for this purpose.

Many ground clamps used in the past were of rather flimsy construction, making their usefulness uncertain. Older codes included a specific subrule on clamps. These issues have been of concern throughout the history of the NESC. The *Official Discussion of the Third Edition* included specific language to address both copper and iron clamps.

The *Discussion of the Third Edition* advised the following:

(1)   When made of copper, clamps should be not less than one-six-teenth inch in thickness, should be provided with strong bolts and lugs for attaching them to the pipe, and should have some means for adjusting them to fit the particular pipe to which they are attached.

(2)   If made of iron, clamps should be galvanized and so made that the protective coating is not broken by bending in putting them on. Whatever the material or operating condition, the NESC requires clamps to be of such substantial construction as to remain in satisfactory service under the conditions of their installation.

095B. Point of Connection to Piping Systems

Grounding connections for circuits preferably should be made immediately at the point where the water-service pipe enters the building or, on a cold-water pipe of sufficient current-carrying capacity, as near as practical to that point. This avoids a possible rise of potential on the building-piping system in cases where disconnections are made for piping repairs.

Wherever practical, the points at which grounding connections are made should be accessible to permit inspection after installation. Such accessibility permits ready detection of corroded or deteriorated connections and of any grounding connection left disconnected following repairs to the piping system.

Where the water meter is within a building, the best place to connect to extensive water-piping systems is on the street side of water meters. Where the meter is at the building but is not within the building, connection may be made on the building side. It is then necessary to shunt the water meter to avoid breaking the ground connection in the event of removal of the meter. However, where the meter is far enough from the building that the piping run to the building is long enough to constitute in itself an acceptable electrode, the shunt is not necessary, although still desirable.

The Fifth Edition required that gas-piping systems could not be used as grounding electrodes if water-piping systems were readily available. Where water-piping systems were a "considerable" distance away, the gas-piping system could be used as the grounding electrode *if* it was well bonded to the water-piping system. This allowance was deleted in the Sixth Edition and such use of the gas-piping system was prohibited.

In the 1977 Edition, installations were further constrained in the area of gas-piping systems operating above 1030 kPa (150 lb/in^2). Under certain conditions, grounding electrodes can be located within 3 m (10 ft) of the gas-piping system, but the gas-piping system cannot *be* the electrode. Calculations or tests are recommended in the 1993 Edition to determine the appropriate separation between the HVDC ground electrodes and high-pressure gas transmission lines.

The Sixth and later editions are silent as to the connection or location near gas-piping systems operating at less than 1030 kPa (150 lb/in^2).

095C. Contact Surfaces

In every case where electrical continuity is desired for the purpose of grounding or bonding, the surfaces of the metals where they come into contact with each other should be carefully cleaned of enamel, paint, rust, or other nonconducting material. In the alternative, special fittings designed to penetrate the coating, or that otherwise do not require removal of the coatings, may be used. The goal is to secure a low resistance of the ground connection.

096. Ground Resistance Requirements

(Prior to being renumbered in the 1993 Edition, the requirements of Rule 096, Rule 096A, Rule 096B, and Rule 096C were contained in Rule 096A, Rule 096A1, Rule 096A2, and Rule 096A3, respectively.)

The desirability of low resistance in ground connections is readily apparent. The lower the resistance of the ground, the less will be the potential difference between the grounded conductor and the earth. In any case, the resistance is required to be sufficiently low to ensure that a faulted circuit is promptly de-energized. This is the overarching requirement of Rule 096 and is to be met in addition to the subrules.

Where secondary distribution circuits are provided with a ground of 25 Ω resistance or less, the current in case of a fault involving the primary distribution circuit will, in general, be sufficient to de-energize the primary circuit at the transformer or elsewhere.

The designer is cautioned that the required resistance to ground of an electrode is a function of the conditions of its service. A ground of 10 Ω resistance that carried 100 amperes would cause a 1000 V drop between the grounded conductor and earth. The power expended through the ground connection would be 100 kW. Even if personnel were not exposed, this is not a desirable situation. The resulting heat could become a fire hazard or could dry out the earth around the electrode and increase the resistance, raising the voltage gradient even higher.

If a secondary circuit is exposed only through transformer windings and not by running under a higher-voltage circuit, it is protected by the primary fuse on the transformer. A *single* ground of 25 Ω could give rise to 250 V between the secondary and ground without enough current flowing to blow a 10 A fuse. While this may not be dangerous, it clearly points out the desirability of having two grounds and two grounding conductors. In the case of two or more services from the same transformer, the multiple grounds are easily obtained.

Rule 096A (Rule 096A1 of the 1990 and prior editions) addresses the step and touch potentials in electric supply stations. The nonmandatory formulas contained in the previous editions were deleted in the 1993 Edition. Satisfactory limitation of step and touch potentials in and around electric supply stations requires complex considerations that are covered in other standards. IEEE Std 80 *IEEE Guide for Safety in AC Substation Grounding* (ANSI) is an excellent discussion of the various problems that may occur and the appropriate methods for grounding supply stations in recognition of the lines and equipment present in the station (see the discussion of Rule 123).

Rule 096B (Rule 096A2 of the 1990 and prior editions) for single-grounded systems, such as unigrounded transmission or ungrounded delta distribution systems, requires a second ground connection if the first exceeds 25 Ω. If the second ground connection brings the resistance down to 25 Ω, no further consideration is required. However, if the resistance is still above 25 Ω, operation of the system-protective devices must be considered (see the main paragraph of Rule 096). If the resistance is low enough to allow prompt operation of the system protection, no further work is needed—otherwise, a new type of grounding electrode(s) must be employed to achieve the overall objective. As a practical matter this means that if two driven rods will not do the job, try plates or some other system; typically this will be required only in areas of high soil resistance. Often one of the biggest problems with achieving the 25 Ω value with two rods is not the soil resistance itself but that the second rod is not installed far enough away from the first to "talk to new earth." Various industry documents recommend separation of the ground rods by an amount equal to or greater than the driven depth.

Under Rules 096B and 097, single-grounded systems are required to have separate grounding conductors from each class of equipment (primary surge arrester and transformer tank versus secondary neutral) to separate electrodes, where each electrode meets the requirements of Rule 096B. Although bonding of the required separate *electrodes* at the ground line (not the grounding conductors) was allowed by the Fifth and Sixth Editions, such bonding has been prohibited since 1977. Rule 096B refers to bonding of electrode elements to form an electrode system, not bonding between the separate electrodes or systems required by Rule 097. In the Fourth and prior editions, a 20 ft separation between primary and secondary electrodes was required. Due to the manner in which the separation requirements changed, and then changed back in subsequent editions, the 20 ft requirement is no longer explicitly stated, although it is accepted as good practice when interconnection of the electrodes is prohibited.

This rule was revised in the 1977 Edition to distinguish between required resistances at these various installations and to require the neutral to be carried throughout the entire system if the system was multigrounded. A *minimum* of four grounds per 1.6 km (1 mi) is required for multiple grounded systems. At least four ground connections **in each** 1.6 km (1 mi) of a multigrounded system, i.e., between transformations, were first required in the 1977 Edition (see IEEE Std 100™ *The Authoritative Dictionary of IEEE Standards Terms, Seventh Edition* [ANSI]) for the definition of *system*. The NESC language was revised in the 1987 Edition to make this requirement more obvious.

IR532 issued May 2003 reinforced the requirement of having four ground connections **in each mile of the entire line** to assure adequate grounding of multigrounded neutral systems. This is not an average of four grounds per mile but rather a requirement to have four ground connections in *each* mile. The intention is to space the ground connections reasonably evenly along the mile at approximately 1/4 mile or smaller increments, not to group them in only one or two locations. Some intervals exceeding 1/4 mile are allowed (see NOTE 2 of the 2007 Edition), which is why the language does not specifically require the maximum distance between electrodes to be 1/4 mile. The rule is based on good engineering practice and is neither arbitrary nor a rule-of-thumb.

In essence, any "sliding 1.6 km (1 mi) segment" of a multigrounded line is required to contain at least four full ground connections. Although Mile A of Figure H096C-1 includes four ground connections, Mile B does not. Figure H096C-2 shows a basic layout of a rural area; distances are shown in tenths of a mile. Numbers in circles represent end poles or tap poles. A practical example is shown in Figure H096C-2 through Figure H096C-4. Mile A of Figure H096C-3 has four grounds in the first 6/10 mile and does not require others to meet in multigrounding requirement. Mile B only has three grounds at transformers; the ideal place for the fourth is at Tap Pole 4 or Tap Pole 6. Mile C needs two at Tap Poles 4 and 6. Mile D also needs two, but one of those would ideally be at Tap Pole 8. As the sliding mile continues up the line, it is easy to see how critical grounds at tap poles become. As a practical matter, placing surge arresters at tap poles can dramatically reduce lightning surge voltages at the ends of radial taps in rural areas. In urban areas, the multiplicity of transformer ground electrodes and surge arresters in close proximity limits the need for additional ground connections or arresters.

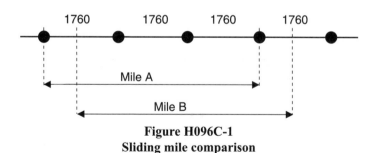

Figure H096C-1
Sliding mile comparison

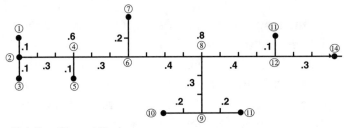

• Existing Ground Rods
at Transformer Locations

Figure H096C-2
Basic layout of rural area

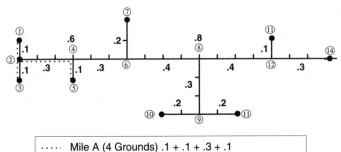

····· Mile A (4 Grounds) .1 + .1 + .3 + .1
• Existing Ground Rods at Transformer Locations

Figure H096C-3
Grounds required in Mile A

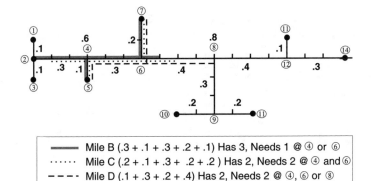

——— Mile B (.3 + .1 + .3 + .2 + .1) Has 3, Needs 1 @ ④ or ⑥
······ Mile C (.2 + .1 + .3 + .2 + .2) Has 2, Needs 2 @ ④ and ⑥
– – – Mile D (.1 + .3 + .2 + .4) Has 2, Needs 2 @ ④, ⑥ or ⑧
• Existing Ground Rods at Transformer Locations

Figure H096C-4
Grounds required in Miles B, C and D

See Rule 094 for discussions of different types of acceptable ground electrodes. Ground electrodes at transformers are allowed to be counted in meeting this requirement, as can electrodes located on attached tap lines (or, if on a tap, the main line). However, ground electrodes at the customer's meter location are not allowed to count. They are not always under utility control and may be adversely affected by customer action.

Additional grounds are required if necessary to "minimize hazards to personnel and to permit prompt operation of circuit-protective devices."

The term *multiplicity of grounding electrodes* refers to multiple points at which grounding is achieved, not to multiple electrodes at a single point. Multiple ground electrodes at a single point used to reduce ground electrode resistance at that point are considered as a single electrode for the purpose of satisfying the four-grounds-in-each-mile requirement of Rule 96C.

Rule 96C applies to both overhead and underground electrical systems, except for underwater crossings meeting the EXCEPTION to Rule 96C. In underground systems, it applies to both cable in conduit and direct-buried cable. Although the four-ground-connections-in-each-mile requirement also applies to long runs of buried, jacketed concentric neutral cable, it does not require a full splice or equipment connection four times in each mile. Instead, the cable jacket can be removed for a few inches at intermediate points to allow connection of a bonding clamp with a flexible cable for attaching to a grounding electrode, and the jacket can be resealed to keep out water.

Rule 096C contains specific requirements for achieving the ground resistance requirements on multigrounded lines. Note that a specific ohm value is not specified for individual electrodes on multigrounded systems. However, it has long been recognized that having the resistance of individual electrodes as low as is practical is highly desirable. Electric supply transmission lines are rarely considered to be multi-grounded wye circuits, within the meaning of the terms used in Section 9. Many transmission-voltage lines will be wye-connected at the station and will carry a grounded overhead shield wire along with the conductors. However, (1) the overhead shield wire is often discontinuous,

(2) the transformer connection to the station ground grid is often a high-impedance connection, and (3) the circuit does not have secondary connections along the line. Thus, the circuit is not constructed similarly to a distribution circuit and is not protected in the same manner. Rule 096C does not apply to or require specific grounding connections or overhead shield wires along electric supply transmission lines. However, if the overhead shield wire does meet the grounding requirements of Rule 096C, Rule 097 will allow connection of a secondary neutral to the shield wire. This was clarified by changes in Rule 096C and Rule 097B of the 2002 Edition.

It cannot be overstressed that the requirements of Rule 096 have been developed carefully over the long experience of the Code. Unfortunately, some of the specifics of Rule 096C (Rule 096A3 of the 1990 and prior editions) are sometimes overlooked. The requirements of Rule 096C are specific, and they are specific for various reasons:

(1) The neutral must be of sufficient size and ampacity for the duty involved. *See previous discussions relating to ampacity, inductance, abrasion, etc.*

(2) The neutral must be connected to made electrodes at each transformer location. It is not generally practical or appropriate to connect surge arrester grounds to *existing* electrodes in areas where the resultant step and touch potentials would be undesirable, or where frequent operation of the arrester might affect the intended operation of the existing electrode. A separate *made* electrode will adequately transfer any small circulating currents that may occur without harming the life or operation of the existing electrode. It is also rare to find acceptable existing electrodes as close as needed to the surge arrester; a made electrode can generally be located to provide a short and straight path to ground (see Rule 093A). An existing electrode may be bonded to a made electrode that grounds a transformer arrester in order to lower the overall resistance. Although the only term specified in Rule 096C for grounding at transformer locations is a *made electrode*, it is apparent, from the consideration of equivalence

between existing and made electrodes by Rule 096 and Rule 094, that an existing electrode could be so used, if desired. Apparently, the term *made electrode* was used specifically to key to the requirements of Rule 094B4 that do not allow the use of a pole-butt plate or wire wrap as the sole grounding electrode at transformer locations.

(3) The neutral shall *also* be connected to ground at a sufficient number of other points in *each 1.6 km (1 mi) segment of line* to total not less than **four** grounds in *each 1.6 km (1 mi) segment* of line. If there are neutral connections to four made electrodes in *each 1.6 km (1 mi) segment* at transformer locations, no others are required; if not, more connections are required. The additional electrodes may be either *made or existing*; see Rule 94. It must be stressed that the effectiveness of the grounding system depends on the number of grounds, the resistance of the grounds, and the size of the grounding conductors (including the neutral). There must be at least four grounds in *each 1.6 km (1 mi) segment* (eight, if only the pole-butt grounds are used). However, **butt grounds may not be the sole electrode at a transformer location** and the neutral must be large enough to appropriately distribute the expected fault current. Grounds at customers' services are not under the control of the utility, are all too frequently damaged or otherwise reduced in capability, and are therefore not acceptable as one of the four required grounds in each 1.6 km (1 mi) segment. **The neutral is required in each span or run of a multigrounded system;** this allows the transformer tank to be connected thereto and lessens problems that may occur if a conductor falls in the line.

For technical guidance in developing suitable grounding systems, safe touch and step potentials, and in determining resistance of electrodes in earth, see IEEE Std 80 *IEEE Guide for Safety in AC Substation Grounding* (ANSI) and IEEE Std 81™ *IEEE Guide for Measuring Earth Resistivity, Ground Impedance, and Earth Surface Potentials of a Ground System* (ANSI). (Although not included in the above, it is noted that the effectiveness of buried, concrete-encased electrodes, per unit, is almost equivalent to that of driven rods under low earth-resistivity conditions [less than 3000 Ω-cm] and is superior to that of rods under average and high soil resistivity conditions.)

The relative problems involved and the practicality of maintaining the specified grounds per segment requirement under water were recognized in the 1990 Edition, and underwater supply cables with a sufficiently sized neutral were exempted. As a practical matter, the longer runs will require a full neutral around each phase to ensure prompt operation of the protective devices and to limit damage to other cables when one cable faults.

096B. Checking

(Requirements for checking the resistance of supply station and distribution grounds were deleted in the 1971 and 1977 Editions, respectively.)

Experience has shown in some areas that one type of electrode system will achieve the desired results; otherwise, individual measurement of electrode resistance may be appropriate. Because of the specifications of requirements in Rule 096 of the later editions, the "checking" rule was eliminated as redundant. Obviously, installers are required to check each ground electrode installation on a single-grounded system to know whether to add the second ground electrode or switch to a different type of electrode, but such requirements do not apply to a multi-grounded system. As a practical matter, many utilities still check the ground resistance of all, or some, electrodes in order to make determinations of grounding needs. See Rule 121, Rule 214, and Rule 313 for present inspection and testing requirements.

The 2007 Edition extensively revised the language of Rule 96D to emphasize the need to have an electrode resistance on a single-grounded system low enough to meet the requirements of Rule 96A, i.e., to have the electrode resistance low enough to permit prompt operation of protective devices and to minimize hazard to personnel. Whereas previous editions allowed the addition of a second ground electrode if the first did not achieve a ground resistance of 25 Ω or less, without specifying the additional measures that might be required to meet Rule 96A, the 2007 Edition requires the use of other methods of grounding if the first electrode cannot meet the 25-Ω limit.

097. Separation of Grounding Conductors

Where the failure of a single grounding conductor might produce undesirable potentials on the equipment or other apparatus, it is advisable to use separate grounding conductors.

One objection to the common use of a grounding conductor by different classes of equipment is the opportunity for a damaged single common conductor to leave a large amount of equipment unprotected. There is also the possibility that, in the event of heavy current flow over the common grounding conductor from one of the connected sources, the impedance of the grounding conductor may be sufficient to give rise to undesirable potential differences between connected equipment and ground.

The failure of a common grounding conductor through mechanical injury may even create a hazard that would not exist if separate grounding conductors were used. For example, frames of equipment are ordinarily at ground potential. If frames are connected to the grounded conductor of a supply circuit, and the ground connection is then lost, an undesirable potential may be imposed on the frames. Connection of separate conductors to the same ground electrode does not involve such potentials during normal operation or many transient conditions, since separate grounding conductors cannot be in electrical connection with each other without being also connected to ground. On this basis, the Fifth and Sixth Editions allow connection of the required separate ground *electrodes* (not the grounding conductors) near the ground level. This is partially because loss of either grounding conductor would not tend to create a hazard to the connected equipment. However, when both grounding conductors were connected, high-magnitude surges were found to have a tendency to run past the connected rods and double up the voltage rise at the end, i.e., on the secondary circuit.

As a result, such bonding between electrodes has been prohibited on *single-grounded* systems since the 1977 Edition.

Where multiple grounds are used, the overall resistance to ground is reduced and the danger arising from the failure of individual grounding connections is minimized, except where one of the pieces of connected equipment is a surge arrester. A surge arrester is basically a voltage-sensitive switch that creates a short circuit to ground when it senses a voltage in excess of its set value. A surge arrester serves the purpose of limiting excess voltage on a circuit during surges caused by lightning or other events. This limits exposure of personnel and equipment to undesirable voltages. However, when a surge occurs and discharges over the arrester, the rate of flow of current is so great that a temporary, but significant, potential difference is set up between the grounding conductor and ground due to the (1) resistance of the ground connection and (2) the effect of the large current pulse (i.e., high frequency) and the inductance of the conductor. In addition, the possibility of arrester failure could introduce fault current into the grounding conductor. If other circuits or equipment have a common grounding connection with the arrester, this potential difference will be impressed on those circuits or equipment as well.

In order to limit voltage potential impression on nonarrester circuits, the NESC requires a direct-earth grounding connection at each arrester location. If the grounding conductor from the arrester is severed between its connection to other circuits and its connection to the grounding electrode, a code violation results. In many situations, supplemental protection from lawn mowers, bush hogs, car bumpers, etc., may be appropriate.

Recognizing the practicability of these factors, the NESC requires that individual grounding conductors be used to ground several different classes of equipment and circuits, either to an electrode or to a qualifying multigrounded cable or bus (see Rule 097A). Under certain conditions, however, the grounding conductors of some of these classes are allowed to be interconnected.

The First Edition of the NESC was issued August 1, 1914. A discussion of this edition suggested that arrester grounds not be used for other purposes.

The Second Edition of the Code (11/15/16) prohibited arrester ground connections to the same artificial ground as other classes and, where practicable, required that they be at least 20 feet from other artificial grounds. The accompanying discussion cited concern over disconnection of the actual common ground connection resulting in circulating currents between different equipment classes. It suggested that the "installation of different ground wires radially to a common ground, or, better yet, to different grounds, offers usually a greater degree of reliability and safety."

The Third Edition of the Code (10/31/20) repeats the language of the Second Edition on arrester grounds. The Discussion (10/31/20) for this edition again points out the concern over loss of a "single wire" resulting in loss of grounding, leaving "a large amount of equipment unprotected." In discussing a common ground wire from an equipment frame and the grounded conductor of a circuit, with regard to the case becoming energized if the common wire is severed, it acknowledged that "connection to the same ground does not create this hazard, since the two cannot be in electrical connection with each other without being also connected to ground." The requirement that "separate artificial grounds shall be used for arresters" is repeated.

The Fourth Edition of the Code (11/15/27) again repeats the language of the Second and Third Editions regarding arrester grounds. Equipment frames, wire runways, and service conduits were permitted at this time to utilize the same grounding conductors, provided the secondary distribution system had multiple grounds to water piping. Water piping at that time was metallic. The Discussion (9/21/28) of this edition pointed out that, "In the case of multiple grounds to water pipe systems, the contingency of losing the grounding connection is rather remote, and an exception is consequently made in this case for a common grounding wire for equipment and secondary circuits." Arrester grounds must be separate to prohibit other circuits or equipment from having a dangerous potential impressed upon them upon discharge of the arrester if a common ground were used.

The Fifth Edition of the Code (8/27/41) for the first time allowed interconnection of the arrester grounding conductor and grounded secondary conductor, *provided there was a direct grounding connection at the arrester* and *the secondary was elsewhere grounded at one point to a continuous metallic underground water-piping system, or the secondary neutral had at least four grounds per mile in addition to service grounds*. Furthermore, the direct-ground connection at the arrester could be omitted if there were at least four water-pipe grounds per mile on the secondary. However, with respect to individual artificial grounds for arresters, the Fifth Edition continued to require that separate grounding conductors and electrodes be used. A new provision was made, though, that "This does not prohibit the bonding together of these separate electrodes near the ground level."

The *Discussion* (7/15/44) of the Fifth Edition states: "Where the failure of a single grounding conductor might produce undesirable potentials on the equipment or other apparatus, it is advisable to use separate grounding conductors. Connection of the separate conductors to the same ground electrode does not involve such potentials, since the separate grounding conductors cannot be in electrical connection with each other without being also connected to ground. Where multiple grounds are used, danger from the failure of individual grounding conductors is eliminated."

The basic concern throughout the first five editions of the Code is essentially to avoid unwanted potential rise in one class of apparatus due to the loss of an effective ground on another. The two main examples of this are impressing a dangerous potential on a secondary circuit due to the discharge of a surge arrester, and energizing an equipment frame through contact with an ungrounded secondary grounding conductor. Initially, separation of the grounding conductors and electrodes for each class of apparatus was presented as the way to solve this problem, since severing any one grounding conductor would not jeopardize others. The Third Edition recognized that the same was true if an equipment frame grounding conductor went to the same grounding electrode as a secondary circuit-grounding conductor. There were still reservations at that time, though, against allowing this for arrester grounds.

The Fourth Edition carried the interconnection of frame and secondary circuit-grounding concept further by recognizing that if you had multiple grounds of low resistance on the secondary, such as to water piping, one grounding conductor could be allowed to ground both the frame and the secondary at the frame, since severing such a grounding conductor still left the secondary well grounded.

The Fifth Edition was the first to allow interconnection of arrester and secondary circuit grounding conductors, recognizing that, with a direct ground at the arrester and with the secondary grounded to the low resistance ground of either a metallic water system or a neutral with at least four grounds per mile, discharge of the arrester upon loss of its direct ground would not be unsafe. Furthermore, if there were four or more water pipe grounds per mile of secondary, even the direct-arrester ground could be omitted. This indicates that if an effective low-resistance ground can be ensured, interconnection is safe; and if multiple such grounds are obtained, no direct arrester ground is even needed. Likewise, it is seen that bonding of two classes of grounding electrodes at the ground line poses little risk to the basic concern, since severing either grounding conductor above this location separates the apparatus.

The Sixth Edition of the Code (6/8/60) is essentially the same as the Fifth Edition.

The 1973 Edition is a reprint of the Sixth Edition grounding rules.

The 1977 Edition includes a significant revision and rearrangement of the wording of Rule 097 from the 1973 Edition. As one of four options, the interconnection of a lightning arrester and grounded secondary conductor was permitted in the 1973 Edition: "In urban water-pipe areas, there are four metallic water-pipe grounds in each mile of secondary and not less than four such ground connections on any individual secondary, in which case the direct-earth grounding connection at the arrester may be omitted." This provision was not included in the

1977 Edition because less reliance can today be placed on water-pipe grounds and the predominant construction is multigrounded wye with an effectively grounded neutral system.

In the 1977 Edition, which was the first revision of grounding rules since the Fifth Edition, the permission to connect separate grounding conductors at the ground line where separate grounding electrodes are required (single-grounded systems) was deleted.

The 1981 Edition of Rule 097 is essentially the same as 1977, with minor editorial changes to parts of paragraph A.

The 1984 Edition includes a substantially modified paragraph D to address means of interconnection through spark gaps or equivalent devices, but made no changes to paragraphs A and B.

The 1987 Edition of Rule 097 is essentially identical to the 1984 Edition. In the 1993 Edition of Rule 097B2, the grounded phase conductor of a corner-grounded delta secondary is allowed to be connected to the grounding conductor for the primary circuit under specified conditions. The 1997 Edition further modified Rule 097B to allow connection to *either* a multigrounded distribution neutral *or* to a multigrounded overhead static wire associated with a unigrounded system (usually transmission or subtransmission). A multigrounded, effectively grounded overhead surge protection wire, commonly called a *shield wire* or *static wire*, above a transmission or distribution line is not a surge arrester. An overhead shield wire may or may not be connected to a distribution neutral, so long as it meets the required multigrounding and effective grounding requirements.

Figure H097-1 summarizes the grounding interconnections that are allowed by the NESC; Figure H097-2 shows interconnections prohibited by the rules.

Table H097-1 shows the evolution of Rule 097 and relationship of these requirements from edition to edition. The letters and numbers within the table indicate the paragraph in that edition that includes the requirement. The letters in parentheses refer to notes that follow the table.

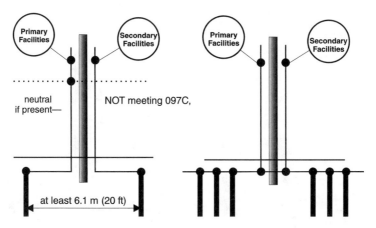

The General Rule: Required for delta, single-grounded, unigrounded, and where the neutral does NOT meet Rule 097C; Interconnection of the grounding conductors or electrodes is prohibited.

Alternate General Rule: Interconnection of grounding conductors is allowed ONLY through a sufficiently heavy, multigrounded ground bus or system ground cable, such as in an electric supply station.

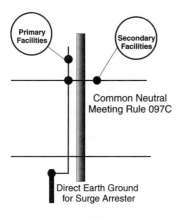

Exception Under Rule 097B: Interconnection of grounding conductors is allowed if the common neutral meets Rule 097C and a direct earth ground connection is maintained at the site for the surge arrester(s).

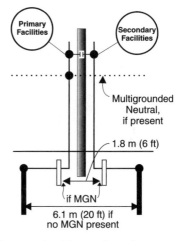

Interconnection of the grounding conductors is Allowed through a spark gap device meeting **Rule 097D**.

Figure H097-1
Connections between primary and secondary installations *allowed* by NESC grounding rules

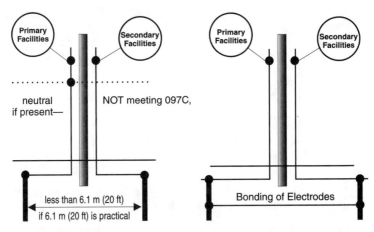

Rule 097D1 requires 6.1 m (20 ft) separation

Rules 097A and B prohibit such interconnection unless the secondary neutral or grounded phase conductor is connected to a primary neutral meeting Rule 097C.

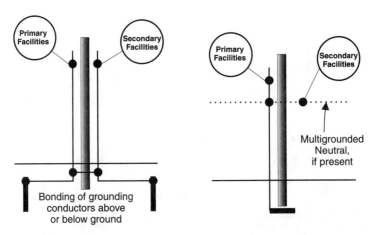

Rules 097A and B prohibit such interconnection unless the secondary neutral or grounded phase conductor is connected to a primary neutral meeting Rule 097C.

Use of butt wrap or butt plate instead of a made electrode (driven rod, etc.) at a transformer site is prohibited by **Rule 094B4**.

Figure H097-2
Connections between primary and secondary
installations *prohibited* by NESC grounding rules

Table H097-1. Grounding Separation Requirements

Rule Language:	Edition					
	3rd	4th	5th	6th	1977–1981	1984–Present
097A. Grounding Conductors						
Grounding conductors from equipment and circuits of each of the following classes, when required by the rules, shall be run separately to the grounding electrode	A	A	A	A	A	A
or to a sufficiently heavy ground bus or system ground cable which is well connected to ground at more than one place	A	A	A	A	A	A
except as provided in the rule and where primary and secondary circuits use a common neutral that is connected to at least four grounds per mile	—	—	A, C	A, C	A, B	A, B
(1) Lightning arresters	A1	A1	A1	A1 (b)	A1 (e)	A1 (e)
(2) Secondaries to low-voltage lightning circuits or equipment	A2	A2	A2(a)	A2(a)	A2 (f)	A2 (f)
(3) Secondaries of CTs and PTs and cases of CTs and PTs	A3	A3	A2(f) A3(e)	A2(f) A3(e)	A1(e) A2 (f)	A1 (e) A2 (f)
(4) Frames of dc railway equipment and of equipment operating above 750 V	A4	A4	A4	A3	A1	A1
(5) Frames of utility equipment or wire runways other than (4)	A5	A5	A5 (a)	A2 (a)	A2	A2
(6) Lightning rods	A6	A6	A6	A4	A3 (g)	A3 (g)

Table H097-1. Grounding Separation Requirements *(Continued)*

097B. Electrodes						
Arrester ground connections shall not be made to the same *made electrode* as circuits or equipment, but should be well spaced and, where practical, at least 6.1 m (20 ft) from other made electrodes.	B	B	—	—	—	—
Where individual made electrodes are used, both separate grounding electrodes and separate grounding conductors shall be used. The separate grounding electrodes may also be bonded together near the ground level.	—	—	B	B (c)	—	—
Grounding conductor(s) for equipment under 750 V may be interconnected with grounding conductor(s) of equipment over 750 V if: 091B1. There is a direct earth connection at each surge arrester location, and 097B2. The secondary neutral (or beginning in 1993, a grounded secondary phase conductor) is common with or connected to a primary neutral meeting Rule 097C, *or*					B	B(i)
(beginning 2002) 097B2. The secondary neutral or grounded secondary phase conductor is connected to a shield wire meeting rule 097C						B(j)

Table H097-1. Grounding Separation Requirements *(Continued)*

Rule Language:	Edition					
	3rd	4th	5th	6th	1977– 1981	1984– Present
097C. Interconnection of Primary Arrester and Secondary Grounded Conductor (h)						
(1) Solid interconnection is allowed if there is a direct ground at the arrester location *and* either:	—	—	C	C	B, C	B, C
the secondary is elsewhere grounded to a continuous underground metallic water system	—	—	C	C	—	—
(if there are enough connections to an extensive water system, the direct ground may be omitted), *or*	—	—	C	C	—	—
the secondary neutral has at least 4 ground connections per 1.6 km (mile) segment *in addition to* a ground connection at each individual service, *or*	—	—	C	C(d)	B, C	B, C
permission for another condition is obtained from the administrating authority	—	—	C	C	—	—
(2) If Rule 097C1 is not met, interconnection may be made through a spark gap meeting the requirements only if the secondary grounding conductor is connected to at least one other grounding electrode that is at least 6.1 m (20 ft) away from the arrester grounding electrode	—	—	C	C	D	D1(k)

Table H097-1. Grounding Separation Requirements *(Continued)*

Rule Language:	Edition					
	3rd	4th	5th	6th	1977–1981	1984–Present
097 D. Multiple-Grounded Systems						
On multiple-grounded systems, the primary and secondary neutrals should be interconnected in accordance with Rule 097B. However, where it is necessary to separate the neutrals, any interconnection of the neutrals is required through a spark gap or equivalent device. Unlike the spark gap installation for ungrounded or single-grounded systems, (1) the spark gap for multigrounded systems is limited to a 60 Hz breakdown voltage not exceeding 3 kV, (2) at least one other grounding connection for the *secondary* neutral shall be located not less than 1.8 m (6 ft) from the primary neutral and surge arrester grounding electrode, and (3) the primary or secondary neutral (or both) must be insulated for at least 600 V.	—	—	—	—	—	D2(k)
097E.						
Where separate electrodes are used for system isolation, separate grounding conductors shall be used	—	—	B	B	E(c)	E(c)
Where multiple electrodes are used to reduce grounding resistance, they may be bonded together and connected to a single grounding conductor [to make a single electrode—*Editor*]	—	—	—	—	E(c)	E(c)

Table H097-1. Grounding Separation Requirements *(Continued)*

Rule Language:	3rd	4th	5th	6th	1977– 1981	1984– Present
Edition						

Rule Language:	3rd	4th	5th	6th	1977–1981	1984–Present
097F.						
Made electrodes used for grounding surge arresters of *ungrounded* supply systems exceeding 15 kV phase to phase should be located at least 6.1 m (20 ft) from buried communication cables	—	—	—		F	F
097G.						
Where both electric supply systems and communication systems are grounded on a joint-use structure, either a single-grounding conductor shall be used for both systems or the electric supply and communication grounding conductors shall be bonded together, except where separation is required by Rule 097A.	—	—	—	—	—	(l)

NOTES:

(a) Where the secondary distribution system has multiple grounds, utilization equipment and wire raceways may use the same grounding conductors.

(b) Except as allowed by Rule 097C.

(c) When a group of made electrodes that have been bonded together are used instead of a single made electrode, in order to achieve the desired level of resistance to ground, the group is treated as a single electrode and a single grounding conductor may be run to the group.

(d) The four grounding connections must also be in addition to the direct-earth grounding connections of arresters.

(e) Over 750 V.

(f) Under 750 V.

(g) Unless attached to a grounded metal structure.

Table H097-1. Grounding Separation Requirements *(Continued)*

NOTES *continued*

(h) Applies only to secondary neutral in 1977–1990 Editions; grounded phase conductor allowed prior to the 1977 Edition and after the 1990 Edition.

(i) The 1993 Edition Rule 097B2 began allowing grounded secondary phase conductors to be connected to primary neutral meeting Rule 097C.

(j) The 2002 Edition of Rule 097B2 allowed connection to a multigrounded overhead static wire associated with a unigrounded circuit (usually above a transmission or subtransmission voltage circuit) as an alternative to a multigrounded distribution neutral.

(k) The 1997 Edition added the specification about the needing to insulate the primary neutral grounding conductor under certain circumstances. The 2002 Edition recognized that both the primary and secondary grounding conductors may need to be insulated on single-grounded and ungrounded systems.

(l) The 2002 Edition of Rule 097G was added as a *should* rule (i.e., mandatory if practical); the 2007 Edition made it a mandatory *shall* rule (i.e., one of the two options must be performed).

Rule 097 carefully delineates what ground may be connected to what other ground when and how. "Single-grounded systems" include delta systems (which are not center-grounded) and unigrounded systems (which are center-grounded at the source through a current-limiting, high-impedance connection and do not carry a neutral meeting Rule 097C along with the circuit), as well as any system having a neutral not meeting Rule 097C.

It should be noted that certain systems are special cases or are neither single-grounded (Rule 096A2) nor multigrounded (Rule 096A3). An example is a case where the secondary neutral and primary neutral of a 7.2 kV distribution circuit are in the vicinity of an HVDC ground and are not interconnected under Rule 097B. To interconnect them could cause current from the HVDC earth return to flow from the customer(s)'s grounding system(s) through the transformer onto the distribution system, saturate the transformer, and cause voltage distortion. Another example is the case where a common secondary and primary neutral is not used. Note that a grounded phase conductor, such as the ground phase of a corner-grounded delta circuit, is not considered a neutral. In these cases, where the primary and secondary neutrals are not interconnected, Rule 097D requires at least one additional secondary ground and requires it to be separated by at least 6.1 m (20 ft) from the primary arrester grounding electrode. The rule allows interconnection of the primary and secondary grounding conductors through a spark gap meeting specified requirements. Interconnection of a grounded phase conductor with a grounded neutral was allowed in the 1993 Edition.

A note was added in the 1987 Edition to Rule 097D2 to recognize that the only way to obtain effective isolation between primary and secondary neutrals often requires the cooperation of *all* utilities involved (including both supply and communication) *as well as* the customer(s), because communication utilities generally utilize a common grounding system along parallel routes. In other words, they often (and are usually required to) bond their messengers to the electric supply pole grounds. This electrically bonds the communication grounding system to the electric primary grounding system. Unless the communication utility has been notified of the isolation of primary and secondary neutrals, they may inadvertently reconnect hose systems at the served structure.

Experience shows that chronic step and touch potential problems in dairy barns, etc., are often the result of violations of the National Electrical Code (NEC) by connections at the customer's facilities and do not involve the incoming utility facilities. Other cases involve a combination of these problems. It also is obvious that in many cases it is not practical to obtain neutral-to-earth voltages as low as might be desired by some operations. In such cases, the most practical method of addressing the exacting needs of the facility may be for the facility to have local gradient control installed at critical locations. This may apply to computer installations, critical areas of hospitals, and critical areas of dairy or swine-farrowing facilities.

It should be stressed that the term *neutral* does not include all grounded circuit conductors; the grounded circuit conductor of a corner-grounded delta circuit is *not* a neutral conductor but a grounded phase conductor. The significance of this is that grounded secondary phase conductors have not been allowed to be interconnected to primary neutrals by the language in the 1977 through 1990 Editions. Before the 1977 Edition, the term *grounded secondary conductor* was used; in 1977, this changed to *secondary neutral*. In the 1993 Edition, Rule 092B2 allows both secondary neutrals and grounded secondary phase conductors to be interconnected with primary meeting Rule 097C *if* a direct-earth grounding connection is present at each surge arrester location.

If the primary and secondary neutrals are separated under Rule 097D, the arrester ground electrode is required to be located at the transformer location under Rule 096C and Rule 092D; it cannot be located remotely and connected by an overhead or underground grounding conductor. This requirement limits the adverse effect on the secondary system of a lightning strike, failure of the high-voltage winding insulation, or a dig-in or other disconnection. The 2002 Edition required the primary grounding conductor or secondary grounding conductor on ungrounded or single-grounded systems to be insulated for 600 V. In practice, both may need to be so insulated because one may be connected to an exposed device, such as a meter base or operating handle on the structure. Exposure of passersby to uninsulated portions of both within reaching distance needs to be avoided, and may require relocation of some devices. The 2007 Edition added the requirement to guard the secondary grounding conductor to meet rule 093D2.

Rule 097G was added in the 2002 Edition to require bonding of the communication system ground to the electric supply primary system ground. When first added, this was a *should* rule, but the 2007 Edition made it a mandatory *shall* rule. Communication utilities were found to be sometimes electing to ground their system separately from the electric supply system in lieu of bonding to the existing grounding electrode conductors, where they exist. In such situations, unless the communications grounding system is either appropriately insulated, or bonded to the electric supply system grounding conductor, voltage potential differences could exist between the two systems which could be a hazard to workers or the public.

098. Not Used

099. Additional Requirements for Communication Apparatus

This rule details the connections that are required for grounding protectors and exposed noncurrent-carrying metal parts in central offices or outside installations. Interconnection of the grounding conductor of this apparatus to grounded conductors or enclosures of supply circuits is required where the supply apparatus is grounded by an acceptable electrode at the site; otherwise, the communication apparatus must, itself, be connected to an acceptable electrode meeting Rule 094. An EXCEPTION allows communication apparatus to be connected to a thinner and shorter driven ground rod than is allowed for a supply circuit.

In concert with the NEC requirements (Article 800—Telephone, Article 820—CATV, and Article 810—Radio and TV Antennas, and NESC Rule 099C, bonding is required between the communication system and radio and TV antenna grounding electrode(s) and the supply grounding electrode(s) with a wire of not less than No. 6 AWG copper where separate electrodes are used at a building or structure (see Figure H099C-1).

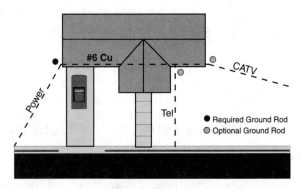

Figure H099C-1
Required bonding of ground rods

A reminder to ensure that the metallic path is continuous when using water piping as a bonding means, along with an NEC reference, was added in 2002. Figure H099C-2 shows a house with two code violations. The CATV does not have the #6 cu bond to the power electrode required by Rule 099C. In addition, the electrical service panel connection to the water pipe is more than 1.5 m (5 ft) from the entrance of the water pipe to the building, thus violating NFPA 70, the NEC. The NEC requires bonds to the water pipe to be within the first 1.5 m (5 ft) to limit the opportunity and need for a plumber to cut the pipe between bonding attachments, thus exposing himself or herself to the voltage involved with any current flowing through that route back to the source.

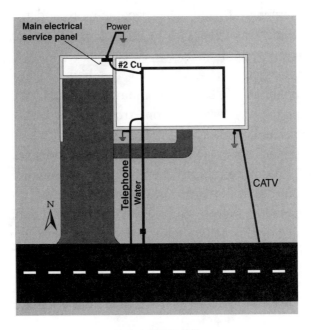

Figure H099C-2
Incorrect Bonding

The tendency today is to connect equipment individually to a common grounding conductor of substantial cross-section and multiple grounds, rather than to run separate ground wires to a single electrode. Cooperation between the customer and all serving utilities may be required to ensure adequate grounding. Both the NESC and NEC require that grounds at the customer premises be tied together, if a single ground electrode is not used by all utilities serving the building. This limits the opportunity for hazardous touch potential and for equipment fire due to lightning surges. If the electric supply primary neutral and secondary neutral are not connected, but communication grounding is connected to the primary neutral, an effective break in the communication grounding system will be required before the customer premises, so that common grounding at the customer premises can be achieved. Rule 99C does not apply on join-use pole lines; see Rule 092-097, Rule 235, and Rule 238 for pole line grounding requirements and methods.

The requirements of Rule 099C become more understandable when considering the consequences of the individual utilities *not* utilizing the same grounding system at the served structure. Without this required bonding, devices such as CATV converter boxes, charging bases for cordless telephones, VCRs and computer modems may contain two separate grounding systems (i.e., one from the electric supply and one from the communication system). This, in effect, is a spark gap (i.e., lightning arrestor) subject to damage for lightning discharge.

The 2007 Edition reinforced the previous language by requiring all separate electrodes at a served installation to be bonded together. This complements the change to Rule 097G to require bonding of supply and communication grounded facilities on joint-use structures.

Since it is the responsibility of all parties to meet the code, it falls upon each utility adding service to install appropriate bonding to existing utility facilities to comply with this requirement.

Part 1. Rules for the Installation and Maintenance of Electric Supply Stations and Equipment

Prior to the 1981 Edition, the term *Electrical Supply Station* was used in the title and the text. The term was changed to *Electric Supply Station* both in Part 1 and in the Definitions (Section 2) to be consistent with IEEE Std 100 *The Authoritative Dictionary of IEEE Standards Terms, Seventh Edition* (ANSI). Part 1 rules are designated by the numeral 1 as the first of the three digits of the rule number. Rules beginning with the numeral 1 do not apply to overhead or underground installations; rules covering overhead or underground installations begin with the numeral 2 or 3, respectively.

Section 10. Purpose and Scope of Rules

(In the Fifth and prior editions, Section 10 was titled Protective Arrangements of Stations and Substations. *In the major revisions of the 1971 Edition, old Section 10 was restructured and divided; much of it, including the title, was moved to Section 11. These movements and changes are detailed in the following sections.)*

100. Purpose

(Rule 100—Scope of the Rules of the Fifth and prior editions was moved to Rule 101 in the 1971 Edition.)

The purpose of Part 1 is to provide *practical* safeguarding of persons performing installation, operation, or maintenance duties in electric supply stations; see the discussion of Rule 010.

101. Scope

*(Rule 100—*Scope of the Rules *of the Fifth and prior editions changed little over the years until it was moved to this position in the 1971 Edition. Significant changes were made in the 1971 and 1981 Editions. Prior Rule 101—* Applications of the Rules and Exemptions *was moved to Rule 102 in the 1971 Edition.)*

In all editions, Part 1 only applies where the covered facilities are accessible to qualified persons. Where the requirements of Rule 110A—*Enclosure of Equipment* are not met, the area is considered to be accessible to unqualified personnel and Part 2 applies. Where the requirements of Rule 110A are met, the area is considered to be accessible only to qualified persons and Part 1 applies.

Part 1 covers electric supply equipment, conductors, and structural arrangements in indoor and outdoor generating stations, switching stations and substations, whether owned and operated by an electric utility or an industrial or commercial complex. Part 1 covers public and private utility systems including utility-interactive generation systems owned and operated by an independent power producer. In the Fifth and prior editions, Part 1 applied to similar equipment, including generators, motors, storage batteries, transformers, lightning arresters, etc., when located in factories, mercantile establishments, vehicles or elsewhere, provided the equipment is in separate rooms or enclosures. Some exemptions were added in the Fifth Edition.

Exemptions were added in the 1971 Edition for (1) installations in mines, ships, railway rolling equipment, aircraft, automotive equipment and (2) conductors and equipment used primarily for the utilization of electric power, except those in electric supply stations. Specifically excluded were industrial and commercial establishments not under the control of, and accessible only to, qualified persons. The definition of *qualified* is included in Section 2 of the Code. Examples given for such exclusions were apartment houses and shopping centers. However, the power delivery systems involved with some of the commercial "megaplexes" of today differ little from a public utility system and are under qualified control, thus allowing the NESC to be applicable. Note also that the NEC contains footnotes for installations above 600 V referencing the user to the NESC requirements.

Electric supply stations owned by, and installed in, an industrial establishment where the facilities are under the control of, and accessible only to, properly qualified persons continue to be covered by the 1971 and later editions. Examples given in the 1971 Edition are paper and steel industries. Clear keys to determining whether the NEC or the NESC is applicable to an industrial installation are (1) does the "electric supply station" have an electricity generation or a delivery function (as opposed to solely a utilization wiring function); and (2) is the facility under the exclusive control of, and accessible only to, qualified persons? If the answer to these questions is "yes," the NESC applies.

The definition of *generating station* was added in the 1993 Edition to make more explicit the type of facility included in the scope of Part 1. In deliberating the 1995 proposal to add the definition, it was explicitly stated that telecommunications central stations were covered by codes other than the NESC. In 1997, the definition of *supply station* was added; *supply station* now includes the subcategories of *generating stations* and *substations*. A generation station includes all facilities, including auxiliary equipment, that are required for the conversion of some form of energy to electric energy. Substations include areas where electricity is switched or transformed, but does not include generation.

It is important to observe the distinction between the requirements of the code for station equipment and for utilization equipment, even when the former is of the same nature as the latter. A somewhat less general use of guards and less complete isolation is allowable with station equipment that is accessible only to qualified persons than is allowable with electrical utilization equipment that is accessible to unqualified persons, as is often the case in workshops, mercantile establishments, and other similar places that are covered by the NEC.

Part 1 is intended to apply to utilization of conductors and equipment by a utility in the exercise of its function as a utility (but not for office buildings, etc., to which the NEC applies).

The addition of grounding requirements for systems over 750 V to Part 1 in the 1993 Edition signals the inclusion of HVDC terminals in the scope of Part 1. DC station clearances were added in Table 124-1 in the 2002 Edition equal to the clearances for ac circuits having the same crest voltage to ground.

102. Application of Rules

*(Rule 101 of the Fifth and prior editions was moved here in the 1971 Edition. Rule 102—*General Requirements *became Rule 110 in the 1971 Edition. Rule 102 of the 1971 Edition was deleted in the 1981 Edition when such rules were consolidated in a new Section 1. Rule 101 of the Fifth and prior editions, which had been deleted in the 1971 revision, is included within Rule 014 of the 1981 and later editions. The present Rule 102 was added in 2007.)*

Rule 013 contains the requirements for application of the current edition to new installations and extensions and application of the current or previous editions to existing installations. Rule 013B1 allows installations of any vintage to meet the current edition without being required to meet previously applicable editions. Rule 013B2 is the so-called grandfather clause that allows previous editions to remain in compliance with the previously applicable edition when the Code changes. Rule 013B3 requires meeting the current edition or the previously applicable edition (which might be either the original one or a subsequent edition with which the installation had previously been brought into compliance) when adding, altering, or replacing components on an existing installation.

102A. Application

*(This rule was moved here in the 1971 Edition from Rule 101A. The rule is included within Rule 013 in the 1981 and later editions.) Rule 102A—*Enclosure of Rooms and Spaces *of the Fifth and prior editions was moved to Rule 110A in the 1971 Edition.)*

102B. Intent of Rules

(This rule was moved here in the 1971 Edition from Rule 101B. The rule is included within Rules 013 and 015 in the 1981 and later editions.) Rule 102B—Rooms and Spaces *was moved to Rule 110B in the 1971 Edition.)*

102C. Temporary Installations

(This rule was moved here in the 1971 Edition from Rule 101C. The rule is included within Rule 014 in the 1981 and later editions. Rule 102C—Rotating Machinery, *which relates to dynamic forces, was moved to Rule 110C in the 1971 Edition.)*

103. Referenced Sections

(Rule 103 was added in the 2007 Edition to cross-reference and formally recognize the application of Sections 1, 2, 3, and 9 to Electric Supply Stations. Section 9 had applied since the inception of the Code; Sections 1, 2, and 3 had applied since their inception in the 1981 Edition.)

Section 11. Protective Arrangements in Electric Supply Stations

(This section comprises rules that were moved here from Section 10 in the 1971 Edition; Section 11 of the Fifth and prior editions was renumbered to Section 12 in the 1971 Edition.)

110. General Requirements

(Rule 102 of the Fifth and prior editions was moved here in the 1971 Edition; previous Rule 110 became Rule 120.)

110A. Enclosure of Equipment

(Rule 102A—Enclosure of Rooms and Spaces was renamed when it was moved here in the 1971 Edition.)

Rule 110A is intended to limit the adverse effect of the unauthorized persons on covered installations, or vice versa. The use of fences, screens, partitions, or walls has proved generally to be an effective means of meeting the intention of this rule, especially when appropriate safety signs are used at entrances to enclosed areas. Such safety signs should be conspicuously located on or near the door, gate, removable barriers, or other entrance area. The sign reference was changed to the generic term *safety sign* from the previous generic term *warning sign* in the 1997 Edition to recognize that ANSI Z535 (which is an update and expansion of ANSI Z35 and ANSI Z53) has three levels of signs involving notification of a personal injury hazard, as well as others denoting equipment damage, safety instructions, safety equipment location, etc.

Under ANSI Z535, the signal word DANGER implies a hazard that has a high probability of causing death or serious permanent injury, if not avoided. The signal word WARNING is also associated with a hazard that could cause death or serious permanent injury, but has a low probability doing so. For this reason, a DANGER sign is appropriate inside an electric supply station or pad-mounted equipment, where there is direct exposure to live parts. A **WARNING** sign is appropriate on a station fence and on the exterior of the cabinets containing energized parts. The signal word **CAUTION** is reserved for association with a hazard that could cause minor or moderate injury and is not appropriate for use on station fences or cabinets containing electrical equipment.

Both ANSI Z535.2 and ANSI Z535.5 require the use of the safety alert symbol with the all three of the previous signal words to alert viewers to potential personal safety hazards for environmental and facility safety signs and for temporary accident prevention tags and barricade tapes, respectively. The safety alert symbol is an equilateral triangle with an exclamation point inside it; this is the global standard to indicate a personal safety hazard.

Prior to the 2006 Editions of the Z535 standards, the signal word **CAUTION** was used without the safety alert symbol to alert viewers to potential equipment damage hazards. The 2006 Editions of Z535 started the process of further differentiating between safety signs alerting viewers to potential personal safety hazards versus potential equipment damage hazards. Previous to 2006, a **CAUTION** sign without the safety alert symbol could be used for equipment damage hazards; the 2006 Edition of ANSI Z535 standards changed to a preference for using *NOTICE* signs for that purpose to better distinguish between personal safety hazards and equipment damage hazards. The **CAUTION** signal word may still be used as an alternate to *NOTICE* without the safety alert symbol to indicate potential equipment damage, but this is only for a one-edition phase-out period; this alternate is schedule to be removed in the 2011 Edition of the Z535 standards (see Appendix B to this handbook).

CAUTION signs may be needed for tripping hazards and similar hazards, if they exist in an electric supply station. *NOTICE* signs may also be appropriate in some stations to alert viewers to potential equipment damage hazards. *NOTICE* signs do not use the safety alert symbol with the signal word.

Beginning in 1997, a safety sign is required on each outward side of supply station fences and ANSI Z535 safety sign standards are referenced. (Special precautions may be required to eliminate confusion when locating warning information around removable barriers). This requirement was discussed in IR 526 issued 21 February 2002. The first sentence of Rule 110A1 requires a safety sign at each entrance to an electric supply station, regardless of whether the station is fenced in whole or part. The second sentence requires a safety sign on each side of a fenced station yard, regardless of whether there is an entrance on that side. A safety sign at an entrance can serve the requirement for both the side and the entrance. Notice that, since viewing distance is limited by letter height and angle of view (see NESC Handbook Appendix B and ANSI Z535.2), stations with long sides may need several signs on the long sides; typically this is both more cost effective and more viewer effective.

Figures H110A1-1 through H110A1-4 illustrate appropriate sign placement on fences enclosing electric supply stations of various configurations. For this purpose, the relatively short panels of recessed entrances are not considered as separate sides

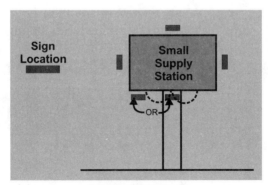

Figure H110A1-1
Signs on small supply station

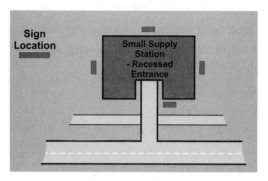

Figure H110A1-2
Signs on station with recessed entrance

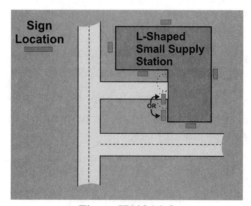

Figure H110A1-3
Signs on L-shaped station

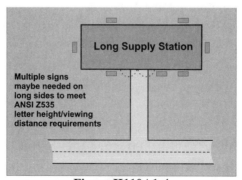

Figure H110A1-4
Signs on long station

Appropriate warning should still exist in an appropriate place when the barrier is removed for access to equipment or parts.

Openings that are not intended as entrances, such as windows or ventilation grills, are not subject to the safety sign requirements. Unauthorized entry through such openings is usually considered "breaking and entering" and subject to the penalties thereof. If an attendant is used to limit entrance to these areas, the attendant should use effective procedures to limit the access of or otherwise warn unqualified persons; otherwise, the entrances are required to be locked.

The 1971 Edition was the first to include specific requirements for fence height and construction. The wording originally proposed for this rule was as follows: "Metal fences and gates, when used to enclose electrical supply stations having energized electrical conductors or equipment that can be reached by trespassers, shall have a minimum of 2.13 m (7 ft) of fabric in height plus a 300 mm (1 ft) extension carrying three or more strands of barbed wire." These words were modified in the final rule to recognize the prohibition in some communities against barbed-wire fences. In the final rule, fences were required to be a minimum of 2.13 m (7 ft) high and to be effectively grounded.

Conductive fences are required to be grounded to reduce the hazard of a failed energized conductor in contact with the fence, the hazard of electromagnetically induced voltage and current on the fence, and the fence touch hazards during fault conditions. IEEE Std 80 *IEEE Guide for Safety in AC Substation Grounding* (ANSI), addresses effective grounding of conductive fences.

A NOTE recognized that some localities restrict the use of barbed wire on fences; it recommended a 300 mm (1 ft) *extension* of barbed wire above the fence fabric if such was allowed.

The intention of the rule was to have generally at least 2.13 m (7 ft) of fabric height *plus* a 300 mm (1 ft) *extension* of barbed wire, for a total of 2.45 m (8 ft). A number of utilities interpreted this rule to allow 1.8 m (6 ft) of fabric with a 300 mm (1 ft) barbed-wire extension; a large number of installations were constructed in this manner. Subsequent data on unauthorized entry indicates that the presence of the fence, with its attendant warning signs at or near gates, is sufficient to deter the general public from unauthorized entrance; whether the fence is 2.13 m (7 ft) high or 2.45 m (8 ft) high is not statistically significant.

The subject of required fence height has been a matter of significant controversy since publication of the 1971 Edition. The initial IEEE Substation Committee vote on the first draft of their proposed 1981 changes barely supported a recommendation of 2.13 m (7 ft) of fence fabric. When the Substation Subcommittee and the Station Design Sub-committee of the Power Generation Committee combined their efforts, the majority vote was to recommend 2.13 m (7 ft) of fabric with an additional recommendation of a 300 mm (1 ft) barbed-wire extension. The comments received from the IEEE Subcommittees along with their proposal to the NESC Subcommittee indicated that the recommendation was far from unanimous.

The NESC Electric Supply Stations Subcommittee, after review of the comments and other data, recommended an overall fence height of 2.13 m (7 ft) to be made up of all fence fabric or 1.8 m (6 ft) of fabric plus a 300 mm (1 ft) extension with at least three strands of barbed wire. That was the recommendation that went to ballot. One of the comments received during the balloting procedure indicated that the NEC required a minimum of 2.13 m (7 ft) of fence fabric. The allowance of only 1.8 m (6 ft) of fabric was rejected in the balloting process because of this inconsistency and the subcommittee was requested to work with the NEC Committees to resolve the problem.

During the period between the 1981 and the 1984 Code revisions, the NESC Subcommittee worked with NEC Committees to develop a joint proposal for submission to both groups. The NEC Task Force declined to consider the recommended change, but the change was adopted by the NESC Committee on the basis of satisfactory operation of existing units in place. The NEC was changed in 1993 to match the NESC requirement.

It was recognized in the 1984 Edition that the existing data on the performance of such fences as effective deterrents to unauthorized entry; 2.13 m (7 ft) *overall* was stated as the required fence height. If

1.8 m (6 ft) of fabric is used with an angled barbed-wire extension, the angled extension must be longer than 300 mm (1 ft) so that the total height of the fence is at least 2.13 m (7 ft) (see Figure H110A1-5).

The 2002 Edition revised Rule 110A1b to remove the specification of a 300 mm (1 ft) extension. This had caused confusion since it did not specifically state that the 300 mm (1.0 ft) was a vertical dimension. The revised rule states the real requirements:

(1) the fence must be not less than 2.13 m (7 ft) high (which effectively means the original fence must be greater than 2.13 m (7 ft) to allow for gravel buildup, etc., during the station life),

(2) the bottom 1.80 (6 ft) must be fence fabric (or a material with equivalent difficulty in climbing), and

(3) if a barbed wire extension is used, it must be at least three strands and may (a) lean in, (b) lean out, or (c) be straight up.

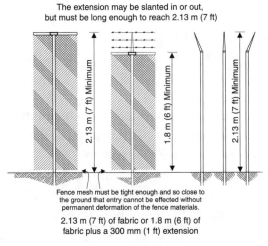

Figure H110A1-5
Fence height requirements

The available data indicates that trespassing in electric supply stations is relatively rare. Accidental entrance into electric supply stations is essentially nonexistent. Furthermore, most fence breaches appear to have occurred from sliding *under* a fence, rather than climbing over it.

Where a chain-link fence is used to meet Rule 110A, the fence fabric is not required to meet the ground, but it is required to be close enough and tight enough to prevent a person from going underneath the fence without permanently deforming the fence. It is not uncommon for utilities to install a concrete curb to control erosion and limit trespassers from crawling beneath the fence. Equivalent barriers were allowed in the 1971 Edition to climbing or other unauthorized entry; in the 1984 Edition, the rule was restructured to detail acceptable fence characteristics.

The main portion of the fence or wall forming the enclosure of an electric supply station should have equivalent difficulty in climbing to that presented by chain link fence mesh. Decorative walls with recesses, holes, or projections that allow easier climbing than chain link fence mesh do not meet the intention of Rule 110A1. A decorative wall like that in Figure H110A1-6 has both handholds and footholds that allow easy climbing. This particular hospital substation also has overhead transformers sitting on a concrete pad; such installations do not meet Rule 124; see further discussion at Rule 124A and 124C3.

Figure H110A1-6
Supply station enclosure not meeting requirements of either
Rule 110A1 requirements for limitation of climbability or
Rule 124 for clearance to energized parts

Gate panels must be kept adjusted so limit the opportunity for unauthorized personnel to slip between the panels. The double-panel gate in Figure H110A1-7 has been thrown open with too much force so many times that the adjustment plates a the fence posts have moved and produced a center gap so large that young children can walk through sideways.

Figure H110A1-7
Supply station fence gate with too large a gap between panels

These rules provide protection for both the general public and authorized workers. The 1984 Edition correctly states that the intention of these rules is "to *minimize the possibility* of entrance of unauthorized persons..." Although in previous editions this was phrased, "to *prevent* entrance of unauthorized persons...," the clarified wording of the 1984 Edition has always been the intention of the Code. The Code has always recognized that, unfortunately, it is not possible to prevent the determined intruder from gaining access to restricted-access areas. To violate the obvious entrance restrictions of the NESC requires conscious action; those who do so are responsible for the consequences of their actions.

CAUTION: If an electric supply station is enclosed on three sides by a fence and on the other by a building that is not restricted to authorized personnel only, the installations within the station do not meet the requirements of Part 1 *unless* exit from the building into the enclosed electric supply station is prohibited. In that case, facilities within the electric supply station are required to meet the clearances of Part 1 *except* that the clearances of Rule 234C are required to portions of the building to which access is not restricted, for example, the roof or a window.

Electric generating stations are often located in thinly populated areas and have perimeter fences that enclose coal storage and handling areas, ash and sludge disposal areas, intake and discharge water ponds, etc. The perimeter fence is intended to prevent intrusion by animals and unintentional trespass. Stranded-metal barbed-wire fencing is a practical alternative to these perimeter fences. This type of fence is not intended to be included in the metal fence requirements of the Code.

The fence requirements of Rule 110A apply to areas containing energized electrical parts, but are not required for contiguous areas. Thus, the utility is provided a choice. Either (1) the entire station can be enclosed in a manner consistent with Rule 110A or (2) the electrical areas can be enclosed in a manner consistent with Rule 110A, and ancillary areas (such as coal piles, disposal areas, storage areas, etc.) can be enclosed in a different manner not meeting Rule 110A.

CAUTION: If a fence meeting Rule 110A is interconnected with a second fence not meeting Rule 110A, (1) the second fence should also be grounded in accordance with Section 9 or be electrically isolated in a manner that would not be hazardous to persons near to or touching the fence(s) when station-protective devices operate, and (2) the second fence should not serve as a climbing step.

A new NOTE was added in the 1990 Edition referencing IEEE Std 1119™-1988 *IEEE Guide for Fence Safety Clearances in Electric-Supply Stations* (ANSI) for guidance in determining appropriate clearances for energized parts from the station fence. The 1997 Edition removed the NOTE and specified clearances for fences from live parts in Rule 110A2—*Safety Clearance Zone*. NESC Figure 110-1 and Table 110-1 define the safety clearance zone. This safety clearance zone will limit the ability of someone carrying a long rod or pipe over their shoulder outside the station to contact energized parts inside the station by either inadvertently passing the pipe or rod above the fence or sticking it through the fence (see Figure H110A2-1). The clearances start at 3.0 m (10 ft) and increase for voltage; they are applied at a 150 m (5 ft) shoulder-high pivot point. Notice that the oil circuit breaker (OCB) of Figure H110A2-1 must be offset to meet the clearance from the fence.

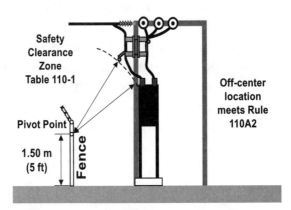

Figure H110A2-1
Required clearance of live parts from fence

Were the OCB to be centered under the disconnect switches, as in Figure H110A2-2, Rule 110A2 would be violated by the closest bushing.

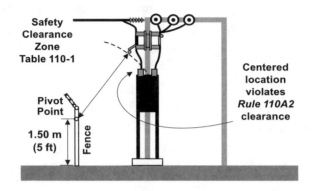

Figure H110A2-2
Fence clearance violation

An alternate is to place a solid wall section beside the OCB extending to the level of the OCB bushings, as in Figure H110A2-3.

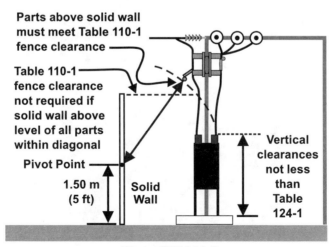

Figure H110A2-3
Using solid wall section to get live parts closer to fence

The entire wall does not have to be solid—only those portions of the wall or fence that would violate Rule 110A2 (see Figure H110A2-4). Note that EXCEPTION 1 to Rule 110A2 allows the exposed live parts to be located within the safety zone only when (1) a solid section of wall, partition, or fence section interrupts the ability of a person outside of the enclosure to insert sticks or other objects through it and (2) the solid section extends to a height not less than that of the energized parts.

Rule 124A and Table 124-1 specify the required height of the energized parts based upon voltage. The literal wording of EXCEPTION 1 to Rule 110A2 requires the fence to be not less than the actual height of the energized parts, which will often exceed the requirements of Rule 124A by several inches. This EXCEPTION does not allow the energized parts adjacent to a perimeter fence to be placed at heights less than required by Rule 124A, nor does it allow reduction of working clearances required by Rule 125. When EXCEPTION 1 to Rule 110A2 is used, (1) appropriate ground clearances must be maintained and (2) appropriate working clearances must be maintained between the solid panel and the energized parts.

Figure H110A2-4
Solid section prevents contact with close
facilities–other areas of fence meet clearances of Rule 110A2

These fence clearance requirements apply to fences around electrical areas that are required to meet Rule 110A and where the public has access to the outside of the fence. The fence clearance requirements do not apply to fences contained within an electric supply station perimeter fence meeting Rule 110A, such as an interior fence guard used to meet Rule 124C3. Note also that a perimeter fence cannot be used to meet Rule 124C3.

110B. Rooms and Spaces

(This rule was 102B in the Fifth and prior editions.)

110B1. Construction

This rule recognizes that supply stations will generally contain combustible materials and liquids, such as insulation, oil, etc. However, to the extent that it is practical, noncombustible materials should be used. For example, wooden floors are prohibited by this rule. IEEE Std 979™ *IEEE Guide for Substation Fire Protection* (ANSI) and NFPA 850 *Fire Protection for Fossil Fueled Steam and Combustion Turbine Electric Generating Plants* are useful references when considering fire safety in electric supply stations. The 2002 Edition clarified that wood poles may be used for supports in stations.

110B2. Use

*(This rule includes the provisions of Rule 102B2—*Storage and Manufacturing Processes *of the Fifth and prior editions.)*

Rule 110B2 prohibits manufacturing or storage inside an electrical supply station. The 2002 Edition specified the conditions under which exception to the storage prohibition may be made, but there are no exceptions to the manufacturing provision.

(1) Material or equipment *essential for maintenance* of the installed equipment may be stored in the station if the material is guarded or separated from line parts as required by Rule 124.

(2) Materials that are related to station, transmission, or distribution construction or maintenance may be stored in a station if located in an area separated from the energized parts by a fence meeting Rule 110A, which includes the distance requirements of Rule 110A2. This effectively cuts the storage area off from the station area and allows access to stored material by nonqualified personnel. This provision is used where the station fence meeting Rule 110A encloses an area much larger than that required for the enclosure of energized parts.

In some cases, some or all of the internal fence may be removable to allow access for replacement of station power transformers, regulators, or other heavy equipment. In that case, materials should be stored in such a manner as to require limited movement to allow such access.

(3) *Temporary* storage in an electrical supply station of material related to work in progress in the station or a nearby transmission or distribution line is allowed only if all five restrictions are met.

Storage of material not only presents a fire hazard, which would be increased greatly by the presence of electrical equipment, but it also creates a life hazard to operators. Exits might be blocked or working space cramped to such an extent that, in case of trouble, an operator could be impeded.

No extended manufacturing process can be performed in the immediate vicinity of supply station electrical equipment without endangering persons engaged in manufacturing, as well as those attending electrical equipment. If the attention of electrical operators is distracted by the presence of other processes, this in itself may present a serious danger. Continuity of service and electrical equipment life may suffer, and the fire hazard is increased by electrical equipment in combustible surroundings.

Arcing at contacts or connections, especially at switches, fuses, and brushes, makes the existence of flammable gas and finely divided combustible material highly dangerous, even where the operator can endure such atmospheric conditions. Substations and generating station areas should be carefully segregated from the dust-producing areas of facilities such as grain elevators, flour and cotton mills, coal-storage plants that are not part of generating stations, and nearly all industrial places where lint, dust, or inflammable gas is customary. Acid fumes should also be avoided because of the deteriorating effect upon insulation and machinery.

Conformance with this rule also reduces the need for personnel to enter areas classified as electric supply stations for purposes not related to the stations. By storing unrelated materials, such as construction components and equipment, in other areas, personnel associated with such activities are not required to be around the station equipment and do not run the risk of contacting live parts or interfering with the operation of station equipment. Where a utility property is intended to house both a station and a construction staging area, the station is intended to be appropriately segregated from the other uses of the property.

This rule was expanded in the 1971 Edition to include the requirements of prior Rule 102B3—*Hazardous Locations*. Battery rooms, and auxiliary equipment in hazardous locations, are covered in Section 14 and Rule 127, respectively.

110B3. Ventilation

(This rule was 102B4 in the Fifth and prior editions.)

Adequate ventilation decreases the accumulation of moisture and dampness on surfaces and prevents the concentration in dangerous quantities of flammable dust and gases. Sanitary conditions are also improved. The primary value of ventilation is, however, a check on the excessive accumulation of heat about electrical apparatus.

In many cases, it is appropriate for the presence of abnormal or dangerous accumulations of explosive gases, acid fumes, etc., to be promptly and automatically indicated by a distinct audible signal. Although early official Discussions of the Code included such recommendations, they were never made a requirement of the Code.

110B4. Moisture and Weather

(This rule was 102B5 in the Fifth and prior editions; it is essentially unchanged from the Third Edition.)

Supply station equipment that is exposed to the weather, or located in wet or normally damp areas, should be designed for the prevailing conditions.

Energized parts should be placed in weatherproof enclosures unless guarded against contact by isolation or elevation in accordance with Rule 124.

When equipment that is not specifically designed for use in damp locations is exposed to moisture, the resulting insulation failure or mechanical failure may endanger personnel. Personnel near energized parts in damp locations are exposed to electrical current leakage over damp surfaces and through deteriorated or moisture-laden insulation. If an accidental contact with an energized part occurs in a damp location, the reduced resistance of the contact tends to increase the seriousness of the result.

110C. Electric Equipment

(This was Rule 102C in the Fifth and prior editions. In the 1981 Edition, the name was changed from Rotating Machinery *and the rule was expanded.)*

Rotating machinery should always be considered as a "live load" in determining the strength requirements of supporting structures or foundations. Adequate support reduces vibrations and the consequent wear on the bearings and insulation. The 1981 Edition expanded the rule to require all electric equipment to be bolted down or otherwise secured to limit movement.

It was recognized in the 1984 Edition that the weight of some equipment, such as a large transformer, may adequately secure it in place in areas of limited seismic activity. The rule was rewritten for clarity in the 1987 Edition to recognize that it is the expected conditions of service that determine required support and security methods.

110D. Supporting Structures and Supported Facilities

(This rule moved to Rule 162A in the 1997 Edition.)

111. Illumination

(Rule 103 of the Fifth and prior editions was moved here in the 1971 Edition; prior Rule 111 became Rule 121.)

111A. Under Normal Conditions

(This was Rule 103A of the Fifth and prior editions; former Rule 111A was moved to 121A in the 1971 Edition.)

It is good practice to provide adequate artificial illumination for all rooms and spaces containing electrical equipment. Both uniform illumination (over working spaces) and special illumination (at points where reading or visual precision is required) are good practice. There should be no deep shadows in working spaces or passageways. Stairs, in particular, should be well-illuminated.

The official Discussion of the Third Edition included a table of illumination levels recommended by the Illuminating Engineering Society (IES). The table included both recommended minimum and "modern practice" footcandle levels. In the Fourth Edition of the Code, specific illumination requirements were adopted (almost all of which were the same as those in the Third Edition Discussion), as shown in Table H111-1.

The Third Edition Discussion recommended, but the Code did not require, provision of moderate illumination over yards, paths, roads, etc., outside of the station, thus ensuring safe access both to and from the station.

The Code of Lighting Factories, Mills, and Other Work Places, A.E.S.C.A. 11-1921, was referenced for more detailed information regarding good lighting practice. Illumination measurements were required on the vertical, horizontal, or intermediate plane, as appropriate. The Fifth Edition repeated the Fourth Edition requirements, except for deleting the "modern practice" values.

Table H111-1 Illumination Intensities

	Minimum (ft-candles)	Modern Practice (ft-candles)
1. Switchboard instruments, gauges, switches, etc.	1	2 to 4
2. Switchboards with no exposed live parts	1/2	1 to 2
3. Storage-battery room	1/2	1 to 2
4. Generating room, boiler room, pump room	1	2 to 4
5.Stairways and passageways where there is moving machinery, exposed live parts, hot pipes, etc. (measured at floor level)	1	2 to 4
6. Any traversed space (measured at floor level)	1/4	1 to 2

NOTE: The above illumination values are to be measured at working surfaces, except as stated.

The 1971 Edition essentially retained the categories of normal illumination requirements of previous editions, but it greatly increased required illumination values. The *IES Lighting Handbook* was referenced, but not required, for nonspecified values. The 1981 Edition revised the rule to specify illumination values for most areas to be found in generating stations. For substation areas not located at a generating stations, the illumination levels should be comparable to those given in the rule. The rules were based upon the minimum illumination levels necessary for safety, rather than levels that might be desirable for efficient visual performance. The rule was also changed from mandatory levels to recommended levels. The reference to the *IES Lighting Handbook* was deleted. The indoor levels are half, and the outdoor levels are generally the same as, those listed in Table 1 of ANSI A11.1-1973. The illumination levels are consistent with the guidance given in ANSI/IES RP-7-1983, *American National Standard Practice for Industrial Lighting.*

Where continuous illumination is not necessary, but may be necessary in case of emergency, a means of illumination is required to be available for use. IR 410 issued 14 February 1989 clarified that (at normally unattended stations) either permanent or portable lighting may be used when personnel work in the station. IR 542 issued 16 December 2005 further clarified that Rule 111A does not require lighting for unattended stations that are not part of a generating station property when personnel are not present. The lighting levels of Table 111-1 apply to generating and service building entrances, including gate houses at such facilities, but not to unattended substation entrances. Neither work areas in unattended stations nor entrances to unattended stations need to be lighted when personnel are not present.

111B. Emergency Lighting

(This rule was numbered 103B in the Fifth and prior editions; former Rule 111B was moved to 121B in the 1971 Edition.)

Operators should not be exposed to the danger of operating switches or performing other operations around live parts in a room that would be suddenly darkened by the failure of the normal power supply.

Emergency lamps are required to be automatically lighted by the failure of the usual energy supply in stations that are attended. Many stations are equipped with a storage battery for the purpose of supplying emergency illumination. In some instances, this battery is especially provided for the purpose; in others, it is used principally to supply energy for operating relay systems and similar equipment. Installation of an automatic relay or other device that will initiate emergency lighting is usually practical and necessary for safety purposes. In certain, rare cases, oil or gas lanterns may provide a sufficient emergency source of illumination; such lanterns are not suitable for battery rooms. The requirement for emergency lighting of exit paths was added in the 1981 Edition. In contrast to Rule 111A, Rule 111B deliberately (1) made the requirements mandatory and (2) did not specify whether the emergency illumination requirements were "minimum" or "average" levels when the 1981 requirement for 11 lux (1 ft-c) in the exit paths was developed. The intention of the rule is to adequately illuminate the areas to be used in an emergency; nearby shadowy areas that do not affect a worker's ability to move about in an emergency situation are not required to receive 11 lux (1 ft-c).

Rule 111B2 includes consideration of the time required for continued emergency lighting for paths. Although the wording of the rule does not state directly that it applies to Rule 111B1 requirements, common sense indicates the need for the same considerations, and such considerations are the intent of the Code.

111C. Fixtures

(This rule includes part of Rule 103C of the Fifth and prior editions. The second paragraph of former Rule 103C was deleted in the 1971 Edition. Former Rule 111C was moved to 121C in the 1971 Edition.)

The use of portable lighting cords in the operation and repair of station equipment should be generally avoided by provision of adequate permanent means for illumination. When the particular arrangement of equipment may necessitate occasional use of portable lamps during station operation, maintenance, and repair, provision should be made for permanently fixed receptacles that are both conveniently located with respect to the equipment and safely accessible to the user. Provision of suitable short cords at convenient points will eliminate (under careful management) the use of long cords attached to distant receptacles and hauled over floors and around equipment.

111D. Attachment Plugs and Receptacles for General Use

(This was Rule 103D in the Fifth and prior editions. Former Rule 111D was moved to Rule 121D in the 1971 Edition.)

Several decades ago, the common use of long cords attached directly to bus bars, switch terminals or blades, and similar makeshifts was frequently the cause of severe burns, eye injuries, and fatalities. As a result, the importance of carefully planning the installations to reduce such accidents became generally recognized in modern practice. It should be rare that such makeshift installations are necessary, and certainly not for normal operation. Special maintenance and personnel security procedures are generally necessary when such temporary power sources are used.

The rule requiring that all poles of cable connectors shall be disconnected by a single operation may be met by the swivel-type connector, but this is not as desirable as the bayonet type. A bayonet-type connector is constructed to disconnect all poles simultaneously; therefore, an interrupted disconnection operation cannot leave a single pole connected.

The fact that tension in the portable conductors tends to disconnect the bayonet-type connector is a desirable feature from a safety point of view.

111E. Receptacles in Damp or Wet Locations

(This rule was added in the 1981 Edition.)

This rule takes advantage of innovations in electrical protection equipment. It recognizes the decreased resistance to ground of wet areas and the need to reduce the potential hazard presented by failure of the insulation of ordinary maintenance tools. The rule recognizes the value of ground-fault circuit-interrupter (GFCI) receptacles as well as GFCI circuit breakers.

112. Floors, Floor Openings, Passageways, and Stairs

(Rule 104 of the Fifth Edition was moved here in the 1971 Edition; previous Rule 112 became Rule 122.)

112A. Floors

(Prior Rule 104A was moved to this location in the 1971 Edition; former Rule 112A was moved to 122A.)

Falls, hitting obstructions, and similar mechanical accidents have been responsible for the greater proportion of all personal injuries in stations over the years. To alleviate part of this problem, it is especially important that spaces about electrical equipment be kept neat and clean and free from extraneous matter.

Unevenness in floors is responsible for many accidents. Iron or tile floors may be dangerously smooth or slippery. Doorway treads, stair treads, and frequently used passageways may develop dangerously slippery surfaces. Footing in these high-traffic areas can be improved by employing antislip treads or antislip materials manufactured for this purpose. Such materials are generally available and can be applied easily after construction.

112B. Passageways

(Rule 104B of the Fifth Edition was moved here in the 1971 Edition; former Rule 112B was moved to 122B.)

Infrequently occupied passageways or spaces, such as busbar or pipe chambers, where the conductors are guarded by insulating coverings, metal sheaths, or barriers, may with reasonable safety have less than 2.13 m (7 ft) of clear head room, since inadvertent movements are less liable to be made in such places. This vertical clearance was raised in the 1981 Edition from 2 m (6.5 ft) to 2.13 m (7 ft) to recognize the increased average height of workers and the required use of hard hats.

Passageways or spaces frequently used by personnel should be of ample dimensions to permit rapid and safe movement. Where the specified requirements for clear passage are not practical, safety signs and proper lighting are required. The reference to ANSI Z535 standards for safety signs was added in 1997.

112C. Railings

Floors should have no abrupt changes of level. Unevenness in floors has been responsible for many accidents. Where a drop in floor level of 300 mm (1 ft) or more exists, such as with an uncovered floor opening or a raised platform (such as that in Figure H112C-1), a railing is required by the 1971 and later editions. Prior editions required railings only for floor openings over 450 mm (18 in) in depth or raised platforms over 1.20 m (4 ft) high. Many companies use railings where the difference in level is much less than that for which a railing is required by the rule.

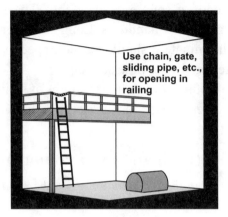

Figure H112C2-1
Elevated platform with railing

Although not specified in the NESC, the standard railing height of ± 1070 mm (42 in) had been in use by many industries before being recognized by OSHA (see Figure H110C-2).

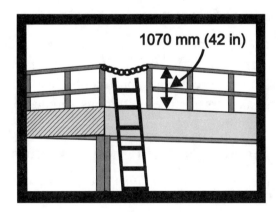

Figure H112C2-2
With ± 1070 mm (42 in) railing with chain across opening

112D. Stair Guards

This rule was numbered 104D in the Fifth and prior editions.)

Stairs are frequently a source of accidents. The seriousness of the hazard is reduced greatly by handrails. Long and steep stairs, especially if not very well-illuminated, should be equipped with guards at the heads, such as gates or sliding pipe sections.

Insurance underwriters have defined a flight of stairs as consisting of "not less than four risers or a series of risers and landings not exceeding one story in height."

112E. Continuity (Fifth and prior editions)

(Rule 104E was deleted in the 1971 Edition when the remainder of Rule 104 was moved to 111.)

There is varied and extensive use of ladders in some stations. Depending upon design, they can be a source of considerable hazard. The hazards involved can be reduced through the use of care and foresight in ladder location. Except on very steep ladders, handrails are desirable.

112E. Top Rails (1981 and later editions)

(This rule was added in the 1981 Edition.)

If railings are to perform their function in emergency situations, they must be easy to grab with a hand. The 75 mm (3 in) required opening will allow the fingers to clear any obstruction and clasp the top rail quickly. Portable ladders for use in stations should always be of the insulating type.

112F. Floor Toe Boards (Fifth and prior editions)

(Rule 104F was deleted in the 1971 Edition when the remainder of Rule 104 was moved to 111.)

Floor toe boards aid materially in preventing tools or other material from falling or being pushed over the edge. Many serious accidents to operators and equipment may be attributed to falling tools, etc.

112G. Stair Toe Boards (Fifth and prior editions)

(Rule 104G was deleted in the 1971 Edition when the remainder of Rule 104 was moved to 111.)

Toe boards on stairways may be attached to the underside of the tread next above and permit a small space above each tread for cleaning away grease and dirt.

113. Exits

(This rule was moved here in the 1971 Edition from Rule 105; previous Rule 113 became Rule 123.)

113A. Clear Exits

(This is Rule 105A of the Fifth and prior editions; former Rule 113A was moved to 123B in the 1971 Edition. References to the exit doors were moved to 113C in the 1987 Edition.)

Exits from rooms and working spaces about electrical equipment should be kept clear of all obstructions, including temporary work equipment. In case of an accident, they provide both a means for escape and ready access for emergency measures.

113B. Double Exits

(This is Rule 105B of the Fifth and prior editions; former Rule 113B was moved to 123A in the 1971 Edition.)

More than one exit is particularly desirable from the back of switchboards, narrow galleries, and long passageways since, in case of arcing, smoke, steam, or other dangerous conditions, a single exit may be shut off. A slight injury to an attendant, such as burns or flashed eyes, or even slight faintness, may make movement through a long passageway to a single exit highly dangerous. Under such circumstances, one is more liable to stumble against live or moving parts adjacent to the passage.

Pipe and cable tunnels can present an increased level of personal hazard because of a conflict between measures for fire prevention and more direct personal safety. Openings at such points assist in spreading fire, but automatic fire doors obviate this. When power distribution cables are run underground, a conduit system is safest from the personnel standpoint. However, the increased ease of inspection and maintenance of cables and piping in tunnels provide strong support for the use of such installations. Tunnels should have barriers to separate the tunnel from other areas of the station. For safety to persons who might be in the tunnel during an accident, a means for rapid exit should be provided at both ends. Small fire doors through the separating walls or manholes readily opened from the inside may be used.

The 1971 and later editions *require* a second exit in many cases; previous editions placed the requirement in what current editions use as a "should" category.

113C. Exit Doors

(Rule 113C of the Fifth and prior editions was moved to 123 in the 1977 Edition, and this rule number was not used until the present 113C was split off from 113A in the 1987 Edition.)

In the 1971 Edition, the requirement was added that locks and latches on doors permit opening by means of simple pressure or torque on the actuating parts under any conditions. The 1984 Edition included the further requirement that doors swing out and limited locking mechanisms to those that actuate on simple pressure. The previous rule had allowed the use of normal doorknobs; such mechanisms have been found to be too hard to actuate and too small to find easily in emergency situations; they were, therefore, specifically prohibited in the 1984 Code. In the 1987 Edition, an exception was made for gates in outdoor fences and certain buildings from the panic hardware requirement.

114. Fire-Extinguishing Equipment

(This is Rule 106 of the Fifth and prior editions. The previous two paragraphs were combined in the 1971 Edition. The rule was generalized in the 1981 Edition. Prior Rule 114 became Rule 124 in the 1977 Edition.)

Electric supply stations vary in type, size, and degree of potential fire hazard, including large generating plants with multiple potential issues to consider all the way down to small industrial substations with little potential fire hazard. This rule is not intended to require permanent fire extinguishers or fire extinguishment systems in all electric supply stations or in all areas of large, complex stations. Where such systems are installed, the type of fire extinguisher or extinguishing equipment must be appropriate for the intended use, must be conveniently located, and must be conspicuously marked.

Useful references for assistance in determining when, where, and what kind of extinguisher or extinguishing equipment may be appropriate include IEEE Std 979 *Guide for Substation Fire Protection*, NFPA 803 *Fire Protection for Light Water Nuclear Power Plants*, NFPA 850 *Fire Protection for Fossil Fueled Steam and Combustion Turbine Electric Generating Plants*, and NFPA 851 *Fire Protection for Hydroelectric Generating Plants*.

Early discussions of the NESC dwelt on the hazards associated with various fire-extinguisher systems, such as soda-acid, carbon tetrachloride, sand, and sawdust-bicarbonate of soda. Today, such systems are generally neither allowed by OSHA regulations nor considered the most suitable for the job at hand. Carbon dioxide, dry-powder systems, and other specialty fire-fighting systems are preferred for portable systems. Permanent, special fire-control systems are appropriate for installations involving large quantities of oil or other combustible materials.

Fire extinguishers should not be installed near steam pipes, radiators, or other heating devices, or in locations subject to high temperatures. In cold climates, provision should be made to protect extinguishers from exposure to low temperature that would render them temporarily ineffective.

114A, 114B, 114C. (Not used in the current edition.)

(Rules 114A, 114B, and 114C of the Fifth and prior editions were moved to 124A, 124B, and 124C, respectively. Rule 124C has been rearranged and parts have been deleted; see the discussion of Rule 124C.)

114D. (Not used in the current edition.)

(Rule 114D of the Fifth and prior editions was merged into Rule 124A in the 1971 Edition.)

115. Oil-Filled Apparatus (1971 and prior editions)

(Rule 115 of the Fifth and prior editions was moved to Rule 125 in the 1971 Edition. The rule discussed below was numbered 107 in the Fifth and prior editions and was moved to 115 in the 1971 Edition. In the 1971 Edition, sub-rules A and B of the Fifth Edition Rule 107 are included within Rules 172 and 153, respectively; Rule 153 was renumbered to 152 in the 1981 Edition. The remaining portion of Rule 115 of the 1971 Edition was deleted in the 1981 Edition.)

Failures of oil-filled apparatus, such as oil switches, transformers, regulators, and lightning arresters, have resulted in personal injuries or damage to property. The severity of the hazard is related to the quantity of oil involved, the capacity of the system to which the apparatus is connected, and proximity to the source of power.

Causes that lead to failures include lightning, short-circuit stresses in equipment, mechanical failure of structural parts, the use of switches and equipment unsuited to the system capacities to which they were connected, and careless or unskilled operation. The nature and causes of the various occurrences of such accidents indicate that no single rule or set of rules could be devised that would adequately meet every condition under which this class of equipment is regularly operated.

The relative degree of hazard presented by oil-filled apparatus is greatly dependent upon the system capacity. Changes in capacity over time preclude one system from being suitable. For example, oil-filled apparatus of standard types may be installed or in use on a system with a generating capacity of 10 000–50 000 kVA with only a remote possibility of serious failure of the equipment. Subsequently, the capacity of the system may be multiplied many times or, through interconnections, the original equipment may become part of a larger system. In either case, the equipment may be entirely inadequate to meet the newer conditions with satisfactory reliability. If the remedy requires the substitution of more modern or substantial equipment, it will be obvious that cost considerations may prevent such a course from being immediately followed.

To design for an ultimate maximum duty in new installations is impracticable when future conditions cannot be determined. It is intended, therefore, that this rule be considered of a general nature and be subject to liberal interpretation under all conditions.

Precautions should be taken to localize trouble that may occur. Failure of cases containing oil should not be permitted to result in spreading burning oil where it places operators in danger. On balconies or upper floors, means should be provided to effectively prevent flowing oil from reaching floors below. A drainage system for oil not only affords added protection, but frequently provides a convenient means for handling the oil during routine operation and maintenance.

115A. Oil Switches or Circuit Breakers (1971 and prior editions)

(When Rule 107 of the Fifth and prior editions was renumbered to 115 in the 1971 Edition, former Rule 107A was merged into former Rule 161—Oil Switches to form 1971 Rule 172—Circuit Breakers, Switches, and Reclosers Containing Oil. This rule never existed as Rule 115A; it is discussed here to keep the discussion of old Rule 107 together.)

Manually operated oil switches, motor starters, and induction regulators in individual locations, or small groups of such apparatus, may involve such a small hazard that the expense of isolating enclosures or oil-drainage systems is not justified. Oil switches and circuit breakers of certain types may be installed safely on the rear of switchboard panels when they are connected to a system of *limited* capacity; if they are connected to a system of large capacity, they should be installed in fire-proof switch cells. Otherwise, they may prove to be entirely inadequate to function properly under the more severe conditions that are occasionally imposed.

Because of the possibility of violent failure of oil switches, they should be located so as to minimize the possibility of damage or injury from scattered burning oil.

Drainage gutters provided for oil switches, instrument transformers, etc., will only contain quantities of oil small enough to permit close approach by attendants during burning; the gutters may lead directly to the sewer pipes in fire-resistive buildings, if local regulations permit. Some apparatus contain enough oil to cause a hazard from explosion and fire by flooding the drainage system with burning and boiling oil. In this case, the drainage gutters should lead directly to a catch basin or other point outside of the building where burning oil can do no damage.

115B, 115C. Transformers, Induction Regulators, etc. (1971 and prior editions)

(Prior Rule 107B of this title was split in the Fifth Edition into two rules, numbered 107B and 107C. Rule 107C included apparatus containing liquids that did not burn. When Rule 107 of the Fifth and prior editions was renumbered to 115 in the 1971 Edition, Rules 107B and 107C were merged into former Rule 143—Location and Arrangement of Power Transformers *to form Rule 153 in the 1971 Edition, using the latter title. Rule 153 was subsequently renumbered to Rule 152 in the 1981 Edition. These rules never existed as Rule 115B and 115C; they are discussed here to keep the discussion of old Rule 107 together.)*

Although modern oil-filled apparatus is improved greatly in safety-related design and construction, too much care cannot be taken in the proper segregation of apparatus containing large quantities of oil.

Transformers and induction regulators, if in large groups or if containing large total amounts of oil, should be installed in fire-resistive enclosures. They should be well ventilated to the outside of the building to prevent dangerous accumulations of oil vapors. The enclosures should be free from apparatus likely to ignite from burning oil, be sub-divided to a reasonable extent, and have doors or windows so located or arranged that burning oil would not be likely to pass through the inflammable material or apparatus outside of the enclosure.

115D. Lightning Arresters (1971 and prior editions)

(This rule was moved from 107C to 107D in the Fifth Edition. It was deleted in the 1971 Edition.)

Lightning arresters containing oil are subject to causing the same problems as other equipment containing similar amounts of oil. Personnel and equipment can be protected adequately by location of the arresters and provision of appropriate oil drainage systems.

116. (Not used in the current edition.)

(Rule 116 of the Fifth and prior editions was moved to Rule 126 in the 1971 Edition.)

117. (Not used in the current edition.)

(Rule 117 of the Fifth and prior editions was moved to Rule 127 in the 1971 Edition.)

118. Shielding of Equipment From Deteriorating Agencies

(This rule was deleted in the 1971 Edition.)

Any hastening of deterioration of electrical equipment by moisture or uncleanliness means greater danger of breakdown of insulation that may fail at the point where the attendant is handling it and cause harmful shock, or may fail near him or her and cause burns or mechanical injuries. The conditions of good contact and cramped surroundings are likely to augment the danger and injury under such circumstances.

The rubber-insulated leads to oil-insulated transformers and switches often deteriorate rapidly and endanger attendants who may come in contact with these leads, relying on their insulating coverings as a guard.

119. (Not used in the current edition.)

(Rule 119 of the Fifth and prior editions was moved to Rule 128 in the 1971 Edition.)

Section 12. Installation and Maintenance of Equipment

(This section was numbered 11 in the Fifth and prior editions. Prior to the 1981 Edition, the title of this section was Protective Arrangements of Equipment.*)*

120. General Requirements

(This rule was numbered 110 in the Fifth and prior editions. Former Rule 120 moved to 130 in the 1971 Edition.)

In general, rules of the National Electrical (Fire) Code[1] that apply to similar equipment and installations serve as good guidelines for reducing accidental fires. By reducing fire danger, those rules also indirectly reduce danger to life.

121. Inspections

(This rule was numbered 111 in the Fifth and prior editions. Former Rule 121 was revised in the 1971 Edition; former Rule 121A was moved to Rules 124 and 131, and Rule 121B was moved to Rule 132 in the 1971 Edition. Both 131 and 132 were deleted in the 1981 Edition; former Rules 121C and 121D were deleted.)

The subcommittee intended these rules to cause the institution and sustenance of inspection and maintenance *programs* that are regular and scheduled. The rules do not specify details of such programs. The operating utility is responsible for using its experience in a forward-looking and prudent manner to design and maintain an inspection and

---

1. Now called the *National Electrical Code* (NEC), this code is the electrical volume of the National Fire Code and has been published by the National Fire Protection Association (NFPA) under various names since 1897, usually incorporating the name "National Electrical Code."

maintenance program of the type and frequency appropriate for the subject facilities.

121A. In-Service Equipment

(Rule 11A—Regular Equipment *was moved here in the 1971 Edition, and was revised for clarity in the 1981 and 1984 Editions. Former Rule 121A was moved to 131 in the 1971 Edition; it was deleted in the 1981 Edition because it also was covered in Rule 122.)*

Systematic inspection and testing of equipment and circuits after operation is valuable from both safety and operations viewpoints. Gradual deterioration of the system will be detected, defective conditions will be avoided by proper repairs and replacements, and injuries will be avoided. Defects that are found should be recorded, especially when they cannot be immediately remedied. A procedure should be in place to track such defects until they are remedied. Records of defects are valuable in improving designs in new installations or extensions. Cleanliness, of course, retards deterioration.

121B. Idle Equipment

(This is Rule 111B of the Fifth and prior editions. Former Rule 121B was moved in the 1971 Edition to Rule 132; those provisions of Rule 132 were deleted in the 1981 Edition and the number was reused.)

Equipment seldom used is more likely to be neglected and, therefore, may be dangerous when placed in service. This can be avoided by periodic inspection or by inspection use.

Where equipment is idle but not permanently removed from service, motors, prime movers, and other rotating apparatus should have rotors rotated periodically. Connections and wiring should be tested to ensure safe and proper operation if the equipment is called into service suddenly.

121C. Emergency Equipment

(This is Rule 111C of the Fifth and prior editions. Former Rule 121C was deleted in the 1971 Edition.)

To ensure the dependability of emergency equipment, especially after long periods of idleness, routine periodic inspections and tests are generally appropriate. The dependability of such equipment is a large factor in providing continuous and safe service. Written reports or logs are good aids in performing periodical inspection of such equipment.

121D. New Equipment

(This is Rule 11D of the Fifth and prior editions. Former Rule 121D—Arcing Shields *was deleted in the 1971 Edition.)*

A thorough initial inspection of each installation of electrical equipment should be made before placing it in service. However carefully the installation has been made, inspection by some person other than one engaged in the work is always desirable with important installations.

An EXCEPTION was added in the 1981 Edition to recognize that it is not necessary or reasonable to test every hardware fitting. The wording caused confusion because of the inclusion of *devices* and *appliances*. Breakers are *devices* that do require testing, and these are industry standards for doing so. This EXCEPTION was revised and added to the rule in the 1997 Edition. The new language requires new equipment to be tested in accordance with *standard industry practices.* The removal of the EXCEPTION is not intended to require testing of fittings, fixtures, and hardware, since much of such testing for which industry standards exist is destructive testing.

122. Guarding Shaft Ends, Pulleys and Belts, and Suddenly Moving Parts

(This is Rule 112A of the Fifth and prior editions. Former Rule 122A was combined with former Rule 113B to form Rule 123A in the 1971 Edition.)

122A. Mechanical Transmission Machinery

(This is Rule 112A of the Fifth and prior editions. Former Rule 122A was combined with former Rule 113B to form Rule 123A in the 1971 Edition.)

While guarding is generally necessary for moving parts near work areas, it may not be necessary to provide guards for very small or very slowly moving parts. However, projecting set screws, for instance, may be dangerous even on such slow or small parts.

ANSI B15.1 *Safety Standard for Mechanical Power Transmission Apparatus* covers the safeguarding of moving parts of equipment used in the mechanical transmission of power. It covers prime movers, intermediate equipment, and driven machines, including connecting rods, cranks, flywheels, shafts, spindles, pulleys, belts, link belts, chains, ropes and rope drives, gears, sprockets, friction drives, cams, couplings, clutches, counterweights, and other revolving or reciprocating parts. It is not appropriate to repeat such detailed requirements in these electrical safety rules.

122B. Suddenly Moving Parts

(This is Rule 112B of the Fifth and prior editions. Former Rule 122B—Coupled Machines was deleted in the 1971 Edition due to the wording of new Rule 123A.)

Small parts that move suddenly, such as circuit-breaker levers and handles, governor levers, and controller handles, have caused several accidents. Mechanical guards may be used; guarding of such parts by isolation is preferred when it is not necessary to make them accessible.

123. Protective Grounding

*(This rule was formed in the 1971 Edition from previous Rules 113, 122, 142, 168, and 175; former Rule 123—*Terminal Bases and Bushings *was deleted.)*

123A. Protective Grounding or Physical Isolation of Noncurrent-Carrying Metal Parts

(This rule was formed in the 1971 Edition from former Rules 113B and 122A; former Rule 123A was deleted.)

The presence of dampness, acid, or acid fumes increases the possibility of good contact between personnel and noncurrent-carrying parts. The danger to personnel of possible electric current leakage to, and then from, ungrounded machine frames is thus greatly increased, and even the lower voltages may become dangerous. Where an explosive atmosphere exists, sparks must be avoided; careful grounding reduces sparking by making frames and surrounding objects the same voltage.

While grounding of noncurrent-carrying metal parts is generally necessary, this is generally understood *not* to apply to such parts as card holders on switchboards and similar parts that are unlikely to become energized by leakage from live parts.

The rule allows protection by either grounding or physical isolation. Some equipment is actually less hazardous if isolated, rather than grounded; examples are dc railway generators, rotary converters, and switchboards, as well as dc arc machines and control boards. This is especially true if ungrounded live parts are exposed as is the case with some so-called "single-voltage" switchboards.

When frames of such equipment are not permanently grounded, they should be effectively insulated from ground by a dielectric that is both suitable for the maximum operating voltage and bonded to neighboring noncurrent-carrying metal parts. Grounded conduit should be kept well away from such insulated frames, so that short circuits or inadvertent personal contact will not occur. Partial and variable insulation between the frame and adjacent grounded parts (such as masonry or concrete usually provides) does not provide suitable protection, either for personnel or equipment.

Small transformers, such as instrument transformers, may be protected to a degree from the hazard of breakdown by leaving the case ungrounded and mounted so that it is insulated from ground. This is permissible only if the hazard to persons is eliminated by isolating the transformer by elevation or by guarding it from persons by barriers or cells or by providing mats.

123B. Grounding Method

(This rule was formed in the 1971 Edition from former Rules 112A and 113B. Prior Rule 123B was deleted.)

Equipment grounding must be thoroughly reliable in order not to give a false sense of security. Obtaining reliable grounding requires careful consideration of the conditions of service. The methods of grounding are, in general, the same whether in stations, on lines, or for utilization equipment and circuits; detailed grounding methods are located in Section 9. IEEE Std 80 *IEEE Guide for Safety in AC Substation Grounding* (ANSI), provides guidelines for the design of substation grounding systems; Rule 96A is based on that standard.

IEEE Std 665™ *IEEE Guide for Generating Station Grounding* (ANSI), and IEEE Std 1050™ *IEEE Guide for Instrumentation and Control Equipment Grounding in Generating Stations* (ANSI) provide information for generating station grounding.

123C. Provision for Grounding Equipment During Maintenance

(This is Rule 113C of the Fifth and prior editions.)

Ungrounded electrical equipment that is separated from its source of electrical energy can, under certain conditions, create an undesirable degree of electrical hazard for personnel working on or near the equipment.

The main part of Rule 123C requires grounding of parts normally energized above 600 V when treating them as de-energized. The 1971 Edition introduced an exception for circuits normally energized up to 25 kV when a visible break switch was present, apparently in anticipation of a similar exception in the work rules expected to be adopted in the 1973 Edition. No such exception to the normal work rules was ever adopted. Such a practice has never been authorized by the NESC work rules, and the language was removed in the 2002 Edition by a Tentative Interim Amendment. TIA 2002-1 deleted Rule 173C and modified Rule 123C to remove the implication that grounding could be omitted on conductors normally operating at 25 kV or less where a visible break exists. Such practice is not allowed by Part 4. The 2007 Edition reaffirmed the TIA and deleted the former language.

123D. Grounding Methods for Direct-Current Systems Over 750 Volts

In the 1993 Edition, dc systems over 750 V were required to be grounded in accordance with Section 9. This edition also revised HVDC requirements in Section 9.

124. Guarding Live Parts

(This rule was formed from Rules 114, 121A, 137, 152, 169, 176, and 184 of the Fifth and prior editions. Former Rule 124—Deteriorating Agencies was deleted in the 1971 Edition.)

124A. Where Required

(This was essentially Rule 114A but includes parts of the above enumerated rules of the Fifth and prior editions; former Rule 124 was deleted in the 1971 Edition.)

The basic rule requires guards around all energized parts operating above 300 V phase-to-phase (150 V phase-to-ground prior to 2007) that are not either insulated or isolated from authorized employees by the clearances of Table 124-1. The 2007 Edition resolved the previous conflict in requirements between the 150 V phase-to-ground used in

Rule 124A1 and the starting point of 151 V phase-to-phase used in Table 124-1. As a result, neither 120-240 V single phase, 120/240 V 4-wire delta with midphase ground nor 240 V 3-wire delta with one leg grounded is covered by Rule 124A1 and Table 124-1. The starting point is 301 V phase-to-phase with a vertical clearance of 2.64 m (8 ft-8 in). However, Rule 124A3 requires parts of indeterminate voltage potential to have not less than 2.6 m (8.5 ft) vertical clearance, which becomes the de facto vertical clearance requirement for the voltages below 301 V.

The 2007 Edition added an exception to Rule 124A1 to remind users of the alternative of providing a separate fence or railing enclosure meeting Rule 124C3 within the outer perimeter fence or enclosure of the supply station.

The intent of the rules is to secure safe installations. Electrical apparatus is manufactured regularly according to commercial designs and generally is not designed for a particular installation. Hence, it may not inherently provide required clearances for safe personnel movement. These rules provide the necessary clearances for guarding and isolating electrical apparatus in its final installation. Table 124-1 is not intended to apply to clearances within or on electrical apparatus as a part of its design and manufacture. The manufacturer is assumed to provide the necessary clearances within or on the apparatus.

In all editions, the guarding provisions of Part 1 only apply where the covered facilities are accessible only to authorized persons. Where the requirements of Rule 110A—*Enclosure of Equipment* are not met, the area is considered to be accessible to unqualified personnel, and Part 2 applies.

NOTE: The clearances of Rule 124A1, Table 124-1, are required to any permanent supporting surface for workers. If a raised concrete foundation extends out far enough for a worker to stand on it without conscious effort, the required clearance is to be measured from the top of the foundation. Otherwise, the height of the foundation above ground contributes to meeting required vertical clearances. The clearance can be achieved by meeting the horizontal clearance, the vertical clearance, or a "taut string distance" equal to the required vertical clearance (see Figure H124-1).

The clearances of Table 124-1 are essentially those of Table 2 of the Fifth and prior editions. The voltage ranges were revised in the 1971 Edition. The increasing average adult height was also recognized in that edition and the lower voltage clearances were increased accordingly. Some of the high-voltage clearances in Table 124-1, Part A were reduced in the 1981 Edition. The tabulations in Part B and Part C were expanded to cover clearances at 800 kV. The initial range was reduced in the 1984 Edition to 151 V. In the 1997 Edition, Table 124-1 Part A clearances were further distinguished based upon differences in BIL levels; the maximum design voltage between phases (from ANSI Std. C84.1 *Electric Power System and Equipment—Voltage Ratings [60 Hz]*) are referenced instead of nominal voltages between phases. Note that these BIL values apply in electric supply stations; they do not apply to overhead lines, except to the extent that such lines are contained in electric supply stations. Table 124-1, Part D was added in the 2002 Edition to define HVDC requirements.

The vertical measurements are intended to be "taut-string' measurements, as shown in Figure H124-1. The combination of vertical and diagonal distances totalling the length required in Table 124-1 effectively provides isolation of the energized parts from workers in the station.

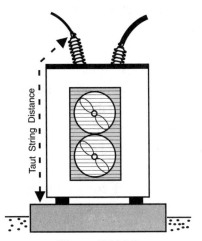

Figure H124-1
"Taut String" measurement

Part 1 does not generally specify requirements for clearances between conductors; it specifies clearances from conductors to other surfaces. Conductors are considered to be in a restricted area, and subject to Part 1 requirements, until they cross over the substation fence; outside of the station fence, they become subject to Part 2 requirements. Inside the substation area, conductor and bus movement generally is controlled more carefully than in typical overhead environments. As a result, the Code does not specify vertical clearances between conductors and any buses or other conductors that they cross inside the fenced supply station area.

The AIEE (predecessor to IEEE) and the IEEE have published guides for phase spacing in substations in the *IEEE Transactions on Power Apparatus and Systems*. These guides are neither required nor recommended by the NESC language, but they are the published guides of the IEEE Substations Committee and are considered to be good practice.

If live parts could always be perfectly guarded when personnel are near them, accidents from electrical shock and burn would cease. Guarding for live parts must, however, be somewhat less than perfect with much station equipment because of the competing requirements for inspection, repair, or adjustment. Quick access, where necessary, should not be obstructed by guards; quick access may be essential in emergencies to avoid unnecessary slowing of operations or partial shutdown of service.

Carefully planned, permanent guards may, however, aid rather than delay service. Carefully planned guards can allow safe repairs near live parts without the necessity of installing temporary protective devices or the de-energizing of adjacent circuits. Guards also tend to prevent accidental short circuits and the spread of short circuits beyond the place of origin.

Guards that protect under all conditions are particularly necessary where, from the nature of the situation, a dry insulating floor or secure footing is not practical.

It is not practical to perfectly guard commutators, brush rings, and other parts of rotating machinery by construction. However, adequate safety may be secured for attendants by supplementing incomplete guarding with safe methods of operation.

It is often difficult to make repairs, replacements, or extensions to existing construction, and to maintain equipment, either existing or new, in safe condition at the same time. Where removal of barriers, or other safeguards, is required to facilitate such construction, unqualified persons such as ironworkers or brick masons may be exposed to unfamiliar hazards. The use of temporary guards or other special care is usually necessary under these conditions.

When new equipment is being installed and is ready for testing, special precautions may be appropriate to guard against (1) accidental contact by personnel in the vicinity, (2) defects, or (3) improper connections or arrangements of apparatus. Installation of temporary guards, barriers, or warning signs may be appropriate, depending on the degree of hazard.

Equipment that is not designed for outside use may become damaged by weather conditions while awaiting installation. This can produce a hazard when the equipment is placed in service. Accumulations of dust, acid, or other deteriorating agencies may have a like effect. It is, therefore, appropriate to protect equipment during processes of construction or reconstruction.

NOTE 1 to Table 124-1 was revised in the 1981 Edition to allow reduced clearances where switching surge factors are known.

Rule 124A3 requires parts of indeterminate potential to be guarded on the basis of the maximum voltage that may be present. When determining the voltage that can be present at any point on an insulator stack, both the insulating material and expected surface contamination should be considered. A minimum vertical distance of 2.60 m (8.5 ft) to the bottom of an insulator stack is appropriate and consistent with the intent of this rule. By subtracting the "guard zone" distance from the vertical clearance required by any Code edition, it can be verified that 2.6 m (8.5 ft) is intended as the closest approach of the guard zone around an energized conductor or part without requiring an appropriate guard (see Figure H124A3).

In recent years, the installation of metering equipment inside bushings tended to make the bushings longer; thus, if the top of the bushing were mounted at the appropriate height, the bottom could be lower than intended by the NESC. Likewise, installation of certain equipment on tall stacks of insulators allowed the insulator bases to be too low. As a result, the intent in 123A3 was clarified in the 1987 Edition by requiring the bottom of an insulator to be at least 2.6 m (8.5 ft) above grade unless it is enclosed or guarded.

124B. Strength of Guards

(This was Rule 114B of the Fifth and prior editions. Former Rule 124B was merged in Rule 123 in the 1971 Edition.)

In earlier years, the strength of guards frequently was given insufficient consideration during design and installation. Prior to the Fourth Edition, the strength of a then-standard guardrail post of 31 mm (1.25 in) wrought-iron pipe 107 cm (42 in) high was questioned. Six posts of 31 mm (1.25 in) iron pipe were attached to flooring in six different ways, including those most frequently used when installing guardrails, and tested. Although the conditions described would rarely be exactly applicable to modern construction, the results are provided below in the interest of those who may be concerned. In all cases, the horizontal force was applied to the post 107 cm (42 in) above the floor.

When the post was fastened to a 50 mm (2 in) maple floor by a standard cast-iron floor flange with four 50 mm (2 in) No. 10 wood screws, the screws started to be stripped out of the floor at a horizontal force of 267 N (60 lbf); the screws pulled completely out with a force of less than 356 N (801 lbf).

When the same type of cast-iron floor flange and post was mounted on 100 mm (4 in) maple flooring with four 75 mm (3 in) No. 14 wood screws, the flange started to deform when the horizontal force reached 400 N (90 lbf); the flange broke and freed the pipe before the force reached 534 N (120 lbf).

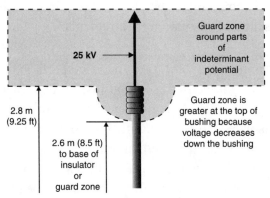

Figure H124A3
Example of guard zone around 25 kV conductor

When methods of holding the 31 mm (1.25 in) pipe post to the floor were used that developed the full strength of the pipe, the pipe took a permanent set at a force of about 125 lb.

When the 31 mm(1.25 in) post was supported by slipping it over a 25 mm (1 in) pipe projecting 150 mm (6 in) above the floor, the post effectively took a permanent set as a result of the bending of its 25 mm (1 in) pipe base at about 400 N (90 lbf).

In a test of horizontal rails, a larger wrought-iron pipe (38 mm [1.5 in]), was supported by posts 2.45 m (8 ft) apart. The rail took a permanent set when a force of 1023 N (230 lbf) was applied in the center of the span.

124C. Types of Guards

(This is essentially former Rule 114C but also includes portions of the above enumerated rules of the Fifth and prior editions.)

124C1. Location or Physical Isolation

(This is Rule 114C1 of the Fifth and prior editions.)

This rule defines the intended use of the terms *isolation* and *location*. The reference to ANSI Z535 safety sign standards was added in the

1997 Edition to complement the previous requirement for safety signs at entrances as a part of isolation requirements.

CAUTION: This rule does *not* mean that live parts in a fenced electric supply station are guarded by isolation just by being within the fence. The further requirements of the rule specify how energized parts can be considered as isolated from the qualified personnel if the *guarding by location* (i.e., clearances) specified in Table 124-1 cannot be met (see Rules 124C2 and C3). The rules of Part 1 apply to areas under the control of and accessible only to qualified personnel. Within such spaces, Rule 124A allows only three alternatives: insulation, isolation *in accordance with Table 124-1*, or guarding; the supplemental fence or railing of Rule 124C3 provides an alternative.

124C2. Shields or Enclosures

(This is Rule 114C4 of the Fifth and prior editions. Former Rule 114C2— Grounded Metal Cable Sheaths *was continued in the 1971 Edition in this number, but it was deleted in the 1981 Edition.)*

Guards inside of the guard zone are permitted by Rule 124A when these guards are located under definite engineering design. This rule provides a general exception to the need for definitive engineering design. Rule 124C allows guards to be located within the guard-zone distance specified in column 4 of Table 124-1 when the guards are of insulating material completely enclosing the live parts and the voltage of the circuit is less than 2500 V to ground (7500 V between phases for the Fifth and prior editions).

This rule provides a second exception to Rule 124A. If the guards are located farther away from the live parts than the radius of the guard zone (column 4 of Table 124-1) plus 100 mm (4 in), then the guards need not extend to the height above the floor specified in column 2 of Table 124-1; they are required only to be a minimum of 2.6 m (8.5 ft) (7.5 ft for the Fifth and prior editions) above the floor (see Figure H124C2).

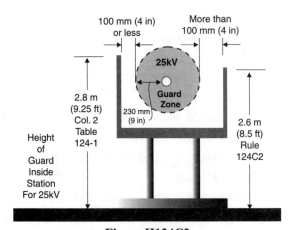

Figure H124C2
Height of guard around 25kV phase-to-phase, 150 kV BIL bus

124C3. Supplemental Barriers or Guards Within Electric Supply Stations

(This was Rule 114C3 in the Fifth and prior editions, which was moved to Rule 124C4 in the 1971 Edition. When Rule 124C2 of the 1971 Edition was deleted in the 1981 Edition, this rule was moved up to 124C3. The name changed from "Railings" in 2002.)

The required location was changed in the 1971 Edition in recognition of the 1971 changes in guard zone heights. The 2002 Edition codified the standard practice, which often used chain link or other fence types in lieu of railings to keep employees out of areas with low energized parts.

The purpose of such a fence or railing is to limit the opportunity for authorized personnel within an electric supply station to inadvertently contact energized parts that are installed below the level otherwise required by Rule 124A (see Figure H124C3-1). No portion of a perimeter fence may be used for such a barrier or guard. This fence or railing is required to be a separate enclosure. This type of low fence or railing is often used to separate rack-mounted capacitor banks from the rest of a supply station area, as in Figure H124C3-2.

This rule traditionally specified (1) a required location of not less than 900 mm (3 ft) and (2) a nonmandated preference for a location not more than 1.20 m (4 ft) from the nearest point of the guard zone that is less than 2.60 m (8.5 ft) above the floor or grade. In the 2007 Edition, an explanatory note was added to (1) recognize that a sufficient clear distance to use the required tools, such as insulated hot sticks, was required and (2) reference the user to Rule 125 (working clearances) and Rule 441 (minimum approach distances to energized parts). Insulated hot sticks sometimes require 10 ft clear distance from the guard zone to limit interference with movement during switching or other energized work.

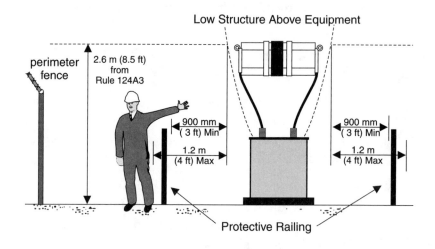

Figure H124C3-1
Location of protective railing inside perimeter fence

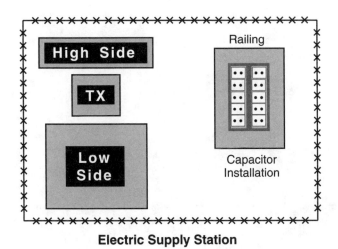

Electric Supply Station

Figure H124C3-2
Example of interior railing or fence around capacitor
bank inside electric supply station

In earlier years, some installations similar to that shown in Figure H124C3-3 were installed as step-down service "substations" behind small industrial or commercial establishments. Such installations typically consisted of overhead transformers sitting on a concrete pad, with energized jumpers dropping down to the high-voltage bushings on the transformers and secondary cables or conductors going to the building to an overhead weatherhead or underground conduit. *Such installations have never been allowed by the NESC.* They have been prohibited by every edition of the Code. Many injuries and deaths have occurred on such facilities from workers contacting the fused cutout switches or jumpers with angle or channel supported on their shoulder as they walked around the corner of the station going from one place to another, from small children walking through a damaged section of fence, etc. This installation does not meet Rule 124C3; it has no exterior fence to keep unauthorized personnel away from this installation. The vertical clearances do not meet Rule 124A1 and Table 124-1. Neither does it meet Rule 110A2; the outer fence (i.e., the only fence) is too close to energized parts for a perimeter fence.

Figure H124C3-3
Pad-mounted substation not meeting Rules 110A2, 124A1, or 124C3

Although no fence or railing height is specified in the NESC prior to the 2002 Edition, OSHA generally uses 107 cm (42 in) as a target height for railings to keep personnel from falling over them. The intention is that a person must be able to fall against the fence or railing without extending an arm or leg into the guard zone. As a practical matter, most utilities used a 1.07 m (42 in) or 1.2 m (48 in) high chain link fence to meet this requirement, if it is needed, because a chain link fence is easier to install than an appropriate fixed railing system. As of the 2002 Edition, the fence or railing height was specified as not less than 1.07 m (3.5 ft). In recent years, such a guard fence has most often been at least shoulder high—often using 1.80 m (6 ft) or 2.13 m (7 ft) fence fabric without a barbed wire extension. Note that, if a railing is to be used, no specification is given for the number of rails required to meet these requirements. Using chain link fence removes the need to consider the number of rails needed to keep someone from falling *through* into the guard zone. However, for structural purposes, to keep a falling person from crushing the fence, a top rail or taut suspension strand is generally required—mesh alone is rarely strong enough to prevent undesired deflection toward the energized parts, unless the span is very short. In essence, the NESC rules are designed to require someone in the station, whether authorized or not, to climb up to or into an area where energized parts are located. Anyone merely walking around inside the station is safe.

124C4. Mats

(This is Rule 114C6 of the Fifth and prior editions. It was moved to Rule 124C7 in the 1971 Edition and to this location in the 1981 Edition. Rule 124C4 of the 1971 Edition was moved to 124C3 in the 1981 Edition.)

124C5. Live Parts Below Supporting Surfaces for Persons

(This is Rule 114C7 of the Fifth and prior editions. It was moved to 124C8 in the 1971 Edition and to this location in the 1981 Edition. Rule 124C5 of the 1971 Edition was deleted in the 1981 Edition.)

The requirement for handrails was added in the 1971 Edition.

124C6. Insulating Covering on Conductors or Parts

(This is Rule 114C5 of the Fifth and prior editions. It was moved to this location in the 1971 Edition.)

This rule provides a general exception to Rule 124A when the insulating covering of the conductors or parts meets the requirements of Rule 124C6. Such insulation may be used in lieu of a guard even though it is within the radius of the guard zone. This exception was enlarged to allow higher voltages under certain conditions in the 1981 and 1984 Editions.

Where mats are used as guards, additional insulating guards may sometimes be necessary. Permanent insulating guards should be provided in addition to floor mats where required so that persons cannot, while touching certain live parts, simultaneously and inadvertently contact other live parts, conducting objects, or surfaces not insulated from ground.

124C7. (Not used in the current edition.)

(Rule 114C7 of the Fifth and prior editions became 124C8 when Rule 114 was moved to 124 in the 1971 Edition. See Rule 124C5. Rule 124C7 of the 1971 Edition was moved to 124C4 in the 1981 Edition.)

124C8. (Not used in the current edition.)

(Rule 114C8 of the Fifth and prior editions was deleted in the 1971 Edition. Rule 124C8 of the 1971 Edition was moved to Rule 124C5 in the 1981 Edition.)

125. Working Space About Electric Equipment

(This rule was formed in the 1971 Edition from Rules 115 and 170 of the Fifth and prior editions. Former Rule 125 was moved to Rule 133 in the 1971 Edition; Rule 133A was retained in the 1981 Edition but the remainder of the rule was deleted.)

If machinery with either live or moving parts is crowded together in a station, a hazardous condition may result. Because of restricted working space and inconvenient access, equipment may suffer from inattention and insufficient cleaning. Consequently, the equipment may deteriorate rapidly to a condition capable of endangering both personnel and continuity of service.

Working spaces about exposed live parts should be accessible only to qualified personnel. Suitable barriers may be used when necessary. Control may also be accomplished (1) through the supervision of an attendant whose duties include restraining the entrance of unauthorized persons or (2) by fencing or otherwise enclosing the area used as a working space. In any case, warning signs should be displayed prohibiting entrance of unauthorized persons (see ANSI Z535.2).

Occasional approach to live parts can be provided in such ways as removal of compartment covers from disconnecting switches. Where live parts are at both sides of a working space, a person may not be able to draw safely away from one side, in case of a slight shock or accident, unless adequate width is provided. Live parts should not be exposed at both sides of working spaces unless there is no practical alternative. The hazard from exposed live parts at both sides is increased greatly in long passageways. Where switches ordinarily are guarded, but work must be occasionally done about them, such as with remotely controlled switches, or where disconnectors and fuses must be handled occasionally, the only feasible safeguard is the provision of adequate working spaces. Then, workers can keep at a suitable distance during inspections and may freely use proper insulating tools to make adjustments. Instances may occur in existing installations where the specified working space cannot be provided. Protection may then be provided by the suitable enclosures or barriers, insulating materials or mats, whichever is most adaptable to the conditions.

Rule 125 was restructured in the 1981 Edition to detail the working space dimensions in areas exposed to 600 V or less. The width of the working space is never less than 750 mm (30 in). However, the depth of the working space varies, depending upon the condition of the wall or equipment surface at the rear of the working space (see Figure H125).

Note that there is no requirement for the access opening into equipment to be 750 mm (30 in) wide. Both the depth and width apply in front of the equipment, not inside it. As a practical matter, when work on parts deep inside narrow equipment is required, the width of the working space *in front* of equipment may need to be wider to allow the worker's body to be positioned for easy access.

Working spaces in areas exposed to more than 600 V were required by Rule 125B in the 1981 through 2002 Editions to be in accordance with the *horizontal* clearances of Table 124-1. The word *horizontal* was removed in the 2007 Edition. The intention of the rule is to provide the appropriate room for the work methods and tools that will be used on the equipment, including insulated hot stick room, if such tools will be used on the equipment.

If the exposed energized parts of conductors or equipment are de-energized before inspection or maintenance, Rule 125A3 does not specify required working space.

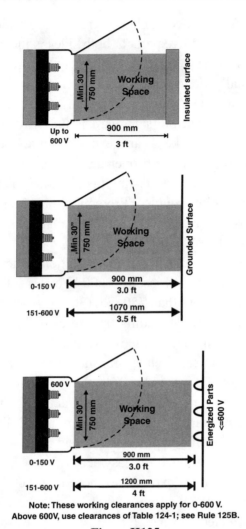

Note: These working clearances apply for 0–600 V.
Above 600V, use clearances of Table 124-1; see Rule 125B.

Figure H125
Requirements of Rule 125A and Table 125-1 for working space
about electric equipment with energized parts of 0–600 V

126. Equipment for Work on Energized Parts

(This is Rule 116 of the Fifth and prior editions.)

Part 4 of the NESC includes required methods for working on live parts and clearances to be maintained between workers and live parts.

In order for workers to comply with Part 4, both working spaces and suitable tools are needed for the work.

127. Classified Locations

(Rule 117 of the Fifth and prior editions was moved to Rule 127 in the 1971 Edition. The rule was completely revised and expanded in the 1981 Edition and in the 1993 Edition. The 1997 Edition removed specific duplication of portions of the NEC and, instead, referred directly to the NEC for guidance in specific classified locations.)

Where equipment will be subject to flammable gases, dusts, lint, or similar materials, special precautions are required to limit the opportunity for ignition of the materials or damage to the equipment. These precautions include limitations on placement of equipment that might be expected to arc during operation (or which might be affected by airborne materials) or the use of special closures for equipment. Note that Section 14 includes special precautions applicable to battery areas.

Rule 127 includes the requirements for electrical equipment and wiring installed, or used, in hazardous areas. The rule was expanded in the 1981 Edition to better define hazardous locations in electric supply stations, and to provide more detail on the requirements. In the 1984 Edition, the word *hazardous* was changed to *classified* because such areas may not be hazardous under normal conditions. The definition of *classified areas* was revised in the 1987 Edition to account for the differences in the electrical resistivities of coal dusts. No changes were made to the rule in the 1990 Edition. Prior to the 1993 Edition, Rule 127A1 classified certain areas to be *both* Class I, Division 1, Group D, and Class II, Division 1, Group F or Group G, as determined by the resistivity of the dust that might accumulate. The definition of *classified areas* in coal-handling systems was revised in the 1993 Edition to eliminate classification based on the electrical resistivity of coal dust. The rule was revised to eliminate the possible misinterpretation that dual classification, i.e., Class I, Division 1, Group D and Class II, Division 1, Group G. Class II, Division 1 or 2, is always necessary for tunnels below stockpiles and surge piles and spaces above, below, and inside coal-storage silos and bunkers.

The classification for preparation plants and coal-handling areas where electrically conductive coal dust might accumulate was eliminated. The classification coal-handling areas on the basis of dust interfering with heat transfer from electrical equipment was also eliminated. However, the Class II definition of NEC Article 500 was retained, and electrically conductive coal dust and interference with normal equipment operation is included in the classification scheme by reference.

Rule 127 specifically incorporates selected articles from Chapter 5 of the NEC. The rule requires that electrical equipment and wiring shall meet the requirements of Article 500—*Hazardous (Classified) Areas.* The NESC therefore includes the definitions of classified areas, i.e., Classes I, II, and III, as they are given in the NEC. However, Rule 127B provides further definition of flammable liquids. Rule 127 also includes the general NEC requirements for Class I, II, and III locations that are given in Articles 501 through 503, and the specific location requirements of Articles 511 through 517. The specific locations, e.g., gasoline dispensing stations, health care facilities, etc., are sometimes located at an electric supply station and they are included in the scope of the NESC. However, these specific locations are also included in the scope of the NEC.

The scope of the NEC does not include electric supply stations, and NEC Chapter 5 does not provide requirements that are specific to supply stations, in particular, coal-handling areas. Rule 127A defines the NEC classification of areas that are unique to electric supply station coal-handling systems. The rule also includes the ventilation requirements, and the requirements for electrical equipment and wiring that is installed, or used in coal-handling areas.

Prior to 1997, Rules 127C, D, E and F included requirements for electrical equipment and wiring installed or used in flammable liquid storage, loading and unloading areas, gasoline dispensing stations, and boilers located in electric supply stations. These requirements were based on NEC Articles 500 and 501, but included additional requirements to reduce the probability of electrical sparks and stray currents. In the 1997 Edition, the previous specific requirements were deleted and users were referenced to applicable sections of NFPA 30 and NFPA 70 (NEC). As a result, Tables 127-1, 127-2, and 127-3 were deleted. Similarly, the *specific* requirements of Rules 127K and 127L were deleted, users were referenced to NFPA 58 and NFPA 70, and Tables 127-4 and 127-5 were deleted. References to NFPA 497M and ARI RP 500 were added to Rule 127L in 1997.

Electrical installations in classified areas must be acceptable for the application. Acceptability is determined by conformance with national standards for the design, manufacturing, and testing of electrical equipment to be used in classified areas. The exceptions and reductions allowed under the conditions specified in NEC Articles 503 through 504 and 511 through 517 are allowed by the NESC if the specific requirements of NESC rules are not violated. Note that an interlocking ventilation system is required when a classification is reduced pursuant to Rule 127A4.

The 2002 Edition modified Rule 127A to limit vapor-air concentration to less than 5% of the lower flammable limit in coal handling areas; previously, the user had been sent to NEC Article 500. The specific ventilation requirements of Rule 127A3 were deleted to eliminate confusion between requirements for equipment and requirements for areas. Where only wet coal is handled, the atmospheric concentration of entrapped volatiles was limited to 8% or less.

It is good practice to consider practical alternatives to the installation of electrical equipment and wiring in areas that are classified.

128. Identification

(This is Rule 119 of the Fifth and prior editions. It was moved to this location in the 1971 Edition.)

The ability to readily identify and trace the connections of equipment, particularly such grouped arrangements as occur commonly at switchboards and in bus chambers, both (1) facilitates repairs and increases reliability of service, and (2) reduces incidences in which workers handle energized parts in the mistaken belief that the parts are de-energized. Labeling frequently provides the best means for identifying switchboard circuits. Sometimes code lettering or a color scheme is used successfully.

Simple and orderly circuit arrangements promote safety, so much so that a multitude of conductors arranged neatly in parallel lines and tagged or labeled have proven to be safer than a lesser number that are crossed and unlabeled.

Parts that are interchangeable, such as some types of switch-compartment doors, should not carry the identification mark. In such instances, a greater hazard may be created than if no identification were used. Care and foresight should be exercised when selecting the methods used for identification.

Individual machines that may be moved and installed in different locations for different uses should have a nameplate giving important operating data. The labels for rotating equipment should specify the capacity rating, speed, voltage and, when necessary, the frequency and condition of operation (such as series or shunt characteristics). Power ratings for both continuous and intermittent loads may be desirable.

129. Mobile Hydrogen Equipment

(This is a part of a new rule that was added in the 1971 Edition as Rule 134. The original rule was expanded and moved to Rule 127 in the 1981 Edition. At that time, mobile equipment was specified in this rule. Rule 132 of the Fifth and prior editions was moved to 142 in the 1971 Edition. In the 1984 Edition, this rule was moved to 132; it was moved to 129 in the 1990 Edition.)

Bonding of mobile hydrogen systems limits potential voltage differences capable of causing sparks.

Section 13. Rotating Equipment

(This is Section 12 of the Fifth and prior editions; former Section 13 was moved to Section 14 in the 1971 Edition.)

130. Speed Control and Stopping Devices

(This is Rule 120 of the Fifth and prior editions; former Rule 130 was moved to 140 in the 1971 Edition.)

130A. Automatic Overspeed Trip Device for Prime Movers

(This is Rule 120A of the Fifth and prior editions.)

Automatic speed-limiting devices are important features of certain types of turbines and engines. Failures of rotating parts by overspeed occur more frequently than do boiler explosions. Speed-limiting devices are particularly needed with steam turbines and belted water turbines, except certain designs of reaction turbines. Even reciprocating engines are frequently fitted with extra valves and independent speed-limiting mechanisms.

Because generating loads may change suddenly from overload to nearly zero, due to the opening of automatic circuit breakers or fuses, the speed-limiting device may need to respond quickly. However, the control devices and piping system must be designed to limit damage to the feeder piping when cutting off the steam or water supply from the engine or turbine.

IEEE Std 502™ *IEEE Guide for Protection, Interlocking, and Control of Fossil-Fuel Unit-Connected Steam Stations* (ANSI) and IEEE Std 1010™ *IEEE Guide for Control of Hydroelectric Power Plants* (ANSI) provide additional information for automatic protection for generating stations.

130B. Manual Stopping Devices

(This is Rule 120B of the Fifth and prior editions.)

Multiple control devices are often desirable for stopping machines that drive electric power generators. In emergencies, this may save valuable time, especially where the control systems for individual equipment are not centrally located. Relay control circuits can allow easy operation of a single valve or disconnecting device from several points.

Control circuits must be properly installed and identified. In emergencies, it is all too common for operators to become confused and control the wrong equipment. It is absolutely imperative, from both safety and operation standpoints, that careful attention be given to the method of identification, type, size, color, wording, and location of labels or markings.

In any case, emergency controls must be located so as to allow the operator to stop the equipment in a timely manner without becoming endangered by the equipment.

The middle paragraph of the former rule was deleted in the 1971 Edition. In some cases, where appropriate secondary control systems are available, it is *not* appropriate to stop the equipment because a control system has failed. This only contributes to discontinuity of service. It may, of course, be appropriate for failure of such control systems to be identified automatically to control operators. The 1990 Edition added motor-generator sets and clarified that the rule applied to machines driving generators, not to motors.

130C. Speed Limit for Motors

(This is Rule 120C of the Fifth and prior editions.)

Separately excited, dc motors are particularly liable to "run away," since their field-excitation current may be greatly reduced while the armature current is still maintained. To a lesser degree, series motors, ac motors of series characteristics, motor-generators operating in parallel, or feeding storage batteries and rotary converters are also subject to runaways. Where such motors are directly connected to mechanical load, dangerous overspeed is not likely to occur. However, where they are belt-connected or have only a generator load subject to the opening of automatic circuit breakers, the danger of overspeeding is considerable. Centrifugal devices are most often used to actuate trip devices to disconnect the source of energy when an overspeed condition occurs. An audible signal that automatically warns of excessive speeds is also advisable.

Motor-generators and converters were deleted from the rule in the 1981 Edition.

130D. (Not used in the current edition.)

(Rule 120D of the Fifth and prior editions moved to 130D in the 1971 Edition; this rule then was added to 131 in the 1990 Edition.)

130E. Adjustable-Speed Motors

(This is Rule 120E of the Fifth and prior editions.)

When the speed of dc motors is adjusted through field control, dangerously weak fields must be avoided to limit dangerous speed. This protection is especially important when the load is (1) not directly connected to the motor or (2) consists of generators subject to release of load by breakers. Release coils may be placed on starting rheostats or other parts through which the field circuit passes to prevent loss of fields during operation. In some cases centrifugal speed-limiting devices are installed.

These control devices should be tested at frequent intervals to ensure proper operation.

130F. Protection of Control Circuits

(This is Rule 120F of the Fifth and prior editions.)

Mechanical protection is essential to ensure reliability of electrical circuits controlling stopping devices.

Circuit configurations that result in stopping the generator or motor when a circuit failure occurs, such as an open circuit, short circuit, or component failure, can enhance the dependability of the protection system. However, there may be some reduction in system availability. Mechanical protection, protection system redundancy, or both, may be used to meet the requirements of this rule where the reduction in availability is unacceptable.

131. Motor Control

(This is Rule 125A of the Fifth and prior editions. It was moved to Rule 133A in the 1971 Edition and to this location in the 1981 Edition; former Rule 125B was moved along with it to 133B in the 1971 Edition, but went into expanded Rule 127 in the 1981 Edition. Part of Rule 121A of the Fifth and prior editions was moved here in the 1971 Edition; that rule was deleted in the general 1981 revision. Rule 131 of the Fifth and prior editions was moved to 141 in the 1971 Edition. Rule 120D of the Fifth and prior editions was relocated to 130D in the 1971 Edition; it was extensively revised to require working and time delay for automatic starting and was added to Rule 131 in the 1990 Edition.)

Motors that automatically start and are not controlled in the immediate vicinity may be a hazard to workers in the area. In such cases, motor controls or the motor power disconnecting device, or both, are required to have provisions that allow tag-out procedures, or the power disconnecting device must have provisions for locking in the disconnect position.

Protection of motors against starting following an interruption of the power supply during low-voltage conditions is a common safety measure. Two forms of this protection are low-voltage protection and low-voltage release.

The motor is automatically disconnected from the electricity source. It will not permit the motor to start again unless a manual starting operation is used after the voltage is restored. Like low-voltage protection, low-voltage release disconnects the motor from the line when voltage falls below specified limits. However, low-voltage release mechanisms allow the motor to restart immediately when normal line voltage is restored.

A requirement that sufficient warning and time delay to allow personnel to take action to prevent injury before automatic restarting occurs was added in the 1990 Edition. The requirement to have the motor control switch less than 15.2 m (50 ft) from the motor was removed in the 2002 Edition. All motor control switches must now have lockout provisions to meet NESC Rule 444 and OSHA 1910.269(d).

132. (Not used in the current edition.)

(Rule 132 of the Fifth and prior editions was moved to 142 in the 1971 Edition. Rule 132 of the 1984 and 1987 Editions was moved to 129 in the 1990 Edition.)

133. Short-Circuit Protection

(This rule was added in the 1990 Edition.)

A requirement was added to provide motor-fault-current protection in the 1990 Edition. This protection may be installed at the motor, or at a remote location, and provide protection for other circuit components in the power supply to the motor. The short-circuit protection must automatically de-energize the motor power supply and it should be selected such that personnel are protected in the event of a fault in the motor.

134. (Not used in the current edition.)

(Rule 134 of the Fifth and prior editions was moved to 143 in the 1971 Edition.)

135. (Not used in the current edition.)

(Rule 135 of the Fifth and prior editions was moved to 145 in the 1971 Edition and to 144 in the 1981 Edition.)

136. (Not used in the current edition.)

(Rule 136 of the Fifth and prior editions was moved to 146 in the 1971 Edition and was deleted in the 1981 Edition.)

137. (Not used in the current edition.)

(Rule 137 of the Fifth and prior editions was moved to 144 in the 1971 Edition and was deleted in the 1981 Edition. See Rule 124.)

138. (Not used in the current edition.)

(Rule 138 of the Fifth and prior editions was moved to 147 in the 1971 Edition and to 145 in the 1981 Edition.)

Section 14. Storage Batteries

Unlike coal-handling areas and other hazardous areas addressed by Rule 127—*Classified Locations*, storage battery areas are not considered as classified locations by the NESC. This section addresses the special concerns relating to ignition of hydrogen, damage to the equipment, and potential personnel-related issues.

140. General

(This is Rule 130 of the Fifth Edition. The rule was generalized in the 1971 and 1981 Editions. Former Rule 140 moved to 150 in the 1971 Edition. The words sealed *and* unsealed *were used in Rule 140 in the 1971 Edition. These words were determined to be obsolete and misleading and were deleted in the 1981 Edition.)*

The requirements of Section 14 provide adequate safety for the maintenance and operation of battery areas. IEEE Std 484™ *IEEE Recommended Practice for Installation Design and Installation of Vented Lead-Acid Batteries for Stationary Applications* (ANSI) is a useful resource for information on methods of meeting Section 14.

141. Location

(This rule was moved from 130 in the Fourth Edition to 131 in the Fifth Edition and to 141 in the 1971 Edition. Former Rule 141 was moved to 151 in the 1971 Edition.)

The presence of electrolyte will decrease the resistance of a personnel contact with a live part and, thus, increase the danger therefrom. Sparks in the gas given off by storage batteries in charging may also be dangerous, especially in rooms with low ceilings. Injury to insulation of other equipment by acid spray may also occur where the battery is not isolated from such other equipment.

For these reasons, battery equipment should be made inaccessible except to qualified persons. It should be placed in a room or compartment away from other equipment.

Extensive editorial changes were included in the 1981 Edition. Rule 141 was restructured in the 1981 Edition to detail the types of battery enclosures that can be utilized.

142. Ventilation

(This rule is numbered 131 and 132 in the Fourth and Fifth Editions, respectively. Former Rule 142 was moved to 152 in the 1971 Edition and deleted in the 1981 Edition; see Rule 123. The intent of Rule 152 was included in Rule 152A1 in the 1981 Edition.)

Rule 142 requires natural or powered ventilation to prevent the accumulation of hydrogen to an explosive mixture. Some types of electric storage batteries generate little or no hydrogen; where such batteries are used, accumulation of significant amounts of hydrogen is unlikely regardless of whether a fan system is installed specifically for the storage battery area.

With large battery equipment, especially in comparatively small rooms, special ventilation by fans may be necessary to reduce hydrogen gas accumulations depending upon the type of battery system used. The battery enclosure and its ventilation should be designed and constructed to limit the opportunity for pockets of high concentrations of hydrogen gas to occur. Gas pockets in ceiling spaces above door and window openings should be avoided. In order to ensure adequate safety, hydrogen content should be limited to less than 2% by volume. Hydrogen-air mixtures are explosive between 4.1% and 74.2% hydrogen by volume of the enclosed space. The requirement to annunciate the failure of continuously operated or automatically controlled ventilation fans was added in the 1981 Edition.

In addition to hydrogen gas, the air from a battery room may contain sulfuric acid spray, small quantities of which have a rapidly destructive effect on both the insulation and metallic parts of electrical apparatus.

The ventilating system should, therefore, be designed to carry air from the storage battery room directly to points outside the building. For the same reason, care should be taken in designing the ventilating system.

Covers or guards arranged to catch the electrolyte spray and return it to the cell are readily devised and applied. Sometimes a beveled edge to each cell is helpful. Sometimes glass plates or other covers placed above the elements prevent the mechanical expulsion of electrolyte, even during violent gassing.

143. Racks

(Rule 134 of the Fifth Edition was moved to this location in the 1971 Edition. Rule 143 of the Fifth and prior editions was moved to 153 in the 1971 Edition and to 152 in the 1981 Edition.)

The racks that support the cells (or support the trays that hold the cells) should be designed to meet expected conditions of service. They should be made or coated with materials that are resistant to damage from electrolytes and should adequately support the cells or trays. Even in areas of reduced seismic activity, floor-mounting is recommended. Dual-mounting to both floor and wall is especially inappropriate in areas of high seismic activity.

144. Floors in Battery Areas

(This is Rule 135 of the Fifth Edition. Former Rule 144 was deleted in the 1981 Edition. See Rule 124.)

Good design for large battery rooms includes a supply of potable water, raised entrance-door sills, and acid-resisting floors (such as vitrified brick set in pitch). Wooden floors are not satisfactory.

Structural steel and other metallic systems under the flooring of battery rooms may require protection from spilled electrolyte. One method is to lay sheet lead so as to interrupt capillary communication between the floor and the steel. Another method is to paint the floor with acid-proof paint.

Steel frames or other metallic fixtures exposed inside the battery room should be protected from the destructive action of acid spray by acid-proof paint. All copper, brass, and iron should be coated.

145. Illumination for Battery Areas

(This rule was developed from Rules 134 of the Fourth Edition, 138 of the Fifth Edition, and 147 of the 1971 Edition. The requirements of the Fifth and 1971 Editions concerning heaters were deleted in the 1981 Edition. See Rule 127. Former Rule 145 was deleted in the 1971 Edition; see Rule 111E. Rule 145 of the 1971 Edition was moved to 144 in the 1981 Edition.)

In order to avoid danger of explosion, no flame devices for illumination are allowed in battery rooms (see Rule 127). Vapor-proof globes are recommended to protect against acid spray. Switches and receptacles and other electrical equipment that may cause an electrical arc during normal operation, e.g., ventilation fan contacts, should be placed outside of the room or in vapor-proof enclosures.

146. Service Facilities

(This rule was added in the 1981 Edition. Rule 146 of the 1971 Edition was deleted in the 1981 Edition.)

This rule was developed in the 1981 Edition to detail the safety items required for personnel protection during battery maintenance and installation. Some eye-protection requirements were slightly relaxed in the 1984 Edition.

Adequate body-care facilities increase the effectiveness of workers and decrease the opportunity and severity of accidents. Water facilities, whether they are portable or stationary, should contain an adequate volume of potable water.

Safety signs are required inside *and* outside of a battery *room*, or in the vicinity of a battery *area* within a room to specifically prohibit smoking, sparks, or flames. The reference to ANSI Z535 safety sign standards was added in 1997 (see NESC Handbook Appendix B).

147. (Not used in the current edition.)

(Rule 147A of the 1971 Edition was moved to 145 in the 1981 Edition; Rule 147B was deleted; see Rule 127.)

Section 15. Transformers and Regulators

150. Current-Transformer Secondary Circuits Protection When Exceeding 600 V

(This is Rule 140 of the Fifth and prior editions; former Rule 150 was moved to 160 in the 1971 Edition.)

The intent of Rule 150 is to provide personnel protection against the effects of accidentally opened or damaged CT secondary circuit conductors *in the vicinity of* primary circuits energized at more than 600 V. Accidentally opened or damaged CT secondary circuit conductors may cause excessively high voltage and arcing that will not be cleared by circuit protective devices. The opening of a current-transformer secondary may cause the insulation to break down or cause serious arcing and danger at the point of opening. If suitable short-circuiting devices are provided, accidental openings are less likely to occur while instruments are being removed or replaced. Because secondary conductors are usually small, with relatively thin insulation, an appropriate means of physical protection is required to ensure safe operation.

Protection in the form of grounded metallic conduit or grounded metallic covering was required until 2002, when nonmetallic ones were recognized, when the CT installation is within an electric supply station (see definition) and the CT or the secondary circuit(s) are *in the vicinity of* a primary conduit energized at more than 600 V (as connected to such a primary circuit). The 2002 Edition requires consideration of circulating currents when nonmetallic conduits or U-guard coverings are used. However, protection is not required for short lead lengths at the secondary terminals of the CT transformer.

The term *primary voltage area* was not defined until 2002, where some help was given, using the term *in the vicinity of*. What is meant by *area* or *in the vicinity of* must still be determined by a competent engineer or designer.

151. Grounding Secondary Circuits of Instrument Transformers

(This is Rule 141 of the Fifth and prior editions; former Rule 151 was moved to 161 in the 1971 Edition.)

The low-voltage and low-current windings of current and voltage transformers used for station metering, relay uses, and control should be effectively grounded. In some cases, such as with voltage-regulator control circuits, proper and reliable operation requires that the entire secondary circuit be ungrounded. As a result, such circuits are subject to leakage or induction of high voltages. Safety requires that such circuits (1) be run in all respects as required for high-voltage circuits and (2) be clearly distinguished by suitable markings from other low-voltage circuits with which it may be associated; see IEEE C57.13.3 *Guide for Grounding of Instrument Transformer Secondary Circuits and Cases* for additional information.

152. Location and Arrangement of Power Transformers and Regulators

(This rule was developed from Rules 107B and C and 143 of the Fifth and prior editions; former Rule 152 was moved to 162 in the 1971 Edition; see also the discussion of Rule 115.)

Rule 152 is intended to minimize the potential conflict between transformers and authorized personnel, whether within a vault or within an outdoor fenced enclosure. Either the energized parts shall be enclosed or guarded, or they shall be isolated in accordance with Rule 124. There is no intended preference between these alternatives. In both alternatives, the transformer case must be grounded in accordance with Rule 123. This changed in the 1981 Edition; only transformers with enclosed live parts were required to be grounded by the 1971 Edition. Rule 110A applies in both cases. Note that a typical underground system pad-mounted transformer meets these requirements by having an outer, grounded case that, in effect, forms a small vault enclosure; within the enclosure, live parts are guarded.

Where transformers are installed with utilization equipment, they are frequently on poles as a part of the yard wiring. If they are not in an area meeting Rule 110A, they must comply with the rules for overhead lines as to clearance from buildings, nonobstruction of climbing space, etc. Even if the installation qualifies as a station under the rules of Part 1, the climbing space requirements of Part 2 should be met, as a practical matter. When transformers are placed against walls of buildings, they should be sufficiently distant from adjacent window openings to ensure (1) that burning oil will not cause a fire hazard, and (2) that persons in or on buildings will not inadvertently contact the frame or high-voltage leads.

Where transformers are placed inside buildings containing other equipment, they may be placed in vaults; these usually will be particularly necessary (1) in buildings that are not used solely for station purposes, and (2) where the amount of oil in the transformer casing is considerable. The wiring and spaces within the vault should comply with the rules for stations or for underground construction, and the interior must be accessible only to authorized personnel.

Rule 152A2 was revised and Rule 152B3 was added to recognize transformer liquids other than oil and recognize various methods of limiting fire hazards associated with liquid-filled transformers in the 1993 Edition.

The 75 kVA value, a lower limit of Rule 152B, matches that of the NEC. See the discussion of Rule 115 for a more complete discussion of oil-filled apparatus.

Oil-filled transformers in pad-mounted enclosures in areas accessible to the public are covered in Part 3; where the requirements of Rule 110A are met, Part 1 applies. The issue of the fire hazard of any contained oil is not addressed in Part 3.

Approved vault construction is specified in Part 3 of the NESC.

153. Short Circuit Protection of Power Transformers

(Rule 153 of the Fifth and prior editions was deleted in the 1971 Edition; see Rules 124 and 127. Rule 153 of the 1971 Edition was moved to 152 in the 1981 Edition. Rule Number 153 was unused from 1981 until the 1997 Edition added requirements for short circuit protection of power transformers.)

If a power transformer suffers a high-magnitude internal fault, the results can be catastrophic for the transformer and/or upstream facilities, if the faulted transformer is not promptly removed from the system. Such protection is especially critical for generator step-up transformers and station auxiliary transformers, in which case the generator electric field and mechanical energy source must be disconnected. These protection requirements apply to power transformers, but are not required for transformers used specifically for control, protection, or metering.

The rule intentionally allows single-phase protection where that is appropriate.

154. (Not used in the current edition.)

(Rule 154 of the Fifth and prior editions was moved to Rule 163B in the 1971 Edition; see Rule 127.)

155. (Not used in the current edition.)

(Rule 155 of the Fifth and prior editions was moved to Rule 163A in the 1971 Edition.)

156. (Not used in the current edition.)

(Rule 156 of the Fifth and prior editions was deleted in the 1971 Edition.)

Section 16. Conductors

(This is Section 15 of the Fifth and prior editions; former Section 16 was moved to 17 in the 1971 Edition.)

The scope of this section is conductors that connect the electric energy sources, such as transmission lines and generators, to power transmission equipment and utilization equipment, such as transformers and motors. This section does not cover conductors that are engineered and manufactured as part of electrical equipment. Conductors that are integral with rotating equipment, storage batteries, transformers and regulators, etc., are covered in other sections of this part of the Code. However, conductors that are integral with metal-enclosed bus are covered by this section. This section covers conductors used for transmission of electric power, control signals, and analog and digital data signals (instrumentation). As used in this section, the term "conductor" includes the devices that connect to electrical equipment, such as connectors and stress cones, as well as equipment such as splices and shield wires.

160. Application

(This rule was created in the 1993 Edition. Rule 160 in the 1990 Edition was moved to 161 in the 1993 Edition.)

The first sentence of Rule 160 of the 1990 Edition was moved to this location in the 1993 Edition, and an ampacity requirement was added.

161. Electrical Protection

(This rule was formed as Rule 160 in the 1971 Edition from Rules 150 and 165 of the Fifth and prior editions; former Rule 150C was deleted in the 1971 Edition. Former Rule 160A was moved to 170 in the 1971 Edition; former Rule 160B was moved to 173B. A new Rule 160C was added in the 1990 Edition to require short-circuit protection on insulated power cable. The rule was

renumbered to Rule 161 in the 1993 Edition when the new Rule 160—Applications, was added. Former Rule 161 was moved to 172 in the 1971 Edition.)

Rule 161 applies in electric supply stations; there is no corresponding rule specifying overcurrent protection for electric supply lines outside of electric supply stations.

Protection of persons in the vicinity of switches or conductors, or operating switches on circuits, requires that live conductors have adequate, automatic protection against currents that are large enough to (1) exert disruptive stresses, (2) cause serious arcing or short-circuits at switches, (3) melt connections or the conductors themselves, or (4) seriously damage insulation.

Electrical protection for conductors is required to limit the potential personal hazards that can result from failure of conductor, conductor supports, and conductor insulation, i.e., conductor faults. For air-insulated conductors, failure of the conductor or conductor support will often result in contact between the energized conductor and personnel, surrounding structures, or earth. For other insulating materials—e.g., oil, gas, or solid dielectrics—failure of the insulating material may have the same result. Contact between surrounding structures or earth and conductors energized from high-energy sources can cause high currents, electrical arcing, and ignition and combustion of material. Electrical shock and severe burns have been caused by conductor and insulation failures.

Another potential benefit of electrical protection for conductor faults is reduction of the degree of personal injury due to unintentional contact with energized conductors. However, conductor fault protection cannot eliminate injuries, and is not a substitute for proper guarding, clearances, and procedures.

Electrical conductors that are connected to a source that can produce high-magnitude current when there is a conductor fault, or conductors that normally carry high-magnitude currents (generally referred to as power conductors) should have automatic trip devices that will interrupt the flow of current in the conductor, or cause a rapid reduction to essentially a zero-current level upon failure of the conductor or the conductor insulation.

Most automatic trip sensors are the type that monitor current and cause actuation of protective devices when a predetermined current level is reached. Because conductors carry load current during normal operation, the trip current level must be greater than normal load current. For this reason, overcurrent trip devices usually do not respond to high-impedance faults of the conductor or insulation. More sensitive automatic trip devices have been developed for high-impedance faults, and they should be applied where it is practical and reasonable to do so. Differential relay schemes, neutral overcurrent relays, and ground-fault interruption (GFI) devices are examples of more sensitive automatic trip devices. All of these devices will meet the requirements of this section of the Code, whereas the NEC permits only the use of overcurrent devices (see NEC Article 240-3).

Some automatic trip devices include both the fault sensor and the circuit interrupter in a single device (e.g., fuses) and some trip devices have the sensors separate from the interrupter. Either arrangement meets the requirements of this section. In addition, the automatic trip sensor, or the circuit interrupter, or both, may be in the electrical circuit at a point that is remote from the conductor for which the fault protection is provided. For example, a differential relay scheme and a circuit breaker on the high side of a transformer can provide protection for faults in the conductors on the transformer secondary.

Devices that can detect conductor faults and reduce the conductor current to essentially zero magnitude also meet the requirements of this section. For example, the conductors that are connected to a steam turbine-generator may have fault protection that includes a differential relay scheme, closure of the turbine steam stop valves, and prompt reduction of the generator field current.

It is also common practice to connect a single device to the conductor that provides both conductor fault protection and equipment fault protection. This practice is allowed by the Code.

A grounded conductor should be neither interrupted nor disconnected from ground by the opening of an automatic circuit breaker or fuse. Such an opening could permit part of the circuit to lose its ground connection and assume the highest voltage of any circuit to which it is exposed or from which leakage might occur.

Equipment and circuits generally should be protected by fuses or automatic circuit breakers. However, where a greater hazard might be caused by the opening of circuits automatically than by overloads and short circuit, other protection methods may be appropriate.

It is generally good practice to provide electrical protection of conductors to limit the magnitude and duration of conductor overloads. Overload of a conductor occurs when the current magnitude and duration exceeds the conductor ampacity or the mechanical strength of the conductor or its supports.

A conductor has a continuous ampacity that is a function of the conductor configuration, insulation, and ambient conditions. The conductor may also have short-duration ampacity that is considerably greater than the continuous ampacity. The mechanical strength of a conductor or its supports is usually described in terms of forces that cause yielding of material or deflection. The mechanical strength is generally the same for continuous or short-time application of the forces.

Above-normal-magnitude conductor current, and above-normal mechanical load on conductor and supports, can occur when there are electrical failures in connected equipment, i.e., equipment faults. Above-normal current levels can also occur because of the electrical system configuration and operation. For example, an outage on a transformer bank may result in a bus-current magnitude that exceeds the continuous ampacity of the bus. Conductor and conductor supports should always be designed to withstand the expected mechanical forces without yielding. However, above-normal current may or may not result in a conductor overload, i.e., current in excess of conductor ampacity.

Conductor overloads cause degradation of the conductor and conductor insulating materials; material degradation can, in turn, lead to failure of the conductor or insulation. Fault protection will provide personnel protection on conductor or insulation failure, but there may be some exposure to energized conductors, arcing, or burning material. For that reason, overload protection to prevent conductor and insulation degradation is considered to be good practice.

The decision to apply a conductor with an ampacity greater than all expected current magnitudes and durations, or to apply or not apply overload protection, should consider the potential level of risk to the general public, and to personnel who may be in the vicinity of a conductor fault.

Overload protection may be in the form of automatic trip devices that sense the current magnitude and de-energize the conductor. This type of overload protection is often combined with conductor fault protection in a single device, e.g., circuit breakers, automatic reclosers, or combination motor starters. Overload protection can also be indicating devices or alarms that notify personnel that the condition exists.

The words of Rule 160A refer to overcurrent, alarm, indication, or trip *devices*. The words "alarm, indication, or trip devices" were substituted in the 1981 Edition for "protection" because automatic overcurrent protection is not necessary in many cases.

With certain types of circuits, the use of fuses or automatic circuit breakers operating on overloads is relatively unnecessary. Resistors, reactors, or suitable regulators may satisfactorily limit the possible currents in circuits from generators or batteries. Series arc circuits that are supplied from special generators are examples of circuits that are limited by their design to a certain maximum current.

Although overloads are imposed upon rotary converters or motor generators from the load side, it is better practice under overload conditions to interrupt the supply circuit rather than the conductors feeding the load. One reason for not opening the dc side of synchronous converters until *after* the ac side has been opened is that synchronous converters will flash if the dc circuit breaker is opened when the current is commutated satisfactorily with the dc circuit closed. DC circuit breakers may, however, be appropriately applied to individual feeder circuits of a group supplied from such a machine; the overload upon an individual feeder can then be interrupted automatically without the need to cut the current off entirely from the machine. In some modern automatic railway substations, the distribution system is being operated with the substation as a single unit and no provision is made for opening outgoing feeder circuits on overload.

When a storage battery is connected in parallel to the dc side of such a machine and the main circuit breaker is on the ac side, the battery should be prevented from backfeeding the machine upon failure of power; the use of a reverse-current or reverse-power relay may be appropriate. However, such equipment cannot be used when power is intended to flow back to the ac side, as in the regenerative control of railway trains.

The secondary current of a series transformer is limited by the current in the primary, and no automatic overload protective device is needed. The interruption of the secondary could cause an abnormal voltage in the secondary and an abnormal reactance in the primary circuit and, hence, would be objectionable.

162. Mechanical Protection and Support

(The mechanical protection part of this rule was Rule 151 of the Fifth and prior editions; it was moved to 161 in the 1971 Edition and to 162 in the 1993 Edition. Former Rule 162 was merged into Rule 173C in the 1971 Edition. The support strength requirements were first added in 1993 as Rule 110D and moved here in 1997.)

Usually the insulating covering of a conductor is not designed also to be a mechanical protection. If it is to retain its insulation function, it must be protected against any mechanical damage so that its value as a dielectric will remain undiminished.

Likewise, the insulation itself must be able to withstand the internal stress placed upon it by conductor conditions.

Where underground cables come out of risers and go up to serve facilities, they must be mechanically protected. The type of protection depends upon the expected type and characteristics of the pedestrian or vehicular traffic and material movement to which it is exposed.

This rule was consolidated and simplified in the 1977 Edition.

The requirement to ensure that structures supporting both ends of spans extending into public areas will meet the strength requirements for facilities in public areas was added in the 1993 Edition. The rule applies only the requirements of Part 2 for grades of construction, strength and loading, and line insulators to these structures. Requirements for clearances, grounding, inspection and testing, arrangement of equipment, and lighting of these structures are included in Part 1, Sections 24–27.

Supply stations are included in the scope of ASCE 7 (formerly A58.1-1982), Minimum Design Loads for Buildings and Other Structures. While it is good practice to design structures within or containing electric supply stations to meet ASCE 7 or similar requirements, no such requirement is explicitly stated in the NESC.

As a practical matter, many supply station structures are designed to equal or greater standards than those contained in Part 2 of the NESC or ASCE 7, because of the overall importance of the station to system operation. Thus, this new rule can be considered a clarification and should not impose additional duties upon the designer.

163. Isolation

(This is Rule 152 of the Fifth and prior editions; it was moved to 162 in the 1971 Edition. It was moved again in the 1993 Edition to Rule 163 when a new Application rule was added. Former Rule 163 was deleted in the 1971 Edition; see Rule 123C.)

When conductors are placed at elevations well above the heads of persons underneath, the conductors rarely will sustain mechanical injury to their insulation, if any, and they will offer little hazard to persons in the vicinity. Adequate elevation provides protection to personnel that is equivalent to actual guarding of the conductors by casing or armor (see Rule 124).

The 1981 Edition recognized limited uses of nonshielded conductors by its additions to this rule and to Rule 124C6.

Insulated conductors with shields that are ungrounded (or upon which there may exist a voltage of indeterminate potential) must be guarded or isolated in accordance with Rule 124.

164. Conductor Terminations

(This rule was formed in the 1971 Edition from former Rules 155 and 153A. It was moved to 164 in the 1993 Edition when a new Application *rule was added. Rule 164 of the Fifth and prior editions was moved to 173 in the 1971 Edition.)*

164A. Insulation

(Rule 155—Taping and Joints of the Fifth and prior editions was moved to 163A in the 1971 Edition and to this location in the 1993 Edition.)

Unless joints of insulated conductors are otherwise guarded, they should be protected with an insulating covering that meets the requirements of Rule 124C6a. Good practice requires either (1) that the joints first be made both mechanically and electrically secure without solder, and then soldered or (2) that some form of approved connector be used. Joints of insulated conductors are, thus, required to be either (1) guarded with insulation meeting Rule 124C6a, (2) guarded with a physical barrier system meeting Rule 124C6b, or (3) isolated in accordance with Rule 124A.

Since shielding is not provided at terminations, insulated conductor equipment at above 2500 V to ground should be guarded on isolation. The guarding requirements apply to the ends of the conductors as well.

164B. Metal-Sheathed or Shielded Cable

(Rule 153A of the Fifth and prior editions was moved to Rule 163B in the 1971 Edition and was moved here in the 1993 Edition; Rule 153B was deleted at the time.)

Metal-sheathed cable affords adequate protection when properly installed. In dry locations, the metal sheath need not be continuous over splices if it is suitably bonded electrically across the splice, such as with a suitable metallic braid. The bond should have a current capacity not less than an AWG No. 6 copper wire. In damp locations, the sheath must not be interrupted.

Both safety and reliability are served by the use of potheads or equivalent methods to protect cables at the ends or outlets from moisture, mechanical injury, and electrical strains.

165. (Not used in the current edition.)

(Rule 165 of the Fifth and prior editions was deleted in the 1971 Edition; see Rule 160A.)

166. (Not used in the current edition.)

(Rule 166 of the Fifth and prior editions was moved to 174 in the 1971 Edition.)

167. (Not used in the current edition.)

(Rule 167A of the Fifth and prior editions was merged into Rule 170 in the 1971 Edition; former Rule 167B was merged into Rule 122B.)

168. (Not used in the current edition.)

(Rule 168 of the Fifth and prior editions was merged into Rule 123 in the 1971 Edition.)

169. (Not used in the current edition.)

(Rule 169 of the Fifth and prior editions was merged into Rule 124 in the 1971 Edition.)

Section 17. Circuit Breakers, Reclosers, Switches, and Fuses

170. Arrangement

(Rule 170 of the 1971 Edition was formed from prior Rules 160, 164, and 167A. Former Rule 170 was merged with 1971 Rules 122, 124, 125, and 180.)

Switches and other control or protective equipment should be very convenient to the operator; no other part of the station installation is used so often during operation and in emergencies. Although accidental operation may cause serious danger to service, to operators, and to equipment, it can be practically eliminated through careful design and arrangement of control and protective equipment. Effective identification of equipment and its operating functions substantially increases the efficiency of well-trained operators.

For station operation, it is not always practical to enclose fuses or circuit breakers in cabinets, as it usually is for utilization equipment. Severe burns and eye flashes are not uncommon and must be considered when handling such appliances. The intensity of the light from the arc may alone cause a severe eye injury. Proper location or shielding tends to overcome such hazards.

Conspicuous markings are required to facilitate identification of devices by employees intended to operate or work on them. The 2002 Edition prohibits duplication of identification within the same supply station.

171. Application

(This rule was formed in the 1971 Edition from the NOTE of Rule 107A of the Fifth and prior editions; former Rule 171 was deleted; see Rule 111.)

Circuit-interruption devices serve various purposes. Both safety and reliability are served by careful choice of devices to match the system capacity and other requirements of its intended service. The stated intention that device capacity should be reviewed whenever significant system changes are considered was added in the 1981 Edition.

Switches are not included in the rule requirement for adequate fault-current-interrupting capability because a switch should not be used to interrupt *fault* current. However, the use of circuit switchers was allowed in the 1997 Edition so long as they are matched in capability with the overall protective scheme requirements for fault current interruption duty.

172. Circuit Breakers, Reclosers, and Switches Containing Oil

(This rule was formed in the 1971 Edition from Rules 107A and 161 of the Fifth and prior editions.)

The wide and varied application of oil switches has resulted in many types of construction and design. In general, oil switches of more than 7500 V are designed for remote control.

Failure of a switch to open a short circuit may result in the explosion of the switch tank, permitting burning oil to escape. If burning oil escapes the unit, it can injure both personnel and other equipment. Such apparatus should be separated by isolation, fireproof surroundings, or fireproof enclosures.

See the discussions of previous Rule 115 and present Rule 152 for a more complete discussion of oil-filled apparatus.

173. Switches and Disconnecting Devices

(This rule was formed in the 1971 Edition from Rules 160B, 162, and 164 of the Fifth and prior editions; former Rule 173 was merged with Rule 170 in the 1971 Edition.)

Note that Part 1 contains no rule forbidding switchblades from being energized when the switch is open.

173A. Capacity

(This rule was formed in the 1971 Edition from Rules 160B and 164A.)

Switches are not *fault*-current-interrupting devices and should not be applied as such. If a switch must be operated under load, its capacity must be adequate for the load that it is required to interrupt. If the switch is intended to interrupt current (e.g., load, transformer magnetizing, or line charging) or might reasonably be expected to be operated as such, the switch should be considered and identified as a current-interrupting switch. A meter should indicate the load carried by such a switch so that the operator will not open it accidentally under loads greater than those that it may safely interrupt. In some cases it will be appropriate to arrange an automatic lock on switches not capable of interrupting currents to which they are subject. Holding such switches or controls in place by a magnetic field that depends upon the current flow through the switch will prevent accidental openings. Such automatic locks are not practical on hookstick-operated switches.

Where switches are to be operated only as disconnectors to open circuits under no load, they (1) only require capability of carrying the full load current and the momentary fault current of the circuit, and (2) should be suitably identified as disconnect switches. Where a number of disconnect switches are placed together, they should be carefully distinguished by suitable markings so that the wrong switch will not be accidentally opened.

173B. Provisions for Disconnecting

(This rule was formed in the 1971 Edition from Rules 160B and 164B of the Fifth and prior editions.)

Except for air-break switches near the equipment controlled, all switches are likely to be operated without full knowledge of the load condition of the equipment. If switches are closed while personnel are working on the controlled equipment, a serious hazard may result. For these reasons, switches are required to be able to be locked or blocked in the open position, where practical, and tagged otherwise.

Mechanical forces due to the magnetic fields around conductors in bus structures have occasionally opened disconnectors. The result has been damage to the equipment, injury to operators, and interruption of service. Locking disconnectors in the closed position is, therefore, strongly recommended. The disconnectors and the supporting parts must be strong enough to resist these mechanical forces. There is at least one case on record in which locked disconnectors have been torn from their supports and the insulators destroyed by magnetic forces.

It is usually practical to install single-throw knife switches to open downward. Double-throw switches should be provided with a proper latch or stop block on one or both sides to prevent the switch from being closed by gravity, unless they are mounted for horizontal throw. Whatever the switch design, if contact positions are not immediately obvious, appropriate marking is required.

173C. Visible Break Switch

(This rule was formed in the 1971 Edition from Rules 161 and 164C of the Fifth and prior editions. A Tentative Interim Amendment for the 2002 Edition deleted this rule and amended rule 123C to remove the implication that, if a visible break (open switch) was available, circuits normally energized up to 25 kV could be worked as de-energized without grounding the circuit. Part 4 does not allow such actions. This deletion was confirmed in the 2007 Edition. See the discussion of Rule 123C.)

The installation of a suitable switch provides the means for disconnecting equipment and circuits entirely from the source of electrical supply. Such precaution may be necessary to safeguard workers or equipment or, in an emergency, to prevent further injury to a person (1) who has been caught in moving machinery or (2) who has come in contact with energized parts controlled by the switches. It is, however, unnecessary to place switches between two pieces of equipment always operated as a single unit. Personnel are not expected to be working on such equipment without special precautions, unless both parts are disconnected from the source of energy.

The location of the switch should be as near to the source of energy as practical. However, the use of conductor leads of moderate length between a generator and a suitable switch is considered to be in compliance with the rule.

Air-break switches usually may be considered free from leakage, but this is by no means true of oil-break switches. Leakage across the gap of oil-break switches may be sufficient to cause dangerous shocks to personnel in contact with circuits supplied through them. Suitable disconnectors should be used to obviate this trouble. Switches that connect buses or are otherwise so located that they can be energized from both sides should be protected by air-break disconnectors at each side.

NOTE: Rule 173C was a companion to Rule 123C. Rule 173C required a visible break *only* if the equipment is to be worked on without protective grounding. Rule 123C implied permission to omit grounding only at 25 kV or less. As a result, visible air breaks were not required to be placed above 25 kV. Since the work rules of Part 4 govern safe work procedures, Rule 123C was never active.

174. Disconnection of Fuses

(This is Rule 166 of the Fifth and prior editions; former Rule 174 was deleted in the 1977 Edition.)

Except for fuses at low voltages, removing fuses from exposed live clips or other contacts generally is dangerous. Even at low voltages, large fuses may present a considerable hazard if they are replaced in a live fuse clip; if there is a serious short circuit beyond the fuse, it may blow and cause burns. The best protection is provided when fuses are inaccessible while their current-carrying parts are energized. This is accomplished in many cases by the enclosure of the fuses such that opening the enclosure disconnects the fuses from the source of energy. While these arrangements generally may be adaptable to industrial uses, they may be impractical for certain parts of station equipment where quick access is needed to minimize service interruptions. In such cases, a second means of protection is preferable if the fuse has to be handled frequently. The operation of a switch in series is more quickly and safely performed than the removal of a fuse from exposed live terminals, by an insulating handle or similar portable appliance.

175. (Not used in the current edition.)

(Rule 175 of the Fifth and prior editions was merged with Rule 123 in the 1971 Edition.)

176. (Not used in the current edition.)

(Rule 176 of the Fifth and prior editions was merged into Rules 124 and 125 in the 1971 Edition.)

177. (Not used in the current edition.)

(Rule 177 of the Fifth and prior editions was merged into Rules 124 and 125 in the 1971 Edition.)

Section 18. Switchgear and Metal Enclosed Bus

(Section 17 of the Fifth and prior editions was moved here in the 1971 Edition; former Section 18 was moved to 19.)

180. Switchgear Assemblies

(In the 1971 Edition, Rule 170—Location and Accessibility *of the Fifth and prior editions was revised and moved here. Parts of that rule were merged into Rules 122, 124, 125, and 180 in the 1981 Edition. The new Rule 180 details requirements in and around switchgear assemblies.)*

180A. General Requirements for All Switchgear

(This rule was codified in the 1981 Edition and revised for clarity in the 1984 Edition.)

Rule 180A addresses the conditions in which switchgear is expected to exist and function. The rule is intended to (1) limit deleterious action by dust, liquids, or harmful gases and (2) limit damage to the equipment from undue stresses or shocks.

It is doubtful that it is practical to keep normal atmospheric dust from settling in or on equipment during operation. However, it is absolutely necessary and practical to limit entrance of significant amounts of dust from construction sources. To do this, either (1) the switchgear location should be completed before its installation, or (2) special precautions should be taken. The intent of this rule is that switchgear should not be located in places and conditions for which it is not rated; nor should it be maintained without the use of methods, equipment, and working spaces designed to limit the opportunity for danger to personnel or equipment.

It was recognized in the 1984 Edition that some installations may be secured in place by their own weight; others will require appropriate support. It was also recognized that excessive settling of pad-mounted or other installations can place undue stress on cable terminations; appropriate support is required.

180B. Metal-Enclosed Power Switchgear

(This rule was codified in the 1981 Edition and revised for clarity in the 1984 Edition.)

This rule carries the provisions of Rule 180A forward with specific applications to metal-enclosed power switchgear. Several of the sub-rules invite careful reading. Rule 180B2 normally requires two exits in switchgear rooms to assure access from all parts of the room, if any particular piece of equipment presents a safety problem (see Figure H180). Rule 180B2 now requires swing-out doors and fire exit hardware; doorknobs do not meet the requirements. The original proposal for the 1981 Edition required the use of panic hardware. This was changed in the final draft for ANSI C2 ballot to be consistent with Rule 113B, which permitted doorknobs; the same change had not been included in the published Preprint for Rule 113B. Both rules were changed in the 1984 Edition to require panic hardware.

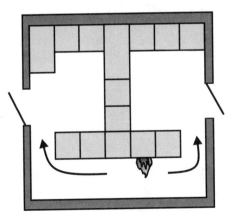

Figure H180
Multiple Exits from Switchgear Room

Rule 180B7 prohibits passing a low-voltage cable from Cubicle A through the high-voltage section of Cubicle B and on to Cubicle C, unless the cable is contained in rigid-metal conduit or is isolated by rigid-metal barriers. Rule 180B8 prohibits bringing a cable from outside the switchgear directly to the current transformers in the high-voltage compartment.

Rule 180B11 requires a safety sign if more than one disconnect device must be operated to de-energize all the conductors in a cubicle, unit, section, etc. The reference to ANSI Z535 safety sign standards was added in 1997 (see Handbook Appendix B). Rule 180B13 was deleted in the 1984 Edition because of the literal impossibility of meeting the rule; certain parts of switchgear are combustible under certain conditions.

180C. Dead-Front Power Switchboards

(This rule was codified in the 1981 Edition. It is a successor to Rule 170 of the Fifth and prior editions and Rule 180 of the 1971 Edition.)

180D. Motor Control Centers

(This rule was codified in the 1981 Edition.)

This rule is intended to flag the necessity of planning for peak let-through currents, rather than limiting currents. A safety sign is required if more than one disconnect switch must be operated to de-energize all of the conductors in the cubicle. The reference to ANSI Z535 safety sign standards was added in 1997(see Handbook Appendix B).

180E. Control Switchboards

(This rule was codified in the 1981 Edition.)

The intention is not to be specific as to limiting the application of this rule; the examples are intended to indicate the characteristics of equipment to which these rules are applicable.

181. Metal-Enclosed Bus

(This rule was codified in the 1981 Edition. Rule 181 of the 1971 Edition, which was Rule 171A of the Fifth and prior editions, was moved to 180B13 in the 1981 Edition; it was deleted in the 1984 Edition. Rule 181 of the Fifth and prior editions was merged into Rule 173C of the 1971 Edition.)

This rule contains the requirements that are particular to this equipment. Again, the intention is to protect the equipment from deleterious action of unrelated forces and to ensure that the equipment is designed properly for its intended function. Rule 181B5, which had been codified in the 1981 Edition, was deleted from the 1984 Edition on the basis that it was related to design, not safety.

Section 19. Surge Arresters

(Section 18 of the Fifth and prior editions was moved to this location in the 1971 Edition.)

190. General Requirements

(This rule was created in the 1971 Edition from Rule 180A of the Fifth and prior editions; it was revised in the 1981 Edition.)

In the large majority of instances, lightning arresters are advisable to protect station equipment supplying overhead line conductors. However, there will be situations where engineering considerations will indicate that installing lightning arresters is not advisable.

Arresters are intended to limit the adverse effects of disturbances in electrical transmission systems that take the form of high voltages. Two sources of such high voltages are atmospheric lightning and internal disturbances originating in the line itself. Although arresters are designed to take care of atmospheric lightning and those internal surges that are transient in nature, they generally are not designed to be operated under continuous surges.

When a discharge from a cloud strikes an electrical conductor directly, it almost always breaks down the insulation at or very near that point; it rarely travels along a transmission line far enough to reach an arrester. If it did, it would probably destroy any type of arrester. Arresters are not designed, therefore, to handle direct lightning strokes. It is usually the line insulators, rather than the station apparatus, that are injured by these direct strokes; the line insulators are best protected by overhead ground wires that are well and frequently grounded, rather than by arresters.

Damage from a surge that is induced by atmospheric lightning is generally caused by either its high voltage, which punctures the insulation to ground, or because of its high frequency, which builds up a high voltage across the end turns of the first inductive winding it strikes, thus causing a breakdown between turns. In either case, the power current flows through the puncture, and a potentially damaging short circuit or an internal surge is started.

Internal surges may be caused by any change in load conditions. They may be transient or continuous.

Transient surges are caused by sudden changes of load, such as those that may be caused by switching, the operation of circuit breakers, etc. They are usually comparatively unimportant, but they may be severe when a very heavy current is broken suddenly.

Continuous surges are caused by arcing grounds, which result in oscillations of larger amounts of power at a relatively stable frequency, usually a few thousand cycles per second. These are very destructive. Arcing grounds frequently result from a breakdown of insulation caused by lightning. Some arrester designs handle continuous discharges such as these for longer periods than others.

Because of their importance as protection equipment, arresters should be installed properly and inspected often; this will ensure continued performance capability.

Placing an arrester on each ungrounded conductor of every open overhead circuit will limit potential damage to connected equipment. In the past, some engineers have preferred to place a single arrester on the station bus rather than separate arresters on each circuit. However, if only one arrester is placed on the bus, the line switch, relays, and transformers are not as fully protected. In addition, should the protected apparatus be somewhat old, its failure is sometimes attributed to the failure of the arrester to function properly, when in reality the fault is more likely to be with the old apparatus than with the arrester.

An important element in the proper protection of circuits against static disturbances is the installation of choke coils and energy-absorbing resistors. Any inductance acts as a partial reflector to high-frequency waves.

CAUTION: Some of these special devices may be supported on special, long insulator stacks. The rules require that parts of indeterminate voltage be protected against the maximum voltage to which the part will be subject. The use of long insulator stacks may require the supported equipment to be raised so that lower areas of the stack will be isolated for the voltage that may be present. Surface contamination and insulation material characteristics should be considered in determination of the subject voltage at any point on the insulator stack.

The 1997 Edition added references to IEEE Std C62.1™ *IEEE Standard for Gapped Silicon-Carbide Surge Arresters for AC Power Circuits* (ANSI) and IEEE Std C62.11™ *IEEE Standard for Metal-Oxide Surge Arresters for Alternating Current Power Circuits* (ANSI).

191. Indoor Locations

Arresters have been frequent causes of fires where located near combustible portions of buildings. It is advisable to locate arresters, especially if oil-filled, outdoors wherever practicable. In some types of arresters, it is impractical to ground their exterior metal frame or case, and in such instances, these parts should be plainly identified and should be guarded as high-voltage parts. This will limit problems resulting from any assumption by attendants or others that these parts were grounded (as would be most exterior metal frames and cases of electrical equipment in the vicinity).

192. Grounding Conductors

(This is Rule 182 of the Fifth and prior editions.)

Appropriate ground connections are absolutely required for satisfactory operation of arresters. For this reason, Rule 191 repeats some of the requirements of Section 9 to emphasize the need for these requirements. It is difficult to overemphasize the importance attached to the need for proper ground connections. Connections should be as short and straight as possible. A poor ground connection can subvert every effort of choke coils and arresters to divert static electricity to the earth. Unfavorable natural conditions should be avoided or satisfactorily circumvented. Many arrester failures are traceable directly to poor ground connections.

193. Installation

(This rule was developed in the 1971 Edition from Rule 184 of the Fifth and prior editions.)

Although arresters generally are located well away from frequently occupied working spaces, they should be treated the same as other live parts of equal voltage for guarding purposes, and their special problems should also be recognized. When arresters are placed inside of buildings, clearances should be allowed above horn-gap arresters, since the arc may be considerable at times. The amount of this clearance is dependent upon the proximity of combustible material and the operating voltage of the arrester.

Live parts in the vicinity of arresters can be satisfactorily protected by location from the action of arresters and associated equipment.

Part 2. Safety Rules for the Installation and Maintenance of Electric Supply and Communication Lines

Part 2 of the NESC was extensively rewritten and updated for the 1977 Edition. This was the first major revision of many of the rules since the 1920s. While most rules retained their original numbering, many changed numbers to improve the readability and usability of the Code. The changes were based upon the need to reflect new materials, construction methods, and types of installation, as well as the additional experience gained in the intervening years.

During the 1980s, an extensive effort was undertaken to improve the readability, understandability, and specificity of the overhead rules, especially in the requirements for clearances. This culminated in an extensive review of the methodology behind the derivation of clearance requirements and resulted in the introduction of a complete system of coordinated clearances.

At the end of the 1980s, a similar review of strengths and loadings requirements was begun. This resulted in the alternate set of requirements for wood structures contained in Section 26. This work is continuing.

In the 1990 Edition, Section 28—*Miscellaneous Requirements* was disbanded; its parts were relocated to, or combined with, other rules with which they were associated in order to increase the understandability of these requirements as well as to expedite their location by the user.

Section 20. Purpose, Scope, and Application of Rules

Section 20 was revised for the 1977 Edition to reflect necessary changes. Because only Part 2 was revised for the 1977 Edition, additional clarifying changes in Rule 200 were delayed until the general revision of the 1981 Edition. At that time, the "general" statements from all parts were revised and moved into a new set of rules numbered 010 through 016.

In the 1990 Edition, Rule 217, Rule 218, Rule 220D and Rule 220E, Rule 223, and Rule 224 were moved here from Section 28.

200. Purpose

In the 1977 and later editions of the NESC, it is made clear by choice of wording that the purpose of these rules is the *practical* safeguarding of persons during the installation, operation, or maintenance of overhead supply and communication lines and their associated equipment (see Section 1).

201. Scope

(See Rule 202 for a discussion of Rule 201B, Rule 201C, and Rule 201D of the Sixth and prior editions. See also Rules 010–016.)

Rule 201 was revised extensively in the 1977 Edition. The present Rule 201 is a clarifying expansion of Rule 201A of the Sixth and prior editions. It is also made clear in the 1977 revision that Part 2 of the Code was not intended to apply to electric supply stations. In essence, Part 2 of the NESC is the general case, with Part 1 and Part 3 (Electric Supply Stations and Underground) as the exceptions to the general case. Although Part 2 contains *no* requirements that apply to electric supply stations, it does duplicate some of the requirements of Part 3 for risers on overhead structures in order to limit the opportunity for code users dedicated to either overhead line work or underground line work to miss the requirements for underground cables as they raise up a structure to connect to an overhead system.

The 2002 Edition added two notes regarding NESC approach distances for communication and electric supply workers and OSHA, federal, state, and local approach distances for nonutility workers. NESC clearance requirements are designed to keep supply and communication lines and equipment at appropriate clearances from activity around the facilities that is normally encountered or reasonably anticipated, even under significant sag changes due to wind, ice, or heating from solar gain and line losses. OSHA regulations and federal, state, and local statutes and ordinances, as well as good common sense, specify that persons other than utility workers should plan and execute their work and other activities to avoid contacting or damaging utility facilities.

202. Application of Rules

Rules are written to cover general cases and, for the described circumstances, are the governing requirements. *EXCEPTIONS* provide for specific conditions under which a rule is not or may not be applicable.

Rule 202 was revised extensively in the 1977 Edition. It encompassed Rule 201B, Rule 201C, and Rule 201D from the Sixth and prior editions. In the 1981 Edition, the general requirements were moved to new Rules 010–016. Rule 202 also serves as a reminder that the requirements of Rule 238C must be met when replacing a structure, even if other changes are not required by Rule 013. Generally, communications facilities are worked on more frequently than power facilities on joint-use lines. It is important to provide adequate working room for communications workers. This requirement is an exception to the general case.

Section 21. General Requirements

(Rule 280 and Rule 281 of the 1987 Edition were moved here as Rule 217 and Rule 218 in the 1990 Edition.)

210. Referenced Sections

(The requirements of Rule 210 of the 1977 and prior editions were moved to Rule 012 in the 1981 revision.)

In order to eliminate repetition of words, Rule 210 references other sections of the Code that apply to Part 2.

211. (Not used in the current edition.)

(This number is not used in the 1981 and 1984 Editions. The requirements of the previous Rule 211 were merged into Rule 012 in the 1981 Edition.)

212. Induced Voltages

(Rule 212—Accessibility of prior editions was renumbered to Rule 212 in the 1977 Edition.)

Since the susceptibility of facilities to induced voltages is so site-specific, no requirements are detailed in the NESC. If induced voltages are to be avoided, an operating cooperative arrangement is necessary between utility organizations locating facilities in the same area or on the same structure. This rule is intended to cover the influence of supply facilities on communications facilities. It does not refer to the induction influence of supply facilities on any other facilities. However, general common sense indicates that application of this rule to pipelines is also reasonable.

IEEE Std 776™ *IEEE Recommended Practice for Inductive Coordination of Electric Supply and Communication Lines,* and IEEE Std 1137™ *IEEE Guide for the Implementation of Inductive Coordination Mitigation Techniques and Applications* may be used to help determine the influence of supply lines on communication lines and the susceptibility of communication lines to induced voltages.

213. Accessibility

(This rule was renumbered from 212 in the 1977 Edition. The previous Rule 213—Inspection and Tests of Lines and Equipment *was renumbered to 214.)*

Although it is necessary to isolate line conductors and equipment for protection of the public, it is essential that such facilities safely be accessible to authorized persons, in order to facilitate adjustment or repairs required to maintain service that is as reliable and safe as is practical. Other rules of the Code, particularly those of Section 23, specify in detail the proper clearances and spacings for conductors, as well as the proper location of the wires and apparatus required to provide safe accessibility for authorized employees.

214. Inspection and Tests of Lines and Equipment

(This rule was renumbered from 213 in the 1977 Edition. Previous Rule 214—Isolation and Guarding *was discontinued in the extensive 1977 revision of Part 2, since the requirements duplicated those in Section 23 in large measure.)*

The NESC recognizes that facilities placed in service may have various opportunities and propensities to wear, break, become damaged, or otherwise be affected adversely by conditions such that continued service in that state would be inappropriate for safety reasons.

As a result there are two sets of requirements for inspections and tests—one for those lines and equipment that are *in service*, and another for those lines or equipment, or portions thereof, that may be *out of service*.

The distinction between being in service or being out of service is not affected by whether customer facilities currently are connected to the utility system; customers might or might not be connected to a utility system or system component regardless of whether it is in or out of service. Rather, the distinction between being in service or out of service hinges upon whether a subject line or equipment is connected to a utility system as an integral, functional part or extension of the system.

Facilities that are out of service include those that intentionally are disconnected from the system, whether by manual operation or disconnection by a worker or by automatic operation of sectionalizing devices, for the purposes of system protection, maintenance, reconstruction, removal, abandonment, etc.

Definitions of *in-service* and *out of service* were added in the 1993 Edition to limit the opportunity for misinterpretation of the requirements of Rule 214A and Rule 214B.

214A. When In Service

It is not intended that new construction shall be inspected by state or city officials before being put into use, or that such official inspections regularly shall be made. The operating utility, or other responsible party if so designated by the operating utility, is required to perform such inspections or practical tests in such a manner, and at such intervals, as experience has shown to be necessary. In general, the "experience" referred to is that of the utility responsible for operation and safety of the facilities in a manner to secure adequate and reliable results. If the responsible utility does not have experience with such an installation under such conditions, and information is available elsewhere, good design practice would suggest that such information should be examined. The utility is responsible for considering the conditions of service to which the installation reasonably can be expected to be exposed. It is not contemplated that provisions must be made for all *possible* occurrences if such occurrences are not also *reasonably expected to occur.* Neither is it expected that all parts and components necessarily will require either inspections or tests, although some parts may require both. The 2002 Edition clarified in a NOTE to Rule 214A2 that inspections may be performed while performing other duties; separate inspections are not required.

The phrase "from time to time" was deleted from the inspection requirements in the 1984 Edition. This language could be misinterpreted to imply that a specific schedule was intended. While schedules may be appropriate for some inspections, they may not be necessary for others.

When inspections or tests identify defects that affect compliance with the NESC, and such defects are not corrected immediately, they are required to be recorded until corrected. Identified defects that reasonably could be expected to endanger life or property are required to be remedied promptly. The intention of the rule is that, when items are identified as needing repair or replacement, either (1) the work will be done at that time or (2) the condition will be recorded to be addressed later. There is no requirement to record items that are addressed initially or to keep records after the work has been done.

Some lines and equipment in some locations may require daily inspections; lines and equipment in other locations may need only annual, or even less frequent, inspections. As a result, this rule could not be made specific. For example, if the concern is only with decay and weakening of pole timber, experience shows that some treated poles have lasted 60 or more years, while others have only lasted half that time—*or less*. Also, there is definite evidence that decay is influenced by the amount of rainfall, and hence moisture, in the soil. This, of course, varies from one part of the country to another. Other factors, such as woodpecker, insect, and lightning damage, vary considerably from one area to another. Salt spray or industrial atmospheres may contaminate insulators or cause accelerated corrosion of guys, hardware, etc. Again, these factors vary from one area to another; inspection procedures and intervals must be tailored to fit the local situation, based upon experience with such installations under such conditions. What is reasonable and necessary in one area may be unsound or unduly burdensome in another area.

214B. When Out of Service

Lines or equipment that are out of service, like idle machinery, may require repair before being fit for active duty; they should not be permitted to deteriorate into an unsafe condition while not in use. Such lines or equipment that may be used infrequently or only in emergency conditions, such as alternative tie lines, may be kept energized (thus considered *in service*) or de-energized (thus considered *out of service*), as desired.

Lines and equipment that are used infrequently are required to be inspected or tested, as necessary, before being placed back into service. Such factors as time of disuse, type of materials involved, and environmental exposure will materially affect the level of inspection or test required before the lines or equipment are placed back into service.

215. Grounding of Circuits, Supporting Structures, and Equipment

215A. Methods

The subject of grounding was studied thoroughly in connection with the preparation of the revision of the rules of Section 9. The methods of grounding are included in Section 9; the requirements to ground-circuit conductors or equipment are included in the rules of Part 2 and other parts. The rules prescribed in the NESC reflect both sound technical analysis and extensive operating experience on this subject.

215B. Circuits

(This rule was extensively revised in the 1977 Edition. Many of the requirements of prior editions were moved into the revised Rule 215C. The requirement of Rule 215C of prior editions is now included within this rule.)

Rule 215B specifies requirements for grounding of neutrals, other conductors, and surge arresters and refers specifically to the requirements of Section 9. In the 1990 Edition, a requirement was also added for nonneutral conductors that are to be grounded, the grounded phase of a corner-grounded delta secondary, to meet the requirements of Section 9; this was not a change in intention but an explicit inclusion of an item that had never been added to the list when the rule was expanded in 1977.

This rule also prohibits the use of the earth normally as the sole conductor for any part of a supply circuit. Prior to the Sixth Edition, this prohibition was required for urban areas and was recommended for rural areas; in the Sixth and subsequent editions, the prohibition was general.

Objections to use of the earth as a part of a supply circuit are made from both safety and service standpoints. Where made electrodes are depended upon for the ground connection, the opportunity for trouble from a high-resistance ground is large. If the resistance is high enough, such as during fault conditions, the voltage potential at a grounding electrode may approach the circuit voltage. In areas where the earth's moisture or water level fluctuates, the impedance of the earth may fluctuate with the water level and so also may voltages fluctuate. Where a vee-phase, or open-wye, earth-grounded circuit is employed, unbalanced voltages and circulating currents resulting from unbalanced loads are frequently encountered. In addition, the destructive nature of current flow through the earth endangers other facilities through electrolysis.

When earth returns were used in some rural areas prior to the 1960s, they became notorious offenders in dairy areas because circulating currents often caused both step and touch potentials. In some cases, these have adversely affected milking operations by shocking the cattle when they were connected to the milking machines, and have affected feeding (see Rule 92D—*Current in Grounding Conductor*). The grounding methods required by the NESC, including the use of a metallic neutral throughout each span of a multigrounded wye system, reduce the opportunity for such occurrences. On modern systems meeting the NESC requirements, alleged dairy farm problems, if found to exist in fact, are usually the result of poor building utilization wiring grounding practices in violation of NEC requirements, or other building-related problems. Note that failure to bond the supply system neutral grounding electrode and a communication system electrode together, when separate electrodes are used, is a violation of both the NEC and the NESC (see NESC Rule 99C).

Where the earth return path is stable and continuous, and of low resistance, connection of the neutral to the earth effectively reduces the impedance of the neutral. This causes overcurrent devices to operate more rapidly and to reduce the rise in step and touch potentials when a conductor falls. However, the earth cannot be depended upon in many areas, such as in sandy areas or rocky areas with high-resistance or discontinuous return paths. As a result, a metallic return is required by the Code (see Rule 96A3 and Rule 97C).

Although the prohibition of using the earth as a normal part of a circuit applies only to supply circuits, it should be noted that such use for communication circuits may suffer from similar objections, but usually to a lesser or negligible degree. Use of the earth as a part of a communication circuit makes the circuit more subject to interference than where a completely metallic circuit is used.

In the 1993 Edition, a NOTE was added to clarify the issue that monopolar operation of a bipolar HVDC system is permissible for short periods of time. This was changed to Rule 215B5b in 1997. Practical requirements necessitate occasional monopolar operation during emergencies and maintenance periods. Such occasional monopolar operation is not considered to violate the rule, but frequent, sustained monopolar operation does not meet the rule.

215C. Noncurrent-Carrying Parts

(The requirements of Rule 215C1 were included in Rule 215B prior to the 1977 Edition. The prior requirements of Rule 215C are now included in Rule 215B. In the 2007 Edition, the requirements for insulators used as an alternative to grounding guys and span wires were moved from Rule 279 to Rules 215C3-7, with the previous 215C3 being renumbered to 215C8.)

The purpose of this rule is to protect persons coming into contact with metal structures, lamp posts, raceways, conduit, cable sheaths, metal frames, cases, etc., by creating an equipotential plane. Such installations are subject to imposition of undesirable voltage potentials through accidental contact with supply circuits. This is one of the most important safeguards in handling supply equipment. The EXCEPTIONS recognize that effective grounding of certain installations is not necessary or is not practical.

Rule 215C1 requires metal structures to be grounded effectively in accordance with Section 9. Rule 96A of the 1990 and prior Editions specified step and touch potential limits to be met within supply stations through the 1990 Edition. Since the 1993 Edition, users have been directed to IEEE Std 80. Neither the previous specifications in Rule 96A nor the ground resistance values of IEEE Std 80 are required for metal structures outside of supply stations. See the definition of *effectively grounded* and Rule 012C.

The requirements for grounding guys were removed from Rule 282H in the 1977 Edition and placed in Rule 215C2. The revision required effective grounding of guys exposed to supply conductors of more than 300 V The 2002 Edition deleted "exposed" and inserted "vulnerable to accidental energization" in its place, to match the 1997 change in Rule 279A2. The alternate requirement "or on a structure carrying more than 300 V" remains. Appropriate exceptions to the general requirement are included. This rule applies both in rural and urban areas to both distribution and transmission structure guys.

The 2007 Edition revised Rule 215C2 extensively and moved the requirements for the use of guy insulators and span-wire insulators as an alternative to grounding the guys and span wires from Rules 279A2 and 279B2 to new Rules 215C3-7. Rule 215C2 continues to specify requirements for grounding or insulating anchor guys and span guys. Rule 215C3 now specifies requirements for grounding or insulating span wires carrying luminaires or traffic signals, while Rule 215C4 addresses requirements for grounding or insulating span wires carrying trolley or electric railway contact conductors. The requirements for

Rules 215C2-4 are essentially the same: either ground the guy or span wire or place one or more insulators at appropriate locations to limit transfer of voltages from one level to another or to the pedestrian area at the ground.

If one or more insulators are used in lieu of grounding the guy or span wire, the insulators must meet the strength requirements of Rule 279A1 or 279B for guy wire insulators and span wire insulators, respectively. Rule 215C5 or Rule 215C6, as applicable, specifies requirements for the placement of the insulators in guys and span wires. The two rules differ as required to meet the special conditions for span wires supporting trolley or electric railroad contact conductors in Rule 215C6. Insulators used only for the purpose of limiting galvanic corrosion have special placement issues that are addressed in Rule 215C7.

The 2007 Edition also removed the former voltage limitation of exposure to more than 300 V in Rule 215C2 for triggering the requirement for grounding or insulating the guy or span wire. Even secondary voltages can be a personal hazard, especially if a line worker is in contact with a grounded neutral or grounded cable messenger at the same time.

The chief reasons for placing strain insulators in guys or for grounding guys are (1) to protect pedestrians and line workers if a guy accidentally contacts, or is contacted by, supply conductors; (2) to minimize the possibility of plant damage that might result in unsafe conditions; and (3) to increase the structure BIL and reduce lightning-caused outages. Either long insulators or multiple insulators may be required if the guy is not grounded (see Figure H215C). Sometimes a combination of insulating upper portions to keep grounded guys out of the work area (see Rule 441A3a) and grounding and bonding lower portions is a practical solution to competing concerns.

Guy-wire insulators are also sometimes used to keep grounded guys from supply-line working spaces, where grounded guys could offer an additional hazard to line workers working in proximity to supply conductors.

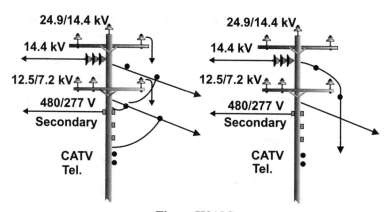

Figure H215C
Multiple guy insulators to limit transfer
of voltage from one level to another

In placing guys, every practical effort should be made to avoid unnecessary crossings or situations involving proximity with power conductors. Where guys must cross or be in located in close proximity to power conductors, as is often the case due to right-of-way or other constraints, it is necessary to provide *both* (1) appropriate clearances between guys and supply wires (see Rule 235E and Table 235-6) and (2) adequate climbing spaces and working spaces for line workers (see Rules 236 and 237). Neither grounding guys nor using the alternative of installing guy strain insulators at appropriate locations is a substitute for providing adequate climbing and working space.

It must be recognized that it is not always possible through the use of insulators to prevent line workers or pedestrians from coming into contact with the exposed parts of guys. For example, in the case of a guy from a joint pole or a supply pole, it is possible for the section of the guy near the pole to be energized by supply wires coming into contact with it regardless of how strain insulators are placed. If the contact is readily visible to a line worker before the pole is climbed, however, the resulting hazard is not great. A similar situation from the standpoint of pedestrians may occur where the section of a guy near the ground is energized by broken supply wires or is cut and sags into supply wires. If the installation meets the requirements of the Code, such situations can only result from the failure of other systems due to so-called *acts of God* and rarely should be encountered. Even if the upper portion of an anchor guy is insulated, as long as the guy is tensioned, the bottom portion will usually have a good enough ground connection through the anchor and rod to operate the protective devices.

Persons come into contact with guys only occasionally and incidentally; contacts between supply wires and guys occur infrequently where ample clearances and proper construction and maintenance have been provided. Consequently, the chance of injury to persons from exposed guys, even without insulators or special grounding, is relatively small in many cases. With the NESC requirements for grounding or insulating guys, the chance of injury is even less. The probability involved is about as remote as the probability of a charged conductor falling directly on a pedestrian, and it is, of course, impractical to provide complete protection against such contingencies.

The requirements for grounding or insulating guys provide the best known practical means of protecting both the public and line workers from conditions that are either normally encountered or reasonably anticipated. It is not practical to protect against *acts of God*.

Where means for adequate grounding are available, grounding often is preferable to the use of guy insulators. An important advantage of grounding is that it facilitates arrangement of protective devices to de-energize the supply circuit promptly upon contact with the guy. Where a guy is attached directly to the pole with a pole plate, and the entire metallic guy wire is grounded to both the system neutral and the earth (via the anchor), the guy wire can be energized only for the short duration required for the protective apparatus to operate.

Rule 215C was extensively revised in the 1977 Edition (when it was located in Rule 279) (1) to better group and more clearly word the requirements and (2) to show the relationship of these requirements and those of Rule 215C2 and Rule 220B2. The 2007 movement of the remainder of the insulator location requirements from Rule 279 to Rule 215C was made to place all of these requirements together in the appropriate section.

The requirement of Rule 215C8 (Rule 215C3 in 2002 and prior editions) to bond all communication messengers on overhead structures together to ensure that communications workers at a given pole location would not be exposed to messengers having different potentials was added in the 1987 Edition. The bonding intervals are specified in Rule 92C, i.e., four or eight ground connections per mile depending upon the size of the messenger. In essence, the smallest messenger dictates the requirements for the frequency of the bonding. Note that when one communication line crosses another on a common crossing pole, the messengers must be bonded at the crossing pole, but the grounding connections may occur elsewhere in the lines (see Rule 92C3).

Section 9 contains the full specifications for the methods of grounding to meet the requirements of this rule.

216. Arrangement of Switches

Inaccessible switches, and switches that do not show *at a glance* whether they are open or closed, such as enclosed switches or switches that may be used at night, tend to increase mistakes in operation and to multiply accidents. This is especially the case in emergencies, when quick action is necessary and time cannot be taken for consideration of unusual connections or arrangement of switches. Careful training of switch operators, when combined with uniform switch arrangements, improves both the safety and reliability of the utility system.

Locking mechanisms serve to limit interruption of operation or introduction of safety problems by unauthorized persons. Locking especially is important where employees are working on a line and a switch control mechanism is readily accessible to other persons.

For these same reasons, switch handles or control mechanisms should have the same position when open and a uniformly different position when closed; if this is not possible, clear marking of the "open" and "close" positions is needed. Uniformity of position and of method of operation within a system makes it easier to avoid mistakes and so promotes rapid and safe operation.

A requirement was added in Rule 216E in the 1993 Edition to provide local provisions on remotely controlled (or automatically controlled) transmission or distribution switches to render the remote or automatic control features inoperable. This rule is intended to limit the opportunity for inadvertent Supervisory Control and Data Acquisition Systems (SCADA) operation to cause a safety hazard in an area being worked by line personnel. See also the changes to Rule 442E in the 1993 Edition.

217. General

(Rule 280 was moved to this location in the 1990 Edition.)

217A. Supporting Structures

(The requirements of this rule have been generally retained from the Fourth Edition. The wording and grouping of the requirements were extensively revised in the 1977 Edition and moved from 280A to 217A in the 1990 Edition. Unless otherwise noted, these references to previous rules refer to the Sixth and prior editions. Previous Rule 280A4 was deleted in the 1977 Edition. Rule 217A1a was renamed from Mechanical Injury *to* Mechanical Damage *in the 2002 Edition)*

Rule 217A1a—*Mechanical Damage* was Rule 280A1a of the 1977–1987 Editions and previous Rule 280A2a of previous editions. The rule recognizes that there are some locations, such as in constricted alleys or parking areas, at which it is impractical (because of building locations or other requirements) to locate structures far enough away from the traveled way to prevent occasional rubbing against the structures by vehicles. This rule has limited application to structures located beside streets, roads, and highways since the 1984 Edition, when structures were required to be located a sufficient distance off the traveled way of streets, roads, and highways that would allow ordinary vehicles to use the traveled way without rubbing against the structures. However, where there are no curbs, the structure must be outside of the shoulder but, in some cases, may need to be located immediately adjacent to the shoulder and, thus, could be subject to this rule. If a structure must be placed in an alley or parking lot or is to be subject to abrasion from such frequent contact by vehicles that the strength of the structure is likely to be significantly reduced, the structure is required by this rule to be physically protected.

Care should be taken in the selection of an appropriate structure guard. Concrete sleeves extending a few feet above and below the ground line, if properly designed and installed, make very effective structure guards and may add to the strength of the structure. However, experience has indicated that, in some cases, such a sleeve or enclosure may actually promote decay in a wooden structure, because it confines moisture in the wood. Steel or iron plates are also used for this purpose; they perform well by allowing the vehicle to slide along the surface of the structure without reducing the diameter of the structure. However, if they are too high or enclose more than half of the structure, they may unduly inhibit line workers when they must climb the structure. Note that some states, such as Pennsylvania, prohibit the use of pole guards in some locations.

Rule 217A1a (1977–1987 Rule 280A2a) does not require protection to prevent collision of a vehicle with the structure; such a requirement would be almost impossible, and certainly impractical, to achieve in most cases. This rule generally applies in alleys, parking lots, and similar tight spaces, where low velocity contact may occasionally occur and rub fibers from wood or dent metal. The practical remedy most often used is to install an oversized pole, so that surface abrasion does not materially affect the strength. Barriers are less often used, because they take up space and may make a tight maneuvering area in an alley, etc., even worse.

Rule 217A2—*Climbing* of the 1997 Edition was formed from the previous Rule 217A1b and Rule 217A2 plus a new rule on standoff brackets.

Rule 217A1b—*Fire* (1977–1987 Rule 280A1c; 1990 Rule 217A1c) replaced previous Rule 280A1—*Rubbish*. The requirement for "guarding" the structure was deleted in the 1977 Edition because it is not a realistic requirement to guard a structure from fire in the same sense as "guard" is used elsewhere in the Code.

The accumulation of brush, grass, and rubbish around the bottom of a pole or tower presents several dangers. It interferes with proper inspection and, with wood structures, it is conducive to decay and increases the fire hazard to the structure. It is advisable that seasonal inspections be made, especially on important high-voltage lines installed on wood supporting structures, particularly during those periods when fires are liable to occur.

Rule 217A1c—*Attached to Bridges* (1977–1987 Rule 280A1d; 1990 Rule 217A1d) is a revision of previous Rule 280A3b. Where water crossings are wide or the earth near the water is substandard foundation material, a practical solution can be to attach supply and communication lines to bridges, as in Figure H-217A1c. Safety signs are required at the attachments of utility supports for supply lines to the bridge structure to warn bridge workers to leave maintenance of the supply line support attachments to the power utility.

Figure H217A1c
Power lines carried on bridge attachments

Rule 217A2a—*Readily Climbable Supporting Structures* (name changed in 2002 from *Climbing*) (1977–1987 Rule 280A1b; 1990 Rule 217A1b) is a more clearly worded rule than Rule 280A2b and Rule

280A3a of the Sixth and prior editions which it replaced. Structures to which the rule applies must be equipped with *either* climbing barriers or appropriate safety signs. The phrase "readily climbable structures" was added in the 1977 Edition; a definition of "readily climbable" was added in the 1981 Edition. The reference to ANSI Z535 safety sign standards was added in 1997.

It should be stressed that the definition of *readily climbable* was changed from "having sufficient handholds and footholds to permit an average person to climb easily without using a ladder or other special equipment" in the 2002 Edition to a definition of "**readily climbable supporting structure**. A supporting structure having sufficient handholds and footholds that the structure can be climbed easily by an average person without using a ladder, special tools, or devices, or extra ordinary physical effort." The 2002 new definition for *readily climbable supporting structure* further requires the 2.45 m (8 ft) section to start not more than 1.8 m (6 ft) above grade. This change codified earlier Interpretations and further experience with various types of structures.

It is recognized that there are people who enjoy the challenge of climbing formidable structures. There is no practical method of guarding a utility line structure from persons determined to climb it. It is, however, practical to design the structure so that the casual observer will be unable to climb the structure without special effort. Experience has shown that such methods as eliminating pole steps lower than 2.45 m (8 ft) above ground, or other accessible flat surface, are practical methods of limiting access to overhead utility facilities by unauthorized persons.

Until 2002, the Code was not specific as to what was considered as "closely latticed poles or towers," except that the structural members of such installations would have to be spaced so closely as to meet the requirements of *readily climbable*. Where handholds and footholds exist, but are not less than 2.45 (8 ft) apart, the structure is not considered to be readily climbable. As a result, riser installations attached to overhead poles are not considered as readily climbable if the supports are at least 2.45 (8 ft) apart. Prior to the 2002 change, the 2.45 m (8 ft) gap could have started above the 1.8 m (6 ft) level. The 1.8 m (6 ft) maximum height for the starting point of the gap required to be considered as not readily climbable is based on recent fall protection restrictions, which require fall protection above that height. Few multileg towers qualify as readily climbable; their support members are spaced so far apart and at such angles as to be difficult to climb. Most of the larger, single-leg towers do not qualify; some of the smaller ones are, however, easy to climb. Anchor guys have neither handholds nor footholds and are, therefore, not easily climbed.

It is left to the discretion of the designer as to whether warning signs or inhibiting barriers are the most appropriate method of limiting climbing by unauthorized persons. Among the matters to be considered are the nature of the area and relative frequency of access to the area by unauthorized personnel, as well as the methods and need for line personnel to work on the structure.

Signs warning against trespass and calling attention to danger are available in durable form to meet most ordinary requirements where metal poles or towers are concerned. On wood poles, stenciled signs are preferable, if the sign must be located in the climbing space, since metal signs may inhibit the movement of line workers. Small individual letters or figures of thin aluminum may be appropriate in many cases. The 1997 Edition added a reference to the ANSI Z535 safety sign standards (see Handbook Appendix B). Rule 217A2b—*Steps* (1977–1987 Rule 280A2; 1990 Rule 217A2) was moved from 280A5 in the 1977 Edition; the required clearance height to the bottom step was raised to 2.45 m (8 ft), as a result of recorded accidents involving unauthorized climbers.

All overhead supply circuits should be inaccessible to unauthorized persons, at least as far as it is practical to make them so. The best method for isolating supply wires is to inhibit climbing of the supporting structures without the use of special means, such as ladders, spurs, or removable steps. Metal steps are not to be installed nearer than 2.45 m (8 ft) to the ground. The use of steps only on the portion of the structure out of reach of the ground is very desirable in most locations.

Some metallic structures require steps in order that they may be climbed easily by authorized persons.

An exception is that the climbing restrictions of Rule 217A2a do not apply when the structure is in a fenced enclosure. The 2002 Edition requires such a fence to meet the height requirements (but not grounding or other requirements) of rule 110A1.

Rule 217A2c was added in 1997 to require standoff brackets to have not less than an 2.45 m (8 ft) clearance between the lowest two brackets or the lowest bracket and the grade beneath.

Rule 217A3—*Identification* (1977–1987 Rule 280A3) was expanded in the 1977 Edition from previous Rule 280A6. It is important that pole or tower structures be identified readily by location, construction, or marking to minimize mistakes by employees working on them or reporting with regard to them. The requirement to record the date of installation was deleted in the 1997 Edition.

Rule 217A4—*Obstructions* (1977–1987 Rule 280A4) was renumbered in the 1977 Edition from previous Rule 280A7.

Obstructions, such as nails, bolts, tacks, or other metal pieces, may keep a line worker's spur from taking hold, thus causing the worker to fall. Mail boxes, street signs, traffic-direction signs, etc., may constitute a serious hazard to workers on poles. Rule 217A4 requires supporting structures to be kept free from climbing hazards, such as vines. See IR 537 issued 3 June 2004.

Rule 217A5—*Decorative Lighting* (1977–1987 Rule 280A5) was added in the 1977 Edition to address the problems caused by the addition of certain types of decorative fixtures to overhead utility structures. If added without proper regard for the function and capability of the utility installation, such decorations can decrease the safety, reliability, and operability of the utility installation. In certain cases, such fixtures create static that interferes with radio and communication system operation. In others, the increased wind loading resulting from such additions may be greater than that which the installation is designed to withstand.

217B. Unusual Conductor Supports

(Rule 280B of the 1977–1987 Editions was moved to 217B in the 1990 Edition. Previous Rule 280B—Crossarms was deleted in the 1977 Edition; see Rule 261D2 and Rule263C.)

Rule 217B applies to unusual conductor supports, i.e., supports other than a conventional pole or tower. The phrase "unusual conductor supports" is not limited to buildings (in fact, the word *buildings* is not used in the rule). Such support(s) could be an outcropping of rock in a mountainous area or a wall around a substation at a conductor entrance or exit. Also, the conductors may be running between poles with intermediate attachments to unusual supports.

Rule 217B is generic with regard to conductors. It covers all electric supply and communication conductors and cables; it does not limit the voltage of electric supply facilities to service drops (see IR 509 issued 6 March 1997).

Note that Rule 217B states that additional precautions may be required to avoid damage to the structures or injuries to the persons using them, and that using roofs or trees as conductor supports should be avoided.

Rule 234C contains specific requirements for electric supply service drops necessary for an entrance to a building while Rule 234C4 permits communication conductors and cables of any type to be attached directly to buildings. Other than for a service entrance, attachments of electric supply conductors and cables to buildings are unusual; specific rules for such situations are not included in the NESC. Also, the NESC does not anticipate attachment of high-voltage electric supply conductors or cables to residential buildings (see IR 500).

It is apparent that when conductors are attached to supports that are not used chiefly for this purpose, such as building or frame structures, different construction techniques should be employed. Masonry and wood buildings are not designed to carry loads that can be placed on a pole and similar structures. Accordingly, designers must pay attention to the weight of the cable, length of span, and tension on messenger at installation. Attachments should *not* be made to deteriorated masonry or wood that shows evidence of serious decay. Note that Grade N Construction is required by Section 42 for service drops. Tree attachments are generally inappropriate because of their displacement due to growth and swaying in the wind. Roof attachments are more subject to interference and they generally involve placing the wires where they are more easily accessible to unauthorized persons than when attached to poles. The 2002 Edition clarified that the restrictions on the use of unusual supports apply to line conductors, not service drops. All voltages are covered.

217C. Protection and Marking of Guys

(Rule 264E of the 1990–2002 Editions was moved to 217C in the 2007 Edition. In the 1987 and prior Editions, this rule was numbered 282E.)

A guy wire is often difficult to see not only at night but also by day in stormy weather. If the guy is in the path of pedestrians, it can cause serious accidents which, in most cases, would be avoided by covering the guy with a conspicuous marker. In earlier years, these markers were made of wood or metal; the newer, colored plastic and related materials have increased the effectiveness of such markers. The rule was revised in the 1977 Edition to recognize these new materials.

The term *guy guard* used in earlier editions is a misnomer in most applications, because the purpose of the marker generally is to call pedestrians' attention to the guy so that they can avoid it. These markers also serve to guard pedestrians from incidental contact with sharp edges of guy-wire terminations or galvanized flashing on bolts or other parts. They are not, however, guards in the sense that they can prevent, or even significantly limit, damage to the guy from contact by an errant vehicle.

It should be noted that, prior to 1997, guy markers were not required unless pedestrian *traffic* was expected. The intention of pedestrian protection was clarified in the 1977 Edition. Neither livestock, road vehicles, nor farm vehicles are mentioned in this rule. Rule 217A1a (Rule 280A1a of the 1987 and prior editions)—*Mechanical Injury* addresses required protection of structures from vehicular traffic; guys are considered to be part of the structure for this purpose.

Nothing in Rule 217C (Rule 264E of the 1990–2002 Editions; Rule 282E of the 1987 and prior editions) prevents guy markers from being used in any place. The use of guy markers may be helpful in areas where not required; they can often serve as a warning to vehicle operators, especially in constrained areas (such as farmyards) where physical protection according to Rule 217A1a (Rule 280A1a of the 1987 and prior editions) would not be required nor be appropriate. However, it is clear that guy markers serve no useful purpose in many locations. They may even be detrimental in some areas, such as where livestock may try to use the marker as a scratching post. In recent years, some municipalities have tried to outlaw guy markers on the grounds that they are unsightly. The 1997 Edition added locations where a guy anchor is in an established parking area, where the guy is not protected against someone backing into the guy.

The 2.45 m (8 ft) length of guy marker has been required since 1941. On a 45-degree-angled guy, the upper end is approximately at eye level; this orientation has served so well that the Strength and Loadings Subcommittee has been reluctant to approve the use of shorter guy markers. Although short guy leads would produce equivalent visibility from a 2.13 m (7 ft) marker, the visibility problems associated with accidental placement of the short marker on a long guy lead require the rule to remain as issued.

218. Vegetation Management

(Rule 281 of the 1987 and prior editions was moved to Rule 218 in the 1990 Edition. The title was changed from Tree Trimming *to* Vegetation Management *in the 2007 Edition.)*

The avoidance of contact of line conductors with trees is a difficult problem in many localities. In some cases, line structures may be high enough to clear the trees without trimming; in other areas, considerable trimming may be necessary. It is important to keep the conductors clear by one method or another to avoid arcing or mechanical damage and to avoid grounds, short circuits, or crosses between circuits, as might occur if two conductors touch the same tree branch. Trees that shed their bark, such as the eucalyptus, or trees that are extremely brittle, such as the poplar, should be avoided if at all practical. Otherwise, they should be trimmed below the level of the supply wires, if permission may be secured. The use of properly designed tools for this purpose is particularly important. Tools designed for the use of orchardists or gardeners are very rarely safe or suitable for use in this connection because metallic connections are often present between the cutting head and the operating handle.

Trees are always a menace to overhead facilities, particularly where they are taller than the supporting structure. If practical, trees should be avoided at crossings, in order to limit the incidence of trees causing one line to contact another during a storm.

This rule has retained its essential requirements since the Third Edition. The requirement concerning crossings was added in the Fourth Edition; the language was revised for clarity in the 1977 Edition; the construction of the rule was further revised for clarity in the 1981 Edition by excerpting the NOTE and breaking the rule into two parts. In the new paragraph 2, the word "insulating" was deleted to recognize that tree wire is considered to be covered, not insulated. It is clearly stated in the 1984 Edition that the major concern is for tree growth getting into energized conductors.

The 2007 Edition revised the rule to reflect modern terminology (*vegetation management* and *pruning*) and to better inform the users of the major concerns and factors to be considered in determining the extent of required vegetation management. The term interfere with ungrounded supply conductors was deleted in 2007, since mere incidental touching of a living leaf with an energized conductor will rarely damage the conductor. The general result is that the leaf will wilt back and kill the branch for a short distance, if it gets close to the conductor (within touching distance). Dry leaves contacting energized conductors can ignite and cause a fire that endangers the lines and structures, if

218. Vegetation Management

they fall on dry leaves or grasses. NFPA Std 1 *The Uniform Fire Code* and the *International Urban-Wildland Interface Code* are fire codes related to electrical apparatus as a source of ignition; if either has been locally adopted, minimum vegetation pruning clearances may be specified for the locality.

Expected experience is a key issue. Both the frequency of pruning and the distance by which vegetation is pruned back from the lines is affected by the line voltage class, the relative growth rates and the failure characteristics of the relevant species, right-of-way limitations, location of the vegetation relative to the conductors, potential movement of conductors and vegetation during routine winds, and the sag of conductors due to elevated temperatures or ice loadings, etc.

Vegetation contact with a grounded neutral or a communication cable in itself is not expected to result in electrical arcing damage to a neutral or cable and is no longer specifically targeted by the rule. However, it is advisable to avoid contact between neutrals and trees, in order to minimize mechanical abrasion. Lashing wires used to attach communication cables to messengers can more easily be broken by mechanical abrasion than a neutral and, if broken, can allow the cable to drop. The main concern with mature limbs growing into grounded neutral conductors and into communication cables is the structural loading that can be transferred to the wires and cables, and ultimately to the structures, during storm loading from ice and/or wind on trees. Significant loads from unpruned trees contacting wires and cables have damaged broken lashing wires and dropped cables, damaged structures, and pulled structures together (or sideways) enough to adversely affect clearances.

It is not required that all lines be maintained with the same vegetation management cycles. Growth rates, species, and even wind and ice loadings may vary across a large system area. Many utilities have such a homogeneous mix of vegetation habits and growth rates that most lines can be served by one cycle length. Many others have service territories with some locations that have (1) vegetation habits or growth rates that dramatically differ from the normal vegetation or (2) limits on pruning distances. Many of the latter utilities have successfully adopted a general vegetation management cycle for all lines and, as a part of the overall program, routinely identify and record particular spans with species that will require one or more spot inspections/prunings between normal pruning cycles. This allows a reasonably timely and economical program to achieve good results without undue impact on the system resources.

Using the *natural pruning method* that is required by Section 5.9.2 of ANSI A300 *Part 1: Tree Care Operations—Tree, Shrub and Other Woody Plant Maintenance—Standard Practices* that has been adopted by the International Society of Arboriculture allows utilities to limit growth under lines and direct future growth away from the utility lines. This is similar to the *Standards for Pruning of Shade Trees* originally distributed by the National Arborist Association (now the Tree Care Industry Association). Both include requirements for crown reduction pruning. The crown reduction method used to prune vegetation around utility lines requires pruning back to a leader that is appropriately sized and located so as to redirect future growth away from the overhead lines.

ANSI Z133.1 *Pruning, Trimming, Repairing, Maintaining, and Removing Trees and Cutting Brush—Safety Requirements* is the industry safety standard for working on vegetation in proximity to energized electrical apparatus.

Rule 217A4 requires supporting structures to be kept free from climbing hazards, such as vines. See IR 537 issued 3 June 2004.

Section 22. Relations Between Various Classes of Lines and Equipment

220. Relative Levels

220A. Standardization of Levels

It is obvious that it is both convenient and simple to have each class of conductor be at a definite level when crossings and joint use of poles are considered. Such situations can then be approached without any change of the levels used at other points, and complicated construction is thus avoided. By permitting the relative levels and required clearances to be readily obtained on jointly or commonly used poles, as well as at crossings and conflicts, this practice (1) facilitates the extension of lines and (2) promotes the safety of the public and workers.

220B. Relative Levels—Supply and Communication Conductors

220B1. Preferred Levels

It is universally recognized that the proper relative positions of supply and communication conductors, considering both public and employee safety, generally place supply conductors above communication conductors.

There was formerly a widespread disposition to run fire-alarm wires at the highest position on a pole with the idea that failure of other wires would not affect such circuits. This policy has now been largely abandoned and fire-alarm conductors are usually below supply conductors.

Supply lines generally use larger, stronger conductors than communication lines, especially if the communication line uses open conductors instead of a cable and messenger; as a result, there is generally less probability of contact between the two if the supply conductors are located in the upper position. This relative location also avoids the necessity of (1) those who work on communication conductors having to pass through supply conductors and work above them, and

(2) increasing the Grade of Construction required for communication conductors.

Special consideration may be appropriate for communication circuits that are used in the operation of supply lines. These circuits are also known as insulated static wires when they both protect the supply conductors from lightning-induced flashover events and provide communication. These circuits generally parallel high-voltage supply circuits for long distances; consequently, high voltages may be induced on the conductors. Because of this, an exception to the general practice of placing communication circuits below supply wires should be made for such dispatching circuits. Where supply circuits customarily employed for distribution purposes are installed on the same poles with dispatching circuits and high-voltage supply circuits, they should be installed beneath the dispatching circuits. The construction of the dispatching circuits is determined by Rule 224A.

220B2. Special Construction for Supply Circuits, etc. (1977 and later editions)

(These requirements were in Rule 220B3 prior to the 1977 Edition. See the previous paragraph for discussion of minor extensions that were covered in this rule number prior to the 1977 Edition.)

For many years, it has been the practice of certain railroads to use the lowest crossarm of the telegraph line to carry circuits that supply the power for operating signal circuits. In view of this successful established practice, this special rule was written to recognize and permit the continuance of such arrangements, but *only* after cooperative consideration with the owners of other circuits that may be involved on the same line. Definite restrictions and limitations are applied to this practice.

The voltage and power limitations were revised in the 1977 Edition to reflect the characteristics of some alarm circuits that generally are not distinguishable from the railroad circuits in actual operation or relative safety. These limitations do not apply to circuits used for signaling purposes, alarm, train control, etc., that meet the definition of communication lines (see Definitions); conductors of these latter circuits are not restricted as to common occupancy of crossarms with those of other communication circuits.

A low-voltage supply cable jumper meeting Rule 230C that extends from a vertical run on a pole to a cable-television (CATV) amplifier is considered as a vertical or lateral communication conductor, (not a line conductor) extending from a vertical communications facility to a horizontal one. No clearances from the pole or the other communications facilities are specified for this case; however, the climbing space rules do apply.

The 600 V reference is intended to be the maximum voltage—not nominal.

The clearance was changed from 600 m (2 ft) to 400 mm (16 in) in the 2002 Edition to coordinate with previous related changes in Rule 235.

220B3. Minor Extensions (Sixth and prior editions only)

(This rule was deleted in the 1977 Edition. The following paragraph refers only to prior editions.)

When the NESC was originated, it would have involved undue expense to specify the immediate standardization of all present construction in conformity with the new rules. This would have been a severe hardship on utilities in localities where it had been the practice to place the communication wires above the supply wires. A gradual change to the preferred type of construction was recommended. Small extensions of the present arrangement of levels were allowed if the construction conformed to the grade required for such arrangements.

Since the NESC has been in effect for many years, the problem of relative levels should be minor if it exists at all. By the same token, the expectation that the necessity or appropriateness of such nonconforming extension should be exceptionally rare led to the removal of this provision in the 1977 Edition.

220C. Relative Levels—Supply Lines of Different Voltage Classifications (as classified in Table 235-5)

220C1. At Crossings or Conflicts

Placing higher-voltage lines at the higher levels allows the lower-voltage lines to be worked without conflict with the higher-voltage lines. Generally, the lower-voltage lines are worked more often than higher-voltage lines.

220C2. On Structures Used Only by Supply Conductors

Several considerations make it desirable to (1) have the circuits of higher voltage on a structure at the higher level, and (2) where there are circuits of a number of different voltages on a structure, arrange them according to the voltage, with those of highest voltage on top. In the latter case, it is preferred that a space of more than the gain spacing between groups of different voltages serve as a dividing line.

From the standpoint of workers, this arrangement is especially desirable. Workers usually will be working on the lowest-voltage circuits more frequently than the higher-voltage circuits. The lower-voltage circuits generally require more frequent addition of equipment, taps, or service connections. The high-low arrangement (1) makes the lower-voltage circuits accessible without requiring line workers to come into proximity with the higher-voltage wires, and (2) necessitates less climbing. In addition, the higher-voltage circuits are usually built to achieve higher levels of service reliability.

Circuits of the higher-voltage classifications are expected generally to provide greater service reliability than circuits of lower voltages; they are expected to be maintained generally in more secure condition mechanically and hence require less attention. These relative levels will often avoid the necessity of increasing the Grade of Construction for crossarms, pins, and conductor fastenings of the lower-voltage conductors.

It is preferred that workers climb through wires operating only at lower voltages. For work on wires operating at extremely higher voltages, conductors are generally de-energized before being worked, bucket trucks are used, or other arrangements are made. A relatively greater amount of power generally moves through the higher-voltage circuits, and thus more people are affected by service interruptions. It is not appropriate to place higher-voltage conductors at lower elevations if they must be de-energized or otherwise worked with in order to gain safe access to lower-voltage conductors placed at higher elevations.

The advantage of having the higher-voltage circuits above the lower-voltage circuits particularly is evident when the types of apparatus that operate on supply lines are considered. The installation and removal of transformers become, at best, rather difficult undertakings when the supply wires are energized, particularly if the transformers must be moved between higher-voltage supply wires. Placing a higher-voltage circuit below a transformer increases the difficulty in providing clearances to secondary conductors.

Where it is not practical to carry the higher-voltage wires at the higher levels, the construction of such lower-voltage circuits as are placed above those of a higher classification must, in general, be made as strong as is required for the higher-voltage circuit in the preferred arrangement.

Note that earlier editions required greater strength for even more classes of lower-voltage conductors than later editions, if the lower-voltage classes of conductors were placed above higher-voltage classes.

220D. Identification of Overhead Conductors

(Rule 285A of the 1987 and prior editions was moved to this location in the 1990 Edition; it has been relatively unchanged since the Fourth Edition.)

Rule 220D and Rule 220E require personnel to be able to identify conductors and equipment (facilities) that each individual worker is authorized to work on. While workers must be trained to identify other facilities as to general type (e.g., electric supply, communication) in order to work safely on the structure, subject rules do not require workers to be able to identify foreign construction by company ownership (see IR 514 issued 20 November 1997).

Joint-use agreements can specify additional requirements for identification of facilities on poles.

In order to safeguard electrical workers, lines should be arranged
systematically by having conductors occupy definite positions throughout a system. Failure to follow this practice may lead to accidents to
persons and to a lessening of the grade of the service rendered. When
arrangements of conductors are not uniform, other means for ready
identification of them should be provided. Diagrams indicating the
position of the various circuits and conductors, especially on the heavy
leads and on corner poles, are valuable aids for the workers and their
supervisors.

Conductors and equipment should not be transferred indiscrimi-
nately from one pin or crossarm position to another. A fixed scheme of
arrangement, such as where series-lighting arc circuits are maintained
on certain pin positions of certain crossarms throughout the system, is
considered an identification. Using characteristic shapes and sizes of
insulators for various voltage classifications frequently secures the
desired result, though too much dependence should not be placed on
this type of identification.

Schemes of line-conductor identification that use insulators of various colors or materials are very satisfactory when properly maintained.
Other systems indicate the character of the conductor according to a
letter or number code on the crossarm opposite the pin position. In
other cases, a colored band or sign placed below any crossarm-carrying
conductors operating in excess of a specified voltage, or a distinctive
color for the crossarm itself, has proven to be a useful identification.
Line workers can, therefore, easily determine the character of the conductors with impunity and choose the appropriate work method. This
rule is essentially unchanged since the Fourth Edition.

220E. Identification of Equipment on Supporting Structures

(This rule has been essentially unchanged since the Fourth Edition. It was moved from Rule 286A to Rule 220E in the 1990 Edition. See the comments on Rule 220D.)

221. Avoidance of Conflict

There are several possible methods of constructing two parallel lines—(1) complete separation, (2) location close enough to have structure or conductor conflict, (3) location of one circuit at a higher level on its structures and in such close proximity as to overbuild the second line, (4) a combination of overbuilding and joint use, and (5) complete joint use of structures. The order of preference is (1) complete separation, (5) joint use, (2) conflict, and then (3) overbuilding.

The best construction is complete separation of lines (see Figure H221-1), but this increasingly is not practical for a variety of reasons: right-of-way is not always available; fewer structures along roads are preferable to more structures; joint-use construction may offer significant advantages in many locations.

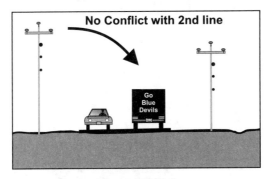

Figure H221-1
No conflict with complete separation of lines

Overbuilding involves most of the disadvantages of joint use of poles, without any of the benefits. Proper clearances are difficult to maintain unless clearance arms are added to the lower structures. The angles in the line and the general impossibility of keeping the structures of each line exactly vertical make such construction difficult at best. This type of construction occurs most often when a transmission circuit must coexist in the same right-of-way with a distribution circuit and the latter requires shorter spans in order to locate poles for service drops or taps. To avoid overbuilding of such facilities, it usually is necessary for them to occupy opposite sides of the road or street. When more than two utilities occupy the same highway, a conflict is almost inevitable unless structures are used jointly. Where conflict (such as shown in Figure H221-2 and Figure H221-3) occurs, Grade B construction may be required (see Rule 241D and Rule 243A4).

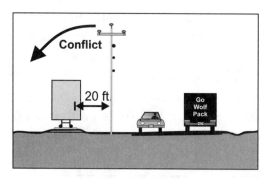

Figure H221-2
Joint-use line near railroad

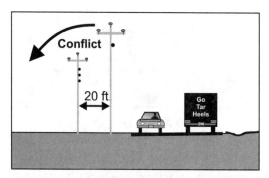

Figure H221-3
Supply line conflict with communication line

The preferable condition is complete separation of the two lines, except as conditioned in Rule 222—*Joint Use of Structures*. Joint use may be preferable to complete separation in locations where significant reductions in cost or environmental factors can be achieved (see Figure H221-4).

Construction of conflicting lines, if they are located as far as practical from one another and meet the Grade of Construction requirements of Section 24, generally is preferable to overbuilding lines.

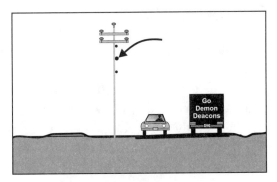

Figure H221-4
Joint use

222. Joint Use of Structures

From a safety standpoint, the ideal condition is for a communication line and a supply line that must follow approximately the same route to be separated adequately. In the case of main toll communication lines and high-voltage transmission lines, the ideal of adequate separation generally can be realized with normal, economical construction methods. Occasions arise, however, when communication and supply lines cannot be so separated.

Where it is impractical to secure separation beyond conflicting distance between the communication and supply lines, a choice must be made between the relative advantages and disadvantages of a joint line and separate, conflicting lines. Both of these types of construction are covered by the rules.

There are cases where one method is to be preferred to the other. Conflicting lines that are not overbuilt naturally offer less opportunity for accidental contact between the conductors of the supply and communication lines, since the likelihood of a broken supply wire falling on a communication conductor is reduced greatly. The possibility of broken poles bringing the two classes of conductors into contact is also perhaps more remote with this method of construction. Such a conflicting line may be preferable to joint use of poles when the supply lines may impress upon the communication circuits a voltage potential against which the communication protective apparatus cannot function reliably; however, if the protective apparatus can function reliably under the circumstances, joint use is preferred. *Consideration* of joint use along highways, roads, streets, and alleys was specified in the 1977 Edition; these areas generally are accompanied by competition for land use. In some cases the issue of consideration is academic, since government authorities sometimes require joint construction unless valid reasons can demonstrate otherwise.

One of the items to be considered is the relative likelihood of (1) the higher-voltage facilities tripping off-line along with lower-voltage facilities when a joint-use line goes down versus (2) the higher voltage lines coming down on, and becoming supported by, a conflicting line without tripping out before damaging the conflicted facilities. On the other hand, from the safety standpoint, a joint line is always preferable to overbuilding, and experience has shown it generally to be preferable to conflicting construction. Another benefit to be derived from the joint use of poles is the reduction in the number of supporting structures on the streets.

Since the available routes for the distribution networks of communication and supply services frequently must coincide, and as the users of both services are, to a large extent, common, the lines of both classes of service will, in general, occupy the same streets or alleys. Since the voltage that such distribution supply circuits may impress upon communication circuits is generally within the limits of reliability of communication protective apparatus, a joint use of a structure line may be a suitable solution. Even when higher distribution voltages are involved, a joint line is usually regarded as safer than separate lines, which must have numerous crossings under or over each other, including service drops to customers' premises. This is especially true when the alternative is either a conflicting line or separate lines on opposite sides of a street which, under the definition, are not in conflict, but yet involve the possibility of mechanical interference.

Where joint use of poles is made by different utilities, there is generally a mutual and reciprocal agreement between them providing for such joint use; a higher degree of cooperation is usually thus obtained than ordinarily is found where the utilities are on separate poles. This valuable spirit of cooperation greatly assists in maintaining a high standard of construction.

In the case of electric railway lines, it is often necessary or desirable to have them on joint poles with communication circuits, but such joint use frequently involves only the attachment of a trolley span wire to poles of the communication line. Where the trolley-contact wires are supported by span wires attached to a double line of poles, it generally is desirable to put the trolley feeders on one line of poles and the communication wires on the other line of poles.

223. Communications Protective Requirements

(Previous Rule 287B—Metal-Sheathed Cable was deleted in the 1977 Edition; it was inappropriate for this location and was superseded by the general revision. This rule was moved from Rule 287 to Rule 223 in the 1990 Edition.)

Most overhead communication lines are, or sometimes will be, exposed to supply circuits exceeding 300 V to ground at some point. Therefore, it is usually advisable to provide one of the methods of protection given in this rule. Typically such measures as those listed in this rule are used in conjunction with bonding of the communication cable messenger(s) to the neutral of the electric supply system, in order to limit the voltage that can be impressed on the communications facilities to a level at which those measures can protect customers' premises.

The wording of this rule was revised for clarity and for consistency with the remainder of the Code in the 1977 Edition. The rule recognizes that qualified persons, because they are aware of the special problems of such facilities and are trained to take appropriate precautions, can safely handle communications apparatus that might be considered unsafe for the general public to contact. Apparatus not subject to contact by the general public, such as aerial splice enclosures and terminals, intentionally are excluded from the requirements.

The 2007 Edition added references to IEEE Std 487™ *IEEE Recommended Practice for the Protection of Wire-Line Communication Facilities Serving Electric Supply Locations* and IEEE Std 1590™ *IEEE Recommended Practice for the Electrical Protection of Optical Fiber Communication Facilities Serving, or Connected to, Electrical Supply Locations* to provide users with guidance on how to evaluate the effects of ground currents on communication circuits near supply stations that have large ground currents.

224. Communication Circuits Located Within the Supply Space and Supply Circuits Located Within the Communication Space

(This rule moved from 288 to 224 in the 1990 Edition.)

In the Fourth Edition of the Code, this rule covered only communication circuits used exclusively in the operation of supply lines. New communications equipment later made it necessary to supply power at points along communications systems; separate requirements were then included in the Fifth Edition for supply circuits used exclusively in the operation of communication circuits.

This rule was reworded for clarity in the 1977 Edition, but it essentially retained the provisions of the Fifth Edition.

With the widespread use of fiber-optic cable in both supply- and communication-line construction in the late 1980s came a series of questions about how to handle fiber-optic cables used in the operation of supply facilities. Rule 230F of the 1990 Edition was the first time that a fiber-optic supply cable was treated as a neutral conductor meeting Rule 230E1 if it met certain criteria.

Neither a fiber-optic cable nor a 230E1 neutral can be located between the supply space and the communication space in the span, nor can either one be supported at a pole between the two spaces.

As written, Rule 224A of the 1990 and prior editions only applies to circuits used exclusively in the operation of supply circuits. Rule 224A was originally written to address communication circuits utilizing metallic signal conductors and is intended to limit the transference of an inappropriate voltage from the supply space to the communication space. Since metallic messengers can transfer voltage, even if the fiber-optic cable cannot, adherence to Rule 224A3 was required if such a circuit that is used exclusively in the operation of supply facilities is to be located in the communication space.

Rule 224A was completely revised in the 1993 Edition to recognize the requirements for work performed in the supply space versus that performed in the communication space. References to the date being transmitted were deleted and the confusing reference to guarding or isolating Supervisory Control and Data Acquisition Systems (SCADA) circuits has been removed. Rule 224A3 has been changed to include the important considerations that must be observed when communication lines make transitions between supply and communication spaces. Previous limitations on the ability of an inappropriate voltage to be transferred to the communication space were retained. The revised rules allow joint use of communication cables located in either space, but they mandate that those working on communication cables located in the supply space be qualified to work in that space (see Section 44). Retention of the visible clearance zone between the supply and communication spaces is ensured by the reference to Rule 230F. That rule covers clearances involving fiber-optic cable used in the supply space and communications facilities.

Where there are hybrid cables containing both 120 V ac power for traffic signals and communication conductors to control the relays, such cables are considered as supply cables and do not meet Rule 224A.

Rule 224B includes special limits on supply circuits used solely to power communication equipment. As of the 2002 Edition, the rule applies to circuits up to 150 V dc, as well as the previously included 90 V ac. The ac voltage limit had been 400 V until the 1997 Edition. As of 2002, an EXCEPTION to the general requirement to meet Rule 224B2 applies when less than 150 W is transmitted.

225. Electric Railway Construction

(This rule was moved from Rule 289 to Rule 225 in the 1990 Edition.)

225A. Trolley-Contact Conductor Fastenings

When a trolley pole slips from the contact wire, it will frequently break the trolley wire loose from its supporting span or bracket suspension wire. It is desirable and reasonable to require that, if the trolley wire becomes loosened from one hanger or if one suspension span fails, no part of the trolley-contact wire or its current-carrying parts may come closer than 3.0 m (10 ft) to any generally accessible place.

225B. High-Voltage Contact Conductors

As voltages become greater, the danger rapidly increases. It is entirely reasonable to require that, for overhead trolley-contact conductors of more than 750 V, the supports should be so frequent that even a break in the trolley conductor itself could not permit its falling low enough to obstruct or endanger either pedestrian or vehicular traffic (see Figure H225B).

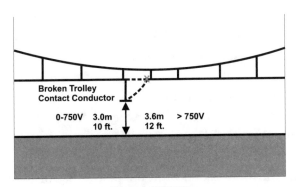

Figure H225B
Broken trolley contact conductor clearance

225C. Third Rails

Third rails are used on many interurban electric railways but, because of the difficulty of providing adequate protection, they are seldom installed in locations that are open to the public. However, where they are so located, the necessity for adequate protection is very evident. This protection is best obtained by the installation of a continuous insulating strip or strips placed above the rail. The construction is, perhaps, the simplest and safest when the underrunning type of rail is employed; the possibility of accidental contact is greatly reduced and the operation of the railway under severe weather conditions is improved considerably. Since it is impossible to prevent contact with live parts under all circumstances, it is recommended that the public be warned by suitable signs and that third rails be eliminated except on private rights of way, elevated, or underground structures.

The previous provisions were retained in the 1977 Edition but the EXCEPTION was added.

225D. Prevention of Loss of Contact at Railroad Crossings at Grade

Because of the greater clearance required for trolley-contact conductors at railroad crossings, the trolley pole at such places takes a nearly vertical position. The trolley springs are adjusted so that they are not in tension when the pole is in the vertical position. Thus, the greater the elevation of the contact conductor, the less effective is the trolley-pole spring and, therefore, the less the pressure of the trolley wheel against the conductor. Because of this reduced pressure, the jar of the car passing over the crossing can easily cause the wheel to slip from the trolley wire. An inverted trough above the trolley wire, if placed in such a manner as to catch the wheel should it leave the wire, will ensure that the car has continuous power and limit the possibility of it being forced to stop on the crossing. Maintaining the trolley conductor in the spans immediately adjacent to the crossing span at the same elevation as the crossing may help to reduce the effect of a sudden change in level. Even in this case, however, the trough is recommended.

Such special construction is not necessary where a pantograph trolley with rollers or shoes is employed.

225E. Guards Under Bridges

The foregoing recommendations also cover the protection of the trolley conductor where the electric railway passes beneath a metallic bridge except that, here, the guard must be of insulating material. If the trolley pole leaves the wire under the bridge, the accidental short circuit produced between the trolley wire and the bridge could burn the wire down.

The provisions of the previous two-paragraph rule were combined into one paragraph in the 1977 Edition.

Section 23. Clearances

Section 23 specifies the clearances of wires, conductors, and cables from each other and from other facilities. It also specifies clearances for structures and structural components. Movement of conductors and cables, as well as flexure of structures, insulator strings, etc., are taken into account to assure that the clearances at the design conditions will be appropriate for the items concerned.

In the 2007 Edition, a concerted effort was made throughout Section 23 to (1) specify categories in all clearances tables for supply neutrals meeting Rule 230E1 (The previous specification of neutrals separately from conductors of 0–750 V in some tables and not in others had caused confusion.), (2) respecify values in clearances tables and rules to consistent decimal places (so that calculation results would be consistent), (3) specify how results of required calculations would be rounded (so that the results would appropriately match values specified in rules and tables), and (4) recognize *overhead shield wires* along with *surge protection wires*.

230. General

230A. Application

The clearances and spacings specified in Section 23 are intended to perform two functions, under the expected conditions of operation:

(1) to limit the opportunity for contact by persons with circuits or equipment, and

(2) to inhibit the covered utility facilities from coming in contact with other utility or public facilities.

In the comprehensive revision of 1990, clearances were adjusted (where necessary) to meet a coordinated clearance system. The 1990 system combines consistent components for reference dimensions of potentially conflicting activity with uniform mechanical and electrical components that recognize the characteristics of the different types and voltages of lines and equipment. A new Appendix A was added to the Code to explain the new methodology and to compare it to that used for previous editions.

In the 1993 Edition, portions of Rule 014 concerning the clearances required for *emergency* installations and for *temporary* installations were expanded and moved to Rule 230A1 and 230A2.

230A1. Permanent and Temporary Installations

(Rule 230A1 was added in the 1993 Edition.)

Note that temporary installations are not required to have the same strengths (see Rule 014) as permanent installations, but temporary installations are required to have the same clearances as permanent installations (see Rule 230A1). Both permanent and temporary installations are planned installations, with temporary installations being planned for shorter durations, and have the same clearances.

230A2. Emergency Installations

(Previous Rule 014A1 was moved to Rule 230A2 in the 1993 Edition)

Rule 230A2 specifies reduced clearances that are applicable only to emergency installations. These clearances particularly recognize the overall safety value of quickly restoring power to traffic signals and street lighting, even if the full permanent clearances will not be in effect during the term of the emergency. Two requirements are worthy of special comment here. First, the allowed reduced clearances for emergency installations are still great enough to be out of contact range by pedestrians (nothing exceeding 2.45 m [8 ft] in height) or trucks (as high as 4.3 m [14 ft]), as applicable. An EXCEPTION to these emergency clearance requirements was added in 1997 Rule 230A2e if the area is made accessible only to qualified personnel. Second, the installation must be put in permanent condition as soon as practical. This rule is typically only used when there is significant, widespread storm damage requiring days or weeks to repair. However, it is sometimes useful to restore safe service quickly while making permanent repairs when only one or a few structures are damaged, such as by vehicle wreck.

It is useful to think of an *emergency* as applying to the clearance of facilities during system reconstruction after an *unplanned* outage and *temporary* as applying to a *planned* installation of short duration.

Rule 230A2d was revised in the 1997 Edition to allow laying secondary service cables meeting Rule 230D (i.e., with a covered neutral) on the ground if properly protected during an emergency. Note that Rule 230A2 only applies to emergency outages, not to temporary installations. If a temporary, above-ground "underground cable" service is needed to a new dwelling or building, and trenching is not practical due to frozen ground, this rule cannot be used to allow placement of the cable on the ground. In such a case, the cable would need to be installed in a conduit and suitably protected to meet Section 32.

230A3. Measurement of Clearance and Spacing

(Rule 230B–Constant-Current Circuits *of the Sixth and prior Editions was moved to Rule 230G in the 1977 Edition. In 2007, Rule 230B*—Measurement of Clearance and Spacing *was moved to 230A3 to allow the use of Rule 230B for specification of required ice and wind loadings used to calculate inelastic deformations used in computing sags under the conditions required by Section 23.)*

This rule was added in the 1977 Edition to clarify where center-to-center and surface-to-surface measurements are applicable (see Figure H230-1).

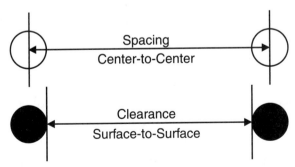

Figure H230-1
Method of measurement

For clearance purposes, live metallic hardware that is connected electrically to line conductors is considered to be of the same voltage classification as the line conductors. Metallic bases of potheads, surge arresters, and similar devices, however, are considered the same as a part of the supporting structure (see Figure H230-2).

The Code is clear in its requirements that parts of indeterminate potential above 150 V to ground, such as portions of insulator bodies, are to be either isolated or guarded on the basis of the maximum voltage potential present.

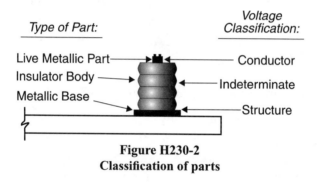

Figure H230-2
Classification of parts

It must be recognized that there will be a significant voltage potential to ground at points near the conductor end of insulating sections. The voltage potential on the surface of an insulator decreases for surfaces nearer to the base end. The maximum voltage potential depends upon the gradient characteristics of the insulator material and construction, and upon the amount of exterior contaminants present on the insulator. Achieving appropriate clearance is rarely a problem with most insulators, since their length is short enough that, if the conductor end meets the required vertical clearance above ground, the portion of the insulating section near the grounded end also is isolated effectively by elevation. However, certain insulator bodies are long enough to require additional care in setting the construction height for the unit. These longer units, in effect, require increased clearance for the conductor end in order to adequately isolate the points of indeterminate voltage potential near the base end of the insulator. While this problem is most often found in electric supply stations and is addressed by Rule 124, the problem can exist with certain installations in areas accessible to the public covered by Part 2.

230A4. Rounding of Calculation Results

(Rule 230A4 was added in the 2007 Edition)

The 2007 Edition added Rule 230A4 to specify how the results of calculations required by rules in Section 23 would be rounded. Unless otherwise specified, the results will be rounded up.

230B. Ice and Wind Loading for Clearances

*(Rule 230B–*Constant-Current Circuits *of the Sixth and prior Editions was moved to Rule 230G in the 1977 Edition. In 2007, Rule 230B*—Measurement of Clearance and Spacing *was moved to 230A3. Rule 230B*—Ice and Wind Loading for Clearances *was added new in the 2007 Edition.)*

This rule was added in the 2007 Edition to bring all of the requirements for clearances determinations into Section 23. Previously, the ice and wind loadings used for structural purposes in Rule 250B had been referenced by Section 23 for the purpose of determining the inelastic deformation of conductors and messengers to be used in determining sags and sag changes that were required to be considered by the clearances rules of Section 23.

Establishing a set of clearance conditions in Section 23 separate from those in Section 25 for structure loading purposes will allow potential differences to be specified in future editions between the sag and tension conditions used for clearance purposes and those used for wire and structure loading purposes. The new rule refers to *clearance zones* instead of *loading districts*.

230C. Supply Cables

Where a supply cable is covered with a continuous grounded metal sheath or armor, high voltage cannot be maintained on the sheath because it is essentially at ground potential. An internal failure of insulation normally will short-circuit the supply system and cause it to de-energize. Similarly, insulated high-voltage conductors that are cabled together and in physical contact with a grounded messenger usually can be expected to de-energize in the event of insulation failure, especially if the assembly is wrapped with metal tape. It is therefore reasonable to make clearances for such facilities independent of, or less dependent on, voltage.

Where cable sheaths or messengers are not grounded effectively, dangerous voltages can be present. Similarly, where insulated conductors are not in physical contact with a grounded messenger, such as when separated from the grounded messenger by insulating spacers, there is no assurance that de-energization will take place in the event of insulation failure. In both cases, the facility is to be classified the same as open supply wires of the same voltage.

For clearance purposes, early editions of the NESC classified "permanently grounded continuous metal sheathed supply cables of all voltages" to be the same as open-supply wires of 0–750 V. The Sixth Edition added "insulated conductors supported on and cabled together with an effectively grounded messenger" and reclassified both groups to be the same as guys and messengers. Among the types of cables determined to be included in this classification were:

(1) self-supporting lashed cable construction (for primary circuits);

(2) conductors lashed to a grounded messenger and treated as a cable (for secondary circuits); and

(3) conductors twisted around the grounded support wire (neutral) (for secondary circuits).

Self-supporting aerial spacer-type construction does *not* meet the requirements of Rule 230C (see Rule 230D). This type of cable uses a grounded messenger to support covered conductors by means of insulated spacers placed at intervals in the span; when the covering material fails electrically, the outside of a phase conductor can exhibit essentially the same voltage as that on the conductor inside the covering.

Rule 230C was expanded in the 1977 Edition to differentiate between the several general types of insulated cable in a manner consistent with their relative potential for creating a safety problem. Metal-sheathed cable, as defined in Rule 230C1 (now Rule 230C1a), offers the greatest protection against abrasion. Also considered as meeting Rule 230C1 was nonmetallic-shielded cable of 8.7 kV or less and meeting the other requirements (now covered by Rule 230C1b). Such cable of voltages higher than 8.7 kV phase-to-ground was then considered as 230C2 cable. The voltage limit on the cables meeting present Rule 230C1b was raised from 8.7 to 22 kV in the 1987 Edition, thus effectively limiting the applicability of Rule 230C2. Cable surrounded by multiple wires forming a concentric neutral was added to the 230C1 category, without voltage limitation, in the 1990 Edition. Previously such cable qualified under Rule 230C1 subject to the 22 kV voltage limit, since it has a semiconducting sheath as well as the concentric neutral conductors.

Nonmetallic shielded cable meeting Rule 230C2 is safe to touch because there will be no capacitive current between the grounded object touching the cable and the energized conductor. The semiconducting shield will drain off any leakage current to the required effectively grounded messenger or neutral.

Nonmetallic unshielded cable has not been practical above 5000 V. When new, leakage currents and capacitive currents of these cables are below the threshold of perception, but the insulation could be damaged without faulting the conductor to ground. If the insulation deteriorates over time or is otherwise damaged, the damaged conductor may flash over to the neutral during a rain, removing the circuit from service and eliminating any potential hazard.

The general revision of the 1977 Edition continued to allow lesser clearances for cables meeting Rule 230C than for open conductors in certain situations. The degree of lesser clearance depended upon the cable classification. The 1984 Edition recognized relative differences by further reducing the clearance requirements for supply cables of

0–750 V meeting Rule 230C2 or Rule 230C3. These reductions were based upon relative potentials for creating safety-related problems.

It should be noted that the types of duplex, triplex, and quadruplex etc., that use an *insulated* neutral do *not* qualify under this rule. The neutral is not bare and cannot function as expected of these installations in the event that the integrity of the insulation around the conductors is violated.

The differences between the treatment of these cables in the coordinated clearance system were further clarified in the 1990 Edition, but the treatment of those differences in the tables may, at first, seem confusing since various references are made to 230C2 cable of 0–750 V. Clearly such cable meets 230C1. However, the Clearances Subcommittee intentionally left the table classifications unchanged in the 1990 Edition while other cable review is being undertaken. This also allows the clearances in the tables to clearly apply to cables that were classified as 230C2 cables at the time of the original installation; these older cable constructions should not now be classified as meeting 230C1b if they were originally classified as meeting Rule 230C2. The clearance tables maintain the 230C2 clearances at lower voltages in order to continue to apply to earlier installations.

In the 1990 Edition, Rule 230C1 was split into 230C1a and 230C1b in order to facilitate proper application of the guarding requirements of other rules, such as near buildings, in the communication worker safety zone, etc. Unfortunately, all of these rules were not revised accordingly until a coordinated effort in the 2007 Edition. Because of the time delay, IR 533 issued 5 November 2003 indicated that neither the 230C1a nor the 230C1b cables were required to be guarded in the communication worker safety zone. That IR answer was negated by the 2007 changes.

In 2007, the rules that specify guarding requirements were revised to differentiate between cables meeting Rule 230C1a and those meeting Rule 230C1b. The older so-called *metal clad* cables fit under Rule 230C1a and may not need to be guarded in specified circumstances. The typical underground residential distribution (URD) supply cables in use today have a semiconducting insulation shield with neutral wires serving as metallic drainage; these cables are covered by Rule 230C1b, regardless of whether they have an insulating jacket over the neutral, and require guarding in those locations. Neither the NESC work rules nor OSHA regulations allow line workers to touch those cables unless they are de-energized and grounded, and it is not appropriate for them to be exposed at public levels. As a result, URD and other supply cables meeting 230C1b are required to be guarded at various specified locations accessible to the public and to communication line workers.

230D. Covered Conductors

(This rule was added in the 1977 Edition. The previous Rule 230D—Neutral Conductors was renumbered to 230E.)

Covered conductors generally are used only when right-of-way constraints require close spacing of conductors or where there is no practical alternative to passing conductors near or through trees. Although covered conductors are not economical or appropriate for normal construction, they may provide a reasonable means of limiting outages caused by momentary contact between two conductors or between a conductor and a noninsulated surface (such as a tree). They also reduce the opportunity for mechanical damage to conductors from such abrasion or arcing. This increases the reliability of operation of distribution lines that face right-of-way constraints.

For clearance purposes, these conductors are treated as bare conductors because they have no specific insulation rating. While covered conductors with the dielectric composition and thickness generally in use today may be touched without receiving a perceptible shock (assuming that the insulation did not become damaged during installation and is new and dry, an assumption that is all too often wrong), there is no industry-wide standard that would prevent covered conductors rated at 5 kV from being operated at 15 kV. When the surface of the dielectric is wet, the capacitance between the energized conductor and a human is increased, perhaps sufficiently to be hazardous. Over time, with exposure to the elements and abrasion, the dielectric can crack or be worn away to effectively expose bare conductor without short-circuiting the covered conductors permanently to ground. In addition, storm damage may be able to bring the line down without tripping it off-line, thus creating a potential hazard.

While in certain situations the installation of covered conductors may reduce the possibility of circuit operations, the use of covered conductors does not allow a reduction in clearances to personnel spaces. See "Design and Application of Aerial Systems Using Insulated and Covered Wire and Cable" and "Safety Considerations of Aerial Systems Using Insulated and Covered Wire and Cable" presented at the IEEE T&D Expo, Los Angeles, 15–20 September 1996 and published in the *IEEE Transactions on Power Delivery*, Vol. 12, No. 2, April 1997 on pages 1006–1011 and 1012–1016, respectively. These papers reported the results of calculations and tests by experts from the utility and cable manufacturing industries. In effect, even when new, such cables can present personal safety hazards if touched under certain circumstances; future degradation in service only serves to increase the level of potential personal safety hazard, if touched.

Covered conductors are considered to have at least the same potential for involvement in a safety-related problem as bare conductors due (1) to the high probability of a covered conductor exhibiting the same voltage potential to ground during its life as a bare conductor (see Figure H230-3) and (2) to the opportunity for an uninformed person to assume that a covered conductor is insulated and approach it in an inappropriate manner because of a false sense of security.

In addition, the covering causes the conductor to be larger and heavier. Sag and wind displacements are greater, especially where ice will accumulate on the conductors. The covering also limits the cooling capability of the conductor, either limiting its current flow or increasing its sag due to the resultant increase in thermal loading. Unless a covered conductor is properly installed, it can be more susceptible to lightning damage than a bare conductor.

Figure H230-3
Covered conductor with deteriorated covering

The rule has traditionally used the term *spacing*, but the term was corrected to *clearance* in the 2007 Edition.

230E. Neutral Conductors

(This rule was added in the Fifth Edition as Rule 230D. It was moved to this number in the 1977 Edition.)

Neutral conductors of multigrounded wye systems are frequently connected to water pipes, metallic conduit, metallic cable sheaths, guys, etc. These objects are contacted by the general public every day. Except under fault conditions (which ordinarily will be of extremely short duration), these conductors have been considered to be relatively harmless; appropriate reduction in clearances is, therefore, justified. Where the neutral conductor of a multigrounded supply circuit in the low- or medium-distribution-voltage range is grounded effectively throughout its length, there is little likelihood of its carrying significant voltage potentials. Accordingly Rule 230E1 recognizes that, for clearance purposes, neutral conductors of such multigrounded systems are considered to be the same as 0–750 V open-supply conductors. Conversely Rule 230E2 recognizes that, if the supply-circuit neutral conductor is not grounded effectively throughout its length, it may carry the phase-to-neutral potential; thus, such neutrals are classified the same as the phase conductors with which they are associated.

The above-voltage limit was originally 15 000 V *between conductors* when the rule was introduced in the Fifth Edition. That limit was determined because it represented the maximum line potential usually employed for such systems and with which there was sufficient experience at that time to justify the reduced voltage limit for the neutral conductor.

As more experience was gained with higher-voltage systems, and as the equipment for grounding and for promptly de-energizing such systems was improved, the voltage limit was raised. The upper limit became 22 kV *to ground* in the Sixth Edition. That edition also reclassified such effectively grounded neutrals to have the same clearances as guys and messengers. Although guys and messengers generally have the same clearance requirements as conductors of 0–750 V, in certain circumstances, this reclassification further reduced the clearance requirements. The limit of 22 000 V to ground was adopted as encompassing the highest voltage to ground of multigrounded wye distribution systems in current use with which there had been adequate experience. This value has been intentionally maintained because (1) it exceeds the voltage to ground of a 34.5/19.9 kV wye line and (2) when subtracted from 50 V it leaves 28 kV which, when multiplied by 10 mm/kV (0.4 in/kV) produces a 300 mm (1 ft) voltage adder from 22 to 50 kV that is used elsewhere in recent editions of the Code.

This rule is particular in the limiting of conditions to which such reductions can be applied. For example, if a 12.5/7.2 kV wye line has a multigrounded neutral, it normally would be eligible for the reduction. However, if that line is joint use with a wye-connected transmission line whose phase-to-ground voltage is greater than 22 000 V, and the neutral of the two lines is common, the neutral is assumed to be associated with the higher-voltage circuit and does *not* qualify for the reduction. This limitation rarely affects joint transmission/distribution construction because most of the transmission voltage circuits in use today are high-impedance grounded, not solidly grounded; the multigrounded overhead surge protection wire that is often used to protect tall transmission lines from lightning strikes is not a neutral—its only connection to the transmission circuit is through a high-impedance connection at the source transformation(s).

It should be noted that such neutral conductors do not behave the same under loading as high-strength guys and messengers. While the same basic clearances may be used, there are additional requirements, such as those in Rule 232B, which affect neutrals but not guys.

Rule 230E determines whether a neutral conductor must meet the full clearance requirements of Rule 234B. If the neutral conductor meets the requirements of Rule 230E1, it is considered to be equivalent to a messenger-neutral meeting the requirements of Rule 230C and, thus, EXCEPTION 1 of Rule 234B applies. If the neutral conductor meets the requirements of Rule 230E2, the full clearance is required as specified in Rule 234B.

NOTE: EXCEPTION 2 of Rule 234B applies only to the vertical clearance requirement and may only be used in lieu of EXCEPTION 1. The clearance reductions of the two EXCEPTIONS are not cumulative.

230F. Fiber-Optic Cable

(The rule was added in the 1990 Edition.)

While fiber-optic "conductors" within a cable are nonmetallic and do not conduct electricity, such cables frequently are carried by conductive messengers or energized phase conductors and may include auxiliary metallic conductors within the cable. Therefore, fiber-optic cables are treated the same as a conductor of the voltage that can be carried on the cable messenger or its interior auxiliary conductors. Such cables must be treated as supply *or* as communication *and be located accordingly.*

Even if carried on a nonmetallic messenger, fiber-optic cables are prohibited from being located between the supply space and the communication space. It must be stressed that such cables cannot be located between the spaces even if they meet Rule 224A3. The safety zone between the lowest facility in the supply space and the highest communication facility by Rule 235 and Rule 238 is intended to be kept clear of such cables.

The 1993 Edition revised the language for easier identification and specifically addressed entirely dielectric cable; such cables are considered the same as a neutral meeting Rule 230E1 for purposes of clearances between supply and communication facilities. Under Rule 235 and Table 235-5, FN 11, fiber-optic cables that are entirely dielectric have no specified clearances from supply conductors and cables; they can be attached directly thereto. However, if not attached, Rule 235G should be considered to limit mechanical interference.

230G. Alternating- and Direct-Current Circuits

(This rule was moved from 230F in the 1990 Edition.)

This rule was added in the 1977 Edition to recognize dc transmission lines in the extra-high-voltage range. It is consistent with the general treatment of dc lines in the Code; they generally are considered to be equivalent to ac lines having the same crest voltage to ground. A NOTE was added in 2002 to remind users that the ratio of peak (or crest) voltage to the normally used term (generally rms) will not be 1.414 for nonsinusoidal wave forms.

230H. Constant-Current Circuits

(Prior to the 1977 Edition, this rule was numbered 230B; it was moved to 230H in the 1990 Edition.)

Where a person may come into contact with a constant-current circuit, the degree of safety problem (assuming the circuit to be intact) depends mainly on the full-load voltage of the circuit. However, in the event of a contact of a constant-current circuit with other facilities, there may be an additional voltage occasioned by either the value of current or the open-circuit voltage. As long as no open circuit occurs in the constant-current circuit, however, the voltage of interest during a contact with other facilities generally would be the full-load voltage of the constant-current circuit.

230I. Maintenance of Clearances and Spacings

(This rule was added in the Fourth Edition as Rule 230D. It was moved to Rule 230E in the Fifth Edition, to Rule 230H in the 1977 Edition, and to Rule 230I in the 1990 Edition.)

It is intended that the clearances required by Section 23 shall be maintained over the life of the installation. If sags change, or if the use or character of the site changes enough to reduce clearance below required levels, action is required. The Code is intentionally silent as to a preferred method of maintaining clearances. It may involve any of the following:

(1) initial installations at such clearances that permanent sag increases during the life of the line will not reduce required clearances,

(2) appropriate resagging of conductors, or

(3) any other mechanism to ensure maintenance of clearances.

It is the responsibility of the utility to perform the necessary inspection and maintenance.

It is not the intention of this rule to place the expense burden of maintaining required clearances on a utility's ratepayers in the case where a line meets required clearances and, later, a private landowner builds something in a place that would violate the required clearances. In many jurisdictions, the Administrative Authority requires utilities, once they have installed facilities designed with good engineering judgment on the basis of reasoned expectations, to seek reimbursement for any cost of relocating facilities required to meet the special needs of site developers and maintain Code compliance. Often, the solution to such conflicts that is best for all concerned is relocation of the recently added private facilities, not the utility facilities.

One purpose of this rule was to limit misunderstandings of the so-called "grandfathering" provisions of Rule 013B. In the 1993 Edition, a reference to Rule 013 was added and the word "currently" was added to

Rule 013B2 so that installations that *currently* comply with a previously applicable edition may continue to do so.

> **Example:** An existing line over a former cow pasture that is now covered by water from a new flood-control reservoir can continue to meet the previously applicable edition if it now meets the newly applicable sailboat clearances of that edition. If the applicable edition is the 1977 or later, sailboat clearances are specified therein. If the applicable edition is the Sixth Edition (all previous installations being required by the Sixth Edition to be upgraded if necessary to meet the Sixth Edition), sailboat clearances were not specified; thus good practice is required by Rule 200C. Because later editions specified "good practice," it is reasonable to use sailboat clearances of the edition in effect at the time of the conversion from pasture to lake as good practice, if the conversion took place after the 1977 Edition went into effect. If the land conversion occurred before the 1977 Edition, the line(s) would have to be located high enough to clear the expected sailboat mast height to meet good practice, but not necessarily located exactly to the specified clearance later. If larger structures would have to be installed to meet the changed condition, the then-current edition would be applicable (see Rules 013A and B).

IR 529 issued 24 July 2002 clarified that abnormal conditions from actions of others or from severe storms are not covered by Rule 230I. Rule 230I was revised in 2007 to ratify IR 529. Application of Rule 230I is limited to the conditions specified in Section 23 of the applicable edition. Such conditions as a structure broken by an out-of-control vehicle, a tree over a line, ice loading beyond the specifications of Section 23, a conductor being burned down by contact from flying debris during a storm, or other activities beyond the conditions specified in the NESC are not considered to be violations of NESC requirements. However, utilities are responsible for making timely corrections whenever they have knowledge of the noncompliant conditions, whatever the cause. The seriousness and immediacy of the potential hazard from such noncompliant conditions affect the required timeliness (see also Rules 214A4 and 214A5). Some noncompliant conditions will require immediate action (Rule 214A5); others may be scheduled for correction at an appropriate future time (Rule 214A4).

231. Clearances of Supporting Structures From Other Objects

231A. From Fire Hydrants

The clearance in Rule 231A between line structures (including guys) and fire hydrants is required to make the latter readily accessible when needed. The clearance of a pole to a fire hydrant was increased from 900 mm (3 ft) to 1.20 m (4 ft) to use the usual style of placing the larger, desired value first. It also recognizes the modern practice of fire-fighters to attach a gate-valve unit to one side of a fire hydrant so a second truck can attach without turning water off to the first truck. In 2007, EXCEPTION 2 was added to allow lesser clearances by agreement between the local fire authority and the pole owner.

Previous to the 1977 Edition, Rule 231B—*From Street Corners* was included to further identify desired structure location on corners where fire hydrants were on the corner. Where hydrants are located at street corners, structures cannot always be placed at the intersection of the lines, and this may make the use of inconvenient flying taps necessary. This type of construction was discouraged because such taps generally are inaccessible from the structure. With the advent of ladder trucks and aerial lift devices such as *bucket* trucks, flying taps became less of a problem, and the rule was eliminated in the 1977 Edition.

This rule does not require clearances between structures and other similar objects, such as telephone pedestals. Recent accidents have indicated that certain cautions may be appropriate in locating such objects near line poles. In one case, a line worker chipped out and fell to the ground with his belt still attached around the pole. He apparently reached the ground safely except that, before his legs and knees could take up the shock of landing, the base of his spine struck a telephone pedestal squarely and caused him damage. It would appear appropriate that any such objects should either meet the requirements of Rule 231A or be located away from the climbing side or corner of the pole.

Although the dimensions of required climbing spaces are covered by Rule 236 and such dimensions start at the ground level, it is not apparent that the situation described here is covered since workers can change quadrants to move around a pole.

231B. From Streets, Roads, and Highways

(Rule 231B—From Street Corners of the Sixth and prior editions was deleted in the 1977 Edition. See the previous paragraph. The prior Rule 231C was renumbered to Rule 231B in the 1977 Edition. Rule 231B does not apply to pad-mounted equipment. Rule 300, Rule 310, and Rule 311 apply but do not specify a distance.)

The 150 mm (6 in) clearance of previous editions was retained in Rule 231B2 of the 1977 Edition as the clearance required in all cases where curbs existed. However, this rule was also expanded in the 1977 Edition to allow for recommended increases in clearances under certain conditions. Some ordinary trucks and delivery vehicles overhang the curb by more than 150 mm (6 in). Superelevated curves and heavily crowned roads further increase this overhang. The 300 mm (1 in) clearance behind the front face of curbs on local streets recommended by Rule 231B2 in the 1977 Edition would place the edge of the supporting structure approximately 150 mm (6 in) back of the curb and minimize conflicts. A clearance of 600 mm (2 ft) behind the front face of the curb on arterial streets was recommended in the 1977 Edition because of the larger traffic volumes, greater truck traffic, and higher vehicle traveling speeds.

The recommended 600 mm (2 ft) offset reflected the current requirements of many state highway departments in their utility accommodation policies. It also matched the design criteria adopted by the American Association of State Highway Officials (AASHO) in their design guide for local roads and streets, *Geometric Design Guide for Local Roads and Streets; Part 2, Urban.* The guide stipulates that a clearance of at least 600 mm (2 ft) generally should be provided on all streets between the face of the curb and the obstruction.

The 4.6 m (15 ft) vertical dimension to which clearances apply relates to attachments and corresponds with the provisions of Rule 232, Table 232-1 (revised Rule 286 of the 1977–1987 Editions).

Early copies of the 1977 Edition contained a gross typographical error in Rule 231B2: the requirements of (a), (b), and (c) were *all* shown to be mandatory; only (a) was intended to be mandatory. This was corrected in the 1981 Edition.

Because of difficulties in determining which roadways met the requirements of (b) and (c), the rule was again revised in the 1984 Edition (see Part A of Figure H231-1). The requirement is now to locate structures a sufficient distance from the curb to avoid contact by *ordinary* vehicles using and located on the traveled way (see Figure H231-2). The discretion and responsibility for proper placement is in the hands of the utility. It is recognized that it is often impractical to locate structures so that oversized vehicles or towed objects that project off the roadway cannot touch the structure if the driver is negligent.

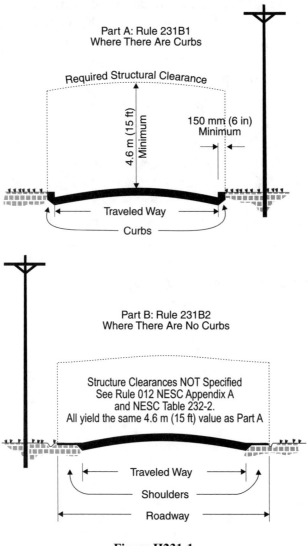

Figure H231-1
Structure clearances from the
traveled way and shoulders of roadways

Clearances for both mountable and swale types of curbs were specified in the 2002 Edition. Poles must now be behind the curb, regardless of curb type (see Figure H231-2).

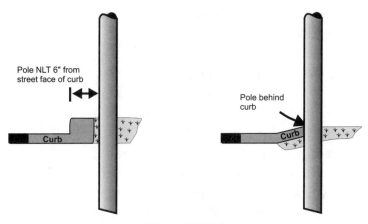

Figure H231-2
Poles back of curbs

The recommendation regarding the location of supporting structures along streets, roads, and highways *without* curbs was included for the first time in the 1977 Edition. Its purpose was to minimize the possibility of vehicle-structure conflicts by placing supporting structures as far from the shoulder line as practical. It was recognized that, in many cases, the most practical location may be directly alongside the shoulder.

The requirement for placement of structures along roadways without curbs was revised in the 1984 Edition. The wording of the requirements of the 1977 and 1981 Editions unfortunately had been misinterpreted, through ignorance or by intent, by some who wished to argue that structures should always be located exactly at the right-of-way line. This was definitely *not* the intention of the Clearances Subcommittee. There are a variety of reasons why it may not be practical to locate a utility pole very near the highway right-of-way line. For example, frequently (1) pole holes need to be sufficiently inside the R/W (by at least half the width of a truck) to allow maneuvering room for the digger truck for both initial installation and replacement, (2) poles need to be far enough from the R/W line that conductors will not hang outside the R/W, (3) room is needed to guy down to the ground to take the conductor forces for tap lines, and (4) poles must be located to dodge ditches or underground facilities located near the R/W line (see Figure H231-3).

It *must* be recognized that, while it is desirable to reduce the possibility of damage to a structure by the action of negligent drivers, it is impossible to limit that possibility to zero.

The Code does not specify whether lines should be located within or without highway rights-of-way. The Code does, however, specify various requirements affecting construction within and without such rights-of-way and, where appropriate, recognizes differences in expected use and restrictions, if any, applicable to each area. For example, Rule 232 requires clearances for a truck over highways as well as adjacent farm land.

The actual placement of a structure depends upon many factors; some affect the reliability of the line, while others affect the ability to construct the line. Because of trees, underground structures, aboveground structures or other installations at or near rights-of-way lines; locations of driveways, parking lots, fences, ditches, and steep slopes; service location requirements; and other factors that are not under the control of the utility, it is often not practical to construct lines at the edge of rights-of-way. The lines and equipment must be located where they are easily accessible for maintenance and emergency operation. The complete line design should be considered when locating structures. It generally is not practical to locate some portions of lines at the right-of-way and other portions closer to the roadway.

The present wording adequately expresses the true requirement to which utilities should and must be held—that of allowing unobstructed use of the traveled way by ordinary vehicles. It also recognizes the practical demands of both roadway use and structure placement. Since many highway shoulders (the outer sections of the roadway that border the traveled way for the purpose of parking or stabilization of the base course) end abruptly, especially beside major ditches or other slopes, it is often not practical to locate utility structures further away than the

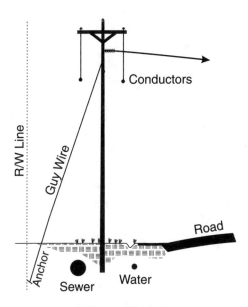

Figure H231-3
Practical location for pole
away from roadway and R/W line

edge of the shoulder. By being located off the roadway, the structures allow full use of the shoulder by most vehicles.

Drivers of vehicles with significant overhangs are, as usual, required to exercise normally expected diligence with movement of those vehicles. To the extent that additional clearance can be obtained without detriment to good design practice or maintenance of lines, such additional clearance is desirable and, obviously, is a part of good design practice, but it is *not mandatory.*

It must be recognized that experience has shown that it is neither possible to predict with certainty the actions of drivers, especially those who are sleepy or otherwise impaired, nor predict the paths of their vehicles if they leave the roadway. It is not clear that there are preferable locations near curves; some errant vehicles go straight off a curve, others partially recover and leave the roadway at an angle. It is also not clear whether the probability of vehicles leaving the road and striking a structure is significantly greater at curves or along straightaways. The fact that a vehicle leaves the traveled way out of control speaks to the driving habits of one or more drivers, not the appropriateness of a pole location; the pole might as well be a tree for all practical purposes in that instance. For example, in one instance, a truck left the roadway of an interstate highway and traveled well over 275 m (900 ft), shattered a utility pole, alongside an underpassing road, and continued across the underpassing divided highway until it hit a bridge embankment.

Guarding structures, either at curves or along straightaways, is not recommended except in special circumstances (usually involving reliability of the line); the guards increase the probability and extent of conflict for moving vehicles. It generally is far better to allow more open space (to allow the errant vehicle to travel in or off the right-of-way and be slowed by natural terrain and growth factors), than to place a wide guard around a pole. This also reduces the opportunity for the errant vehicle to be flipped into the structure even though it otherwise would have missed.

For similar reasons, a cleared utility right-of-way along a roadway is generally safer as a vehicle recovery area, even though structures are located at intervals within the cleared right-of-way, than are uncleared woodlands alongside the roadway; the vehicle has more opportunity to recover where trees have been cleared from the utility rights-of-way. Statistics on what errant vehicles hit when they leave the roadway (and do not recover) indicate a large proportion of the items struck are trees and utility poles (see Table H231B-1). Pole strikes are in the same general frequency range as guard rails, embankments, and culverts/ditches and about half the frequency of the tree hits. Given that such statistics do not include the number of vehicles leaving the road that do not hit the trees that were cleared off the utility rights-of-way (and, therefore, recover), it should be obvious that the only practical difference between the side of the road with a utility line, versus one with trees, is that there is a greater opportunity to recover if a vehicle leaves the road on the utility line side.

Table H231B-1
First Harmful Event—Fixed Object Fatalities by Object Type

Fixed Object	1988	1989	1990	1991	1992
Tree/Shrub	3329	3296	3252	3236	3053
Utility Pole	1476	1418	1277	1329	1129
Culvert/Ditch	1473	1349	1501	1401	1363
Guardrail	1385	1288	1249	1204	1139
Embankment	1360	1332	1334	1187	1138
Curb/Wall	891	860	843	774	752
Other Fixed Objects	682	696	731	634	619
Sign of Light Support	576	594	578	528	488
Bridge/Overpass	553	519	582	545	545
Other Pole/Support	501	449	450	411	388
Fence	482	448	505	500	455
Concrete Barrier	201	249	236	217	214
Building	106	98	97	111	89
Impact Attenuator	15	20	28	16	20
Totals	**13 030**	**12 616**	**12 663**	**12 093**	**11 392**

From *Roadside Design Guide*, Copyright 1996 by the American Association of State Highway and Transportation Officials, Washington, D.C. Used by Permission.

Rule 231B3 recognizes that roads, streets, and highways with narrow rights-of-way or closely abutting improvements can provide significant challenges to the line designer to provide appropriate clearances to both the roadway and the adjacent land or improvements. The 2007 Edition made it clear that this rule was not intended to distinguish between roads, streets, or highways, urban or rural, for application of this rule. Restricted rights-of-way for overhead lines are special cases and may require special attention, regardless of whether the setting is urban or rural.

Many jurisdictions allow or require utilities to place structures within highway rights-of-way. Rule 231B4, added in the 1987 Edition, recognizes that governmental bodies having jurisdiction over structure locations may choose to require or approve *specific* locations of utility structures within a highway right-of-way. Rule 231B4 does not apply to the mere allowance or requirement to place structures within a road right-of-way; it only applies when specific structure locations are required or approved.

231C. From Railroad Tracks

(This rule was numbered Rule 230D prior to the 1977 Edition.)

The 1977 Edition revision increased the clearances at sidings from 2.13 m (7 ft) to 3.6 m (12 ft) (see Figure H231C-1), unless there is another controlling obstruction present. This change is a result of the difficulty in unloading on the siding because of the obstruction. The difficulty in providing 3.6 m (12 ft) in some areas was recognized in the 1984 Edition. An EXCEPTION was added for industrial sidings to allow them to return to the previous clearances. Industrial sidings may now continue to use 2.13 m (7 ft) horizontal clearances *where space is available for loading and unloading cars.* Note that, if a controlling obstruction (i.e., obstruction closer than 3.6 m (12 ft) from nearest rail) exists on both sides of a potential utility structure area, such as Building A and Building B in Figure H231C-2, then the obstruction which allows the greater room for something to get by the building (in this case, Building A) is the controlling obstruction. It should be noted that, where large cars must traverse sharp curves, the clearances *may* need to be increased appropriately; such increase is, however, not required by the Code.

It was made clear in the 1981 Edition that the intended measurement was from the *nearest* track rail.

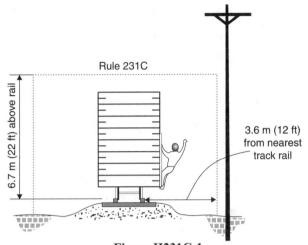

Figure H231C-1
Structure clearances from railroad tracks

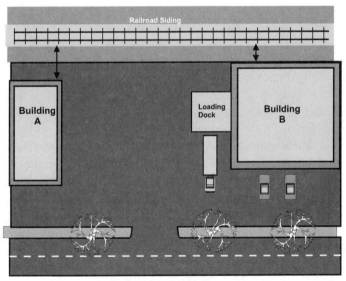

Figure H231C-2
Building A is controlling obstruction

232. Vertical Clearance of Wires, Conductors, Cables, and Equipment Above Ground, Roadway, Rail, or Water Surfaces

A Historical Note on Early Code Development

Prior to the codification of the NESC, the clearances of line conductors above railroads, roadways, and footways had been specified at widely different amounts by different states in their statutes and commission orders. Local variations in practice existed even where no rules were in effect. In general, no variation in traffic existed that would justify the varying requirements. The establishment of higher clearances in one community than in others tended to encourage the local use of high vehicles, such as hay derricks, well-drilling outfits, furniture vans, etc., which, when carried into the neighboring lower-clearance communities, caused serious safety problems.

The NESC was codified and uniform clearances were established in consideration of accidents due to insufficient clearance or extra high loads on vehicles, the current general practice, and the advantages of a more nearly uniform practice.

Railway freight cars were not expected to exceed a height of 15 feet. In most communities, freight cars of greater height were eliminated already by low highway bridges; they were often much lower than the wire clearances specified.

The basic clearances of 27, 28, and 30 ft were required for open conductors in crossings over railways where men were permitted on tops of freight cars; the clearance requirement depended on the voltage of the line. Later editions considered 25 feet to be an adequate clearance for guys and for cables carried on messengers, since this clearance would not be reduced appreciably by temperature changes or ice loading. Clearances of 18, 20, and 22 ft were required for supply conductors crossing over railways not included above. These were intended to be used, in general, in connection with electric and steam roads operating only passenger trains and where men were not permitted on the tops of cars while the cars were in motion.

For wire clearances above highways, the traffic under consideration varied more in its clear-height requirements, although the ordinary roadway vehicles were much lower than freight cars. The higher vehicles considered at that time were hay wagons, box loads, moving vans, etc. It was very rare that the height of such vehicles above ground exceeded 12–14 ft; it was quite practical to restrict ordinary traffic to vehicles not exceeding such a height. The National Bureau of Standards, the original codifier of the NESC, considered that those responsible for the traffic of vehicles more than 12 or 14 ft high could reasonably be expected (1) to know that obstructions exist along highways that would prevent riding on the tops of such vehicles (such obstructions included overhead bridges, branches or trees, trolley, and other wires) and (2) to know also that contact with overhead wires frequently is dangerous, either to men or to the wires, and should always be avoided.

The movement of such devices as hay stackers, well rigs, and derricks along highways was always considered as extraordinary traffic that was subject to the necessity of observing special precaution against contacts with overhead constructions of all kinds. Otherwise, such vehicles could endanger the community by injuring overhead structures. It frequently was practical to reduce the height of such vehicles, but this was often neglected. Low wire elevations were sometimes blamed for avoidable accidents arising out of culpable negligence of the operators of the vehicles.

For conductors of all types less than 300 V to ground, the height above pedestrian thoroughfares was increased from 10 ft to 12 ft in the Third Edition because an average person could, with an umbrella, reach wires having only a 10-ft clearance, such as when a person raises his umbrella at arm's length above his head to avoid hitting that of another person when passing. This clearance applied only where footways or spaces were provided for pedestrians as a thoroughfare. An EXCEPTION was made to the rule in the case of signal wires of less than 150 V to ground, where a 10-ft clearance was permitted.

Supply service leads of less than 150 V to ground were allowed a clearance of only 10 ft at the entrance of the service to the building. This exemption was made because it was often not practical to give a greater clearance.

*The original basic system of clearances remained in effect, with various additions and revisions, until the **fundamental change in the 1990 Edition in the methodology of specifying clearance requirements**. Prior to the 1990 Edition, vertical clearances generally were required when the conductor, wire, or cable was at 15 °C (60 °F). The clearance values specified in the tables applied to spans of varying limits, depending upon the ice loading district, and included expected increases in sag due to ice* *loading or thermal loading up to 50 °C (120 °F), generally 450 mm (18 in) (see Figure H232-1). These values were required to be increased for long spans or high temperature operation. For long spans not exceeding 50 °C (120 °F) operation, the vertical clearance was increased by 30 mm (0.1 ft) per 3 m (10 ft) of span length in excess of the basic span length of Rule 232A (see Figure H232-2).*

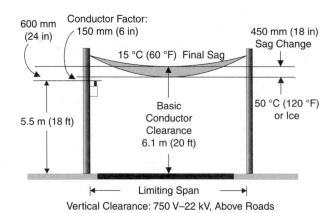

Figure H232-1
Example of 1987 method—Basic clearance

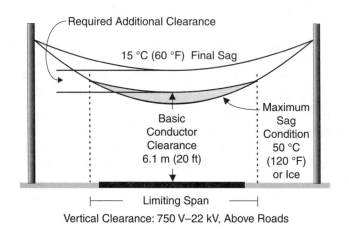

Figure H232-2
Example of 1987 method—Long span clearance

Above 50 °C (120 °F), the actual sag change was used regardless of span length for the 1987 Edition only (see Figure H232-3). For the 1977–1984 Editions, the actual sag change from 15 °C (60 °F) to maximum conductor temperature was added to the table value, thus double counting the sag change component already included. In the 1987 Edition, the sag change component included in the table value was removed and actual sag change from 15 °C (60 °F) to maximum conductor temperature was used in its place, regardless of span length. For all practical purposes, this became the required system for all operating conditions in the 1990 Edition.

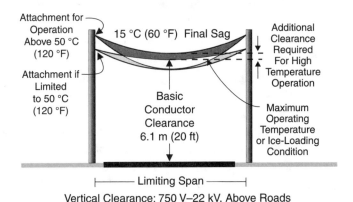

Figure H232-3
Example of 1987 method—Operation above 120°F

*As of the 1990 Edition, the NESC no longer included any sag change within the vertical clearance values (see Figure H232-4). As a result, the values for vertical clearances in the 1990 tables are generally 450 mm (18 in) less than in the 1987 tables (unless increases were required for conformance to the coordinated uniform clearance system instituted in the 1990 Edition). The vertical clearance values specified in the 1990 Edition are to be met under **all** conditions of service, emergency or otherwise; it is the responsibility of the installer to attach wires, conductors, and cables at such locations that, under the conditions resulting in the greatest sag, the clearances will not be less than the specified values. The installer has both complete flexibility and complete responsibility for sag and tension control.*

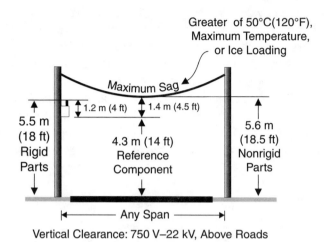

Figure H232-4
Example of 1990 method—Any span

*A brief discussion of these changes is presented as Appendix A located at the rear of the 1990 and subsequent editions of the NESC. Appendix A is based upon discussions included in the predecessor to this document and upon papers presented for public review at an IEEE/PES meeting in Vancouver, British Columbia, Canada, and an IEEE/IAS meeting in San Antonio, Texas, by the Chairman of the NESC and supplemented with new figures. The figures in Appendix A will not be presented here, but the discussion will be elaborated upon. The 2002 Edition added a new Footnote 26 to Table 232-1 to inform the user how to calculate a vertical clearance **if** a line was to be designed for travel of known-sized overheight vehicles under the line. In essence, the known oversized height of the expected vehicle is substituted for the 4.3 m (14 ft) truck height included in Rule 232 clearances.*

The horizontal clearance requirements were also coordinated with the uniform clearance system in the 1990 Edition and, in some cases, table values were changed with the conductor to apply at rest, rather than under wind displacement conditions. Use of the greater of the "at rest" clearances or the "wind displaced" clearances is now required.

232A. Application

(Prior to the 1990 Edition, this rule was called Basic Clearances for Wires,
Conductors, and Cables. *Included was a table—Basic Span Lengths, which is
no longer required by the coordinated clearance method of 1990. Rule 232
was also reorganized in the 1990 Edition, when the conductor clearance
requirements were moved to Rule 232B. See the discussion of Rule 232B for
comparisons between the 1987 and 1990 clearance values. Rule 232A now
contains only application rules.)*

In order to eliminate confusion as to how clearance requirements are
applied, the NESC specifies a set of basic conditions at which the
requirements must be met. The specified clearances have been developed by considering the maximum height of the usual traffic passing
underneath, the normal use of the ground beneath the overhead lines,
and the physical and electrical characteristics of the conductors.

There have been two sets of basic conditions used during the history
of the NESC. From its inception through the 1987 Edition, the specified
vertical clearances were required to exist when the conductor, wire, or
cable was at 15 °C *(*60 °F). Recognizing the sag change effects of ice
and thermal loading, the clearance value included 450 mm (18 in) to
allow for increased sag during the operating life of the facilities. Starting in the 1990 Edition, the vertical clearance values specified in the
clearances tables were changed to reflect the clearances that must be
maintained under the conditions that produce the greatest sag in the
conductor wire or cable. In other words, the system changed from a
15 °C (60 °F) system to a "closest vertical approach" system in the
1990 Edition.

It is recognized, for example, that conductors strung from one support to another undergo changes in sag as a result of ice loading, wind
pressure, or conductor temperature changes. The latter may be caused
by changes in ambient temperature, supply conductor losses, or the
cooling effect of wind. Experience has shown that many conductors
commonly incur a permanent increase in sag as a result of these loadings, unless special measures are taken to prevent it. Thus *final* sags are
considered for clearances *above* an installation, and initial sags are

considered for clearances below an overhanging installation (see Rule 233 and Rule 234).

Prior to the 1990 Edition, clearances were specified as applying at 15 °C (60 °F), without wind or ice loading, and with the conductors either at final unloaded sag (or at initial sag only if they are so maintained). The combined effect of both storm loading and long-term creep was and is to be considered in developing the final unloaded sag. It was recognized that reduced clearances would be experienced under ice loading (Rule 251) or temperatures above 15 °C (60 °F). In the 1977 Edition, it was made clear that the 15 °C (60 °F) referred to *conductor temperature, not ambient* (see Rule 232B2c(3) of the 1987 Edition). The term *no wind* was intended to mean "no horizontal displacement of the conductor," and the term *no wind displacement* is now used. The cooling effect of a nominal 0.6 m (2 ft)/s wind may, however, be assumed in determining sags at various conditions. The basic condition of fixed supports for the vertical requirements was also deleted in the 1977 Edition; appropriate movement of suspension insulators under changes in loading is to be considered.

Prior to the 1990 Edition, the NESC generally added 0.6 m (2 ft) to the vertical clearance required of rigid live parts to obtain the clearance required of conductors. Most conductor installations will not experience more change in sag under the basic conditions that were then used. The original vertical clearances for conductors were calculated using the sags of commonly used conductors. The basic spans were chosen to limit the sag to 450 mm (18 in), and another 150 mm (6 in) was added to allow for uncalculated variances. This made up the 0.6 m (2 ft) difference between the clearances for rigid live parts and conductors of the same voltage in editions prior to the 1990 Edition.

It should be appreciated that some installations, such as spacer cable, may experience greater sag under ice loading than allowed for in the specified clearance requirements for conductors. If so, they may require special clearance calculations. In such cases, it is reasonable and consistent with the Code to use the sum of 150 mm (6 in) plus the clearance required for rigid live parts of the same voltage classification as the minimum clearance required under the maximum sag condition. This is exactly the method used in the coordinated clearance method of the 1990 Edition.

Some utilities make a practice of prestressing conductors in order to avoid differences between initial and final unloaded sag. Also, some utilities maintain their conductors essentially at original sags by pulling slack. This is most commonly done with soft-drawn bare open wire used in communication service (in order to reduce swinging contacts between adjacent conductors). These rules require the clearances to be maintained; if that means pulling slack out of conductors after installation, then such pulling is required. The alternative is to install enough clearance to allow for continued sag creep during the life of the conductor (see the discussion of Rule 230I).

The use of very low conductor tensions (slack-span construction) tends to reduce the amount of sag increase under storm loading. It minimizes, or may eliminate, differences between initial and final unloaded sag. The requirements of Rule 232A apply to slack-span construction, however, just as they apply when normal conductor tensions are employed. Note that larger sags (lower tension) produce larger horizontal displacement under storm wind loadings than smaller sags (higher tension).

Prior to the 1990 Edition, very short spans and slack spans were penalized by the system that required the specified clearance values to exist at the specified conductor temperature of 15 °C (60 °F), regardless of small changes in sag actually expected in such situations.

Clearly the 1990 method of specifying the closest-approach clearance, rather than a clearance required at 15 °C (60 °F), neither penalizes short-span or slack-span construction nor underspecifies clearances needed for nonstandard, special constructions.

Prior to 2007, Rule 232A3 referred users to Rule 250B for the loading conditions to be assumed to produce inelastic deformation in conductors and messengers that must be considered when determining sags for clearance purposes. As of 2007, the rule references new Rule 230B which brought all of these specifications into the Clearances section.

232B. Clearances of Wires, Conductors, Cables, and Equipment Mounted on Supporting Structures

(Prior to the 1990 Edition, Rule 232B contained the requirements for additional clearances for higher voltages for conductors, etc., that have now been combined with additional clearance requirements for rigid live parts in a revised Rule 232C. In the 1990 Edition, the clearance requirements for conductors, etc., that were formerly in Rule 232A were moved to Rule 232B1; the clearance requirements for equipment that were formerly in Rule 232C and Rule 286E were moved to Rule 232B2 and Rule 232B3 and combined into Table 232-2. Also in the 1990 Edition, the street light requirements formerly in Rule 286G were moved to Rule 232B4; this was an inappropriate location for the portion of the rule referring to clearances from the pole itself. As a result, that part of the rule was again moved in 2002 to Rule 236D2, when luminaire clearances to ground were added at Rule 232B4a. See the discussion under Rule 232C for comments on the additional clearance requirements formerly contained in Rule 232B1 and Rule 232C2 of the 1987 and prior editions. Rule 232B2—Sag Increase was deleted in the 1990 respecification of clearances at maximum sag conditions.)

The requirements of Rule 232B cover vertical clearances above ground, roadway, rail, and water surfaces, including both publicly and privately owned areas.

Railroad crossing clearances were originally based on the premise that workers walk on top of box cars and may signal each other by raising their arms. Box cars were considered as being 4.6 m (15 ft) high. Although higher equipment was in use, it was generally of types that did not permit workers to walk on top of them. Today, similar clearances are in use, although for slightly different reasons, as will be discussed later.

Clearances for conductors crossing and overhanging highways, forests, cultivated lands, etc., are intended to ensure free passage of vehicles underneath the line. The maximum legal height of vehicles normally permitted on highways varies somewhat from one state to another; 4.1 m (13.5 ft) is the limit in some states, but the long-established maximum of 4.3 m (14 ft) is the basis for the Code.

Other types of activity considered in the development of these clearance requirements include the activities of pedestrians and sailboat traffic.

During the early life of the NESC, a large proportion of the circuits were delta-configured, with the working voltage potential being between phase conductors. Consequently, the NESC voltage classifications to which clearance requirements applied were based on phase-to-phase clearances. This caused several problems as the electric supply industry evolved. The predominant construction of distribution-level facilities has been of the wye configuration in recent years. Since most accidental contacts with supply conductors have involved only one energized conductor and ground, rather than two energized conductors, the voltage of major interest has become phase-to-ground voltage.

The method of classifying voltages was changed from "conductor-to-conductor" to "phase-to-ground" in the Sixth Edition, and the table headings were changed accordingly. Ungrounded wye- and delta-connected circuits are still considered on the basis of the voltage between conductors. However, where a protection system that will promptly de-energize the circuit under a ground fault creates a reference voltage to ground, the voltage to ground may be used. In later editions, information under the table headings will assist in making the appropriate choice of voltage. This change caused a number of problems when the values within some tables were not also changed. In effect, the clearance requirements for many wye-connected circuits were reduced. Unfortunately, the effects of some of these changes were not identified for several years. Although some requirements were not affected appreciably, others, like some clearances to buildings, were affected adversely (see Rule 234C) to the extent of requiring significant change in the 1977 Edition.

IR 159 issued 7 October 1974 addressed the applicability of the NESC clearances for lines that were (1) specifically to be constructed for construction power at a building site and (2) *expected* to have cranes and large equipment moving beneath them. The rules of the Code do not specify clearances for such lines installed in construction areas because (1) the requirements vary widely and are site-specific and (2) any rule that provided for worst-case conditions would be unduly restrictive in most cases.

In the 2002 Edition, Footnote 26 was added to Table 232-1 to specify how to determine such clearances if a line was being designed specifically for transit of nonstandard height equipment beneath the line. The methodology follows the same methodology used to produce all of the specified vertical clearances that is shown in NESC Appendix A. Road clearances are based upon transit of a 4.3-m (14-ft) vehicle under the line. If a line is to be designed to allow nonstandard equipment under it, the clearances specified in Table 232-1 are to be increased by the difference between the height of the desired equipment and 14 ft. This effectively substitutes the height of the desired nonstandard equipment for the 14 ft already included in the clearance.

Nothing in this rule is intended to infer a duty on a utility to design a line for transit of nonstandard equipment of which it has no knowledge (see Rule 012C). Clearly a utility constructing and maintaining an overhead utility line has a duty to maintain clearances that are appropriate for the activity expected under the line. Also clearly, the utility cannot be expected to be clairvoyant; there is no duty to plan for activities that are not expected under the line.

It should be noted that *the NESC ground clearances in Table 232-1 are required to be met when the conductor or cable is at maximum sag* under the specified ice loading or the maximum thermal loading that will be allowed on the line, whichever produces the greater sag. Often the load on a line and resulting thermal loading will be less than the current-carrying capability of the conductors; this is typical for distribution lines. For transmission lines, load flows are generally controlled so as to neither (1) exceed the thermal capability of the conductors (and, thus, anneal the conductor and reduce its strength) nor (2) create excess sags (reduce clearances) beyond that for which the line was designed. In any case, the maximum thermal loading produced by, or allowed for, the particular line as a result of its loads and/or load controls is the conductor temperature to be considered, not the maximum rating of the conductor itself.

Different lines or line segments using the same conductor will have different maximum loads and, thus, different maximum thermal loadings. The fact that another line with the same conductor can have a greater load and greater thermal (or ice-loaded) sag does not require that the greater sag be used for a conductor that does not have that load—it is the *actual loading conditions for the particular line that must be considered* for that line.

To meet the NESC clearance requirements, *the conductors and cables must initially be installed with enough vertical clearance to allow for both (1) changes in sag from initial sag with unstretched wire to final sag after long-term creep and (2) changes in loading from ordinary conditions to maximum ice loading or maximum thermal loading,* whichever produces the greatest sag. Thus, *under ordinary operating conditions* without ice and less than maximum thermal load, *it is both expected and intended that the cable or conductor have less sag (more ground clearance) than required at maximum sag.* This is part of the NESC clearance system to provide appropriate clearances for expected occurrences.

The NESC vertical clearance above ground includes 1.5 ft for neutrals and cables and 4.5 ft for primary conductors above the built-in 14-ft truck height. As a result, under ordinary operating conditions a neutral or communication cable meeting the NESC vertical clearance requirements will accommodate a vehicle height in excess of 16 ft. Vehicles of this height usually require special permits for movement over roads. As a part of the permitting process, most jurisdictions require those who desire to move vehicles or equipment of this height to notify utilities and work with them to accommodate the move.

The regulations of the Occupational Safety and Health Administration (OSHA) of the U.S. Department of Labor, especially 29 CFR 1926, as well as related state regulations, specify clearances that operators of cranes, dump trucks, backhoes, and similar vehicles/mechanized equipment must maintain between the vehicles/equipment and existing, energized power lines. This is clearly pointed out in the answer to IR 159. There is no intention that lines be built higher than tall cranes, for example, when cranes are not expected to be operated erect under the lines. There is no intention that existing lines are to be moved when portable cranes are to be used in the area.

Under OSHA 29 CFR 1926 Subparts O and N, it is the responsibility of those responsible for cranes, vehicles, and mechanized equipment operating in the vicinity of energized power lines to do two things: (1) utilize a spotter to assist a crane operator in locating the position of the crane and its cables relative to power lines, and (2) stay 3 m (10 ft) or more away from such power lines. It is the responsibility of the equipment operator, who is the only one with control and knowledge of intended equipment activity, to keep the equipment away from energized lines or to notify and work with the utility to produce a combination of equipment activity and line location/protection that meets the OSHA requirements.

Further, under OSHA 29 CFR 1926.416(a), it is the responsibility of the employers of crane operators, etc., to inspect job sites specifically for the location of power lines, whether open or concealed. Where power lines are in a position to be contacted by workers at the job site, their employer is required to inform workers of the location of the lines, put up appropriate warning signs, inform employees of the consequences of hitting the lines, and inform employees of what to do to stay out of the lines. Various OSHA regulation changes have extended these requirements and the OSHA "10-ft rule" to all employers and employees.

NOTE: When vehicles and mechanized equipment to energized power lines are transiting in the over-the-road mode (with booms, buckets, etc., knuckled down), the OSHA clearance to power lines is only 4 ft.

Under OSHA regulations, the only way that an equipment operator is allowed to have the equipment approach within the proscribed 10-ft distance to power lines is if the operator's employer is able to arrange with the utility to insulate, move, or both de-energize the lines and ground the lines within sight of the operator. Sometimes it is practical for the utility to do one or more of these remedies. However, it is often not practical to do any of them. On radial lines with critical loads (or where the line would have to be de-energized for long periods), it will not be practical to de-energize the lines. With small conductors or long spans, the lines may not be structurally able to support insulating line hose. Line hose are not expected to be left up for long periods, due to ozone tracking that will occur with moisture.

In many cases, the only practical answer to the dilemma is for the employer to select different equipment or a different route to accomplish the required work. For example, when constructing on- and off-ramps for an elevated road crossing and rock is encountered, contractors often like to use a 30-ft rock drill. When the ramp goes under an existing utility line, they often substitute a 10-ft rock drill and use multiple explosions to do the work within 30 ft or so on each side of the line—and then continue to use the 30-ft drill outwards from that point.

OSHA does not specify a clearance to be maintained by equipment operators from overhead communication lines. However, it is well known that snagging a communication line on a joint-use supply and communication line can bring the electric supply lines down to the ground or near enough to be contacted by vehicles or personnel. Under the so-called *general duty clause*, Section 5(a) of the Occupational Safety and Health Act of 1970, as amended, employers have a duty to provide both safe work and a safe work place. Section 5(b) of the Act requires employees to adhere to OSHA Regulations. As a result, operators of equipment have a duty to maintain a safe clearance from utility structures and supported lines and equipment so that neither a mechanical hazard nor an electrical hazard will be created by their actions. Spotters are often required to assure safe movement of tall vehicles and equipment near overhead utility lines.

Clearances to *completed* buildings are covered in Rule 234C.

The vertical column headings of the clearances tables were revised in the 1977 Edition to reflect more accurate and concise definitions. Voltages exceeding 50 000 V were then covered under Rule 232B1. After consideration of recent operating experience and relative electrical hazard, the 1984 Edition moved the 15 kV break point under Open Supply Line Conductor clearance headings up to 22 kV. The required clearances above 22 kV were decreased by 1 ft in the 1984 Edition (except for the clearance over nonsailboating water areas, which was already at the lower value).

The 1984 change was the beginning of the switch to the closest-approach, coordinated clearance system that occurred in 1990. The break point in the voltage values was moved from 15 kV to 22 kV for three reasons:

(1)   the safety considerations of a 24.9/14.4 kV wye line are not appreciably different from those of a 34.5/19.9 kV wye line,

(2)   22 kV is above the value for the phase-to-ground voltage of commonly used distribution voltage systems, and

(3)   when 22 kV is subtracted from the old maximum table value of 50 kV, it leaves 28 kV. When this value is multiplied by the 10 mm (0.4 in)/kV voltage adder value, a 300 mm (1 ft) adder is calculated for a 50 kilovolt-to-ground line above a 22 kV base.

Thus, when changes to the 1990 Edition caused all table values above 22 kV in Rule 232 to be removed, and the normal 10 mm (0.4 in)/kV voltage adder above 22 kV was applied, the total required vertical clearances for voltages above 50 kilovolt-to-ground, including the base table value plus the appropriate voltage adder, did *not* change appreciably.

It should be noted that the specified clearance values are *vertical* clearances; there are no specified *diagonal* clearances. The Clearances Subcommittee considered changing the rule to include diagonal considerations to specify clearances to hillsides with slopes small enough to allow vehicular or pedestrian traffic. It was recognized that, since conductors effectively rise vertically when displaced by wind, the clearances over shallow slopes would not be significantly changed. The intermediate scopes that are not steep enough to prevent vehicle travel, but that are steep enough to cause a reduction in clearances, are considered *special cases* to be resolved in consideration of the local conditions. The revision under consideration was rejected because of the difficulty of specification for the general case and the present applicability of Rule 012—*General Rules* to all particulars not specified in the rules.

The heading of the first column of numbers has historically included *surge protection wires*. This was changed in 2007 to *overhead shield/surge protection wires* to better fit normal terms of use. Definitions were also changed in 2007 to recognize the use of the term *overhead shield wires* to designate a number of wires for which various names are used throughout the country, including *overhead ground wire*.

The tables that are shown at the end of the following discussions of categories of Table 232-1 summarize the vertical clearance requirements of Rule 232. They do not illustrate each allowed clearance reduction. Values are given for both the pre-1990, 15 °C (60 °F) system and the 1990 closest-approach system. It will be noted that certain clearance values prior to the 1990 Edition adhere to general relationships; others represent special cases developed over the years from experience with actual construction. A complete review of these relationships was undertaken after the 1984 Edition. This resulted in the complete redevelopment of the clearances rules into a coordinated clearance system in the 1990 Edition.

In general, vertical clearances *prior to 1990* had the following relationships. A 300 mm (1 ft) increment was added to the height of the conflicting activity under the line to obtain the required structure clearance; another 300 mm (1 ft) was added to obtain the required clearance to rigid live parts of 0–750 V. A 600 mm (2 ft) increment was added to obtain the required clearance to live parts of 750 V–22 kV; another 300 mm (ft) was added to obtain the required clearance for live parts of 22–50 kV. If the live parts were not rigid, i.e., if they were open conductors, another 600 mm (2 ft) was added to the clearance required for rigid live parts to obtain the clearance required for the conductors.

The additional 600 mm (2 ft) for conductors included a 150 mm (6 in) factor applied to nonrigid items and an allowance for up to

450 mm (18 in) for sag increase from the conductor position at the 15 °C (60 °F) basic measurement condition to the sag under (1) ice loading or (2) thermal loading up to 50 °C (120 °F) conductor temperature plus a 150 mm (6 in) grace factor. The Basic Spans were developed to limit conductor sag to 450 mm (18 in). Adders were applied for span lengths exceeding the Basic Span lengths. The effect of ice on the sag of smaller conductors is quite apparent from the decreases in the Basic Span length allowed where more ice is expected. A number of special cases did not follow this general pattern; usually these installations were in areas with limited vehicle activity and where two types of conflicting activity may be expected. These considerations were retained as a part of the clearances coordination of the 1990 Edition (see the Reference Components in NESC Appendix Table A2, Item 5). Different installations may present different degrees of hazard under these conditions, and the code clearances reflect the experience with these differences.

The basic difference between the 1990 closest-approach clearance specification (illustrated in Table H232-1) and the prior 15 °C (60 °F) system is that the 1990 closest-approach clearance no longer included the 450 mm (18 in) of expected further sag change previously included, since wires, conductors, and cables are no longer allowed to go below the values listed in the present tables. The 1990 system did, however retain the 150 mm (6 in) difference between nonrigid, open conductors and unguarded rigid live parts of the same voltage level.

Additional clearances were required by Rule 232B of the 1987 and prior editions for high voltages, long spans, and high-temperature operation. In the 1990 Edition, additional clearances were still required for higher voltages, but not for long spans or high-temperature operation, since it is now the responsibility of the installer, regardless of the span length, to initially attach wires, conductors, and cables at such levels that they will not sag below the table values (plus any required voltage adders) at their condition of maximum final sag, whatever that is expected to be.

Table H232-1
Classification of Conductors, Wires, Cables, or Parts

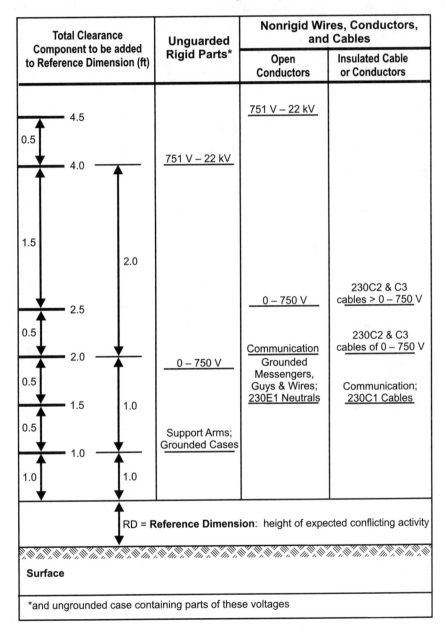

Total Clearance Component to be added to Reference Dimension (ft)	Unguarded Rigid Parts*	Nonrigid Wires, Conductors, and Cables	
		Open Conductors	Insulated Cable or Conductors
4.5		751 V – 22 kV	
0.5			
4.0	751 V – 22 kV		
1.5			
2.0			
2.5		0 – 750 V	230C2 & C3 cables > 0 – 750 V
0.5			230C2 & C3 cables of 0 – 750 V
2.0	0 – 750 V	Communication Grounded Messengers, Guys & Wires; 230E1 Neutrals	
0.5			Communication; 230C1 Cables
1.5	1.0		
0.5			
1.0	Support Arms; Grounded Cases		
1.0	1.0		

RD = **Reference Dimension**: height of expected conflicting activity

Surface

*and ungrounded case containing parts of these voltages

The 1997 Edition clarified the intention of treating ungrounded overhead span guys as a conductor of the voltage to which they are exposed by adding ungrounded guys in the appropriate column headings of Table 232-1 and by adding a new Footnote 14. Anchor guys were exempted from this requirement by new Footnote 15, so long as one or more insulating sections are placed at appropriate places in accordance with Rule 279.

Jumpers are considered to be rigid live parts for clearance purposes. In the 2002 Edition, clearances for support arms, platforms, and braces extending beyond the surface of the structure were added to Rule 232B3 and Table 232-2. As a result, it is clear that structure clearances also apply to *external* braces for platforms, etc. These clearances do not apply to internal structural braces for latticed towers and X-braces between H-frame pole structures. A pole-type push brace must meet the same clearances as an anchor guy; both can (and must) go to the earth, but neither is allowed to obstruct a vehicle or personnel passageway.

Rule 232B4a of the 2002 Edition is new. The clearance requirements of grounded and ungrounded luminaire cases and brackets above ground are now specified in Table 232-2. The clearance for ungrounded luminaire brackets to the pole that was formerly in this rule was moved to Rule 236D2.

Two rules must be met in order to determine vertical clearance above ground for overhead electric supply or communication wires, conductors, and cables. First, the clearance must be not less than that shown in Table 232-1 (see Rule 232B1). Second, this clearance requirement applies with the wire, conductor, or cable at the largest final sag condition outlined in Rule 232A1, Rule 232A2, or Rule 232A3. In other words, lines must be constructed and maintained so that the required clearance will be obtained when the lines are at their largest final sag condition.

For example, consider clearance for a communication cable over a road subject to truck traffic. Table 232-1 requires not less than 4.6 m (15.5 ft) clearance. Rule 232A requires consideration of three conditions. Assume that the installation is in a loading district where icing is a factor and that the Rule 232A3 icing condition produces a larger final sag than either of the Rule 232A1 or Rule 232A2 conditions. Icing becomes the limiting condition and the cable must have at least 4.6 m, (15.5 ft) clearance at the Rule 232A3 icing condition. Obviously, it will have more clearance at other times without the ice load. In this example, the communication cable may have 4.9 m (16 ft) clearance under some operating conditions. However, the cable is not in compliance *at any time* if it will have less than 4.6 m (15.5 ft) clearance under the Rule 232A limiting condition.

The same philosophy applies to sags and tensions. Wires, conductors, or cables are not in compliance with NESC requirements *at any time* if any one of the stated tension limits will be exceeded at the applicable condition. See also Rule 014 and Rule 230A for emergency and temporary installations.

Table 232-1, Category 1. Clearances over track rails of railroads (except electrified railroads using overhead trolley conductors)

The railroad representatives indicated during the 1977 revision that certain utilities had taken the prerogative of reducing clearance height over rails to a minimum of 5.5 m (18 ft) where men are not allowed to ride on the top of cars. This created a hazard in handling certain types of new cars, such as the "Hy Cube" boxcar (5.2 m [16 ft], 11-7/8 in) and the three-deck auto carrier (around 5.8 m [19 ft] with station wagon on top).

On several occasions the Association of American Railroads (AAR) requested the Clearances Subcommittee's consideration of a proposal for greater basic clearances over railroads. Their reasons for greater clearances were as follows:

(1) To provide for the safe operation of certain types of railroad maintenance and construction equipment.

(2)   The height of railroad cars is increasing. Some new types of cars are approaching 6.1 m (20 ft) in height, with a trend towards even taller cars.

(3)   A man standing on top of the newer, tall cars could come in contact with lines installed under the present Code.

After consideration of the AAR's request, the majority of the Clearances Subcommittee determined that, for the following reasons, there should be no changes made to the basic clearances over railroads:

"(a) The clearances have already been raised by eliminating the 18 ft clearance over railroads in the Sixth Edition where men are not permitted on top of cars. The minimum clearance is now 25 ft.

"(b) U.S. DOT regulations (FRA Regulations for Safety Appliance and Power Brakes) prohibited men from the top of most railroad cars beginning in the year 1974.

"(c) Accident statistics included in the 1910 Railroad Reports Act data show that, during the five years investigated, a very low number of reportable accidents involved men on top of cars being struck by overhead lines.

"(d) In light of the present A.R.E.A. highway-railway bridge clearance standards of 23 ft, the height of any future railroad cars will have a practical limit."

The clearance required over rails by the 1977 and later editions of the NESC is based upon a rail car height of 6.1 m (20 ft), the highest car currently in service, and a bridge height of 6.7 m (22 ft). All reference to men walking on top of the cars is eliminated. These clearances were revised in the 1990 Edition to match the coordinated clearance system and eliminate inclusion of any sag change within the table values (see changes to Table 232-1). The various clearances were also coordinated according to the formal system or relationship between clearances required of the various classes of conductors and cables in the 1990 Edition.

Table 232-1—Category 1.
Clearances Over Tracks of Railroads

Potential Conflict Dimension of Conflict:	passage of rail car

car height of 6.1 m (20 ft) (1987)
bridge height of 6.7 m (22 ft)

Basic Clearances (m/ft)

	1987 System: at 15 °C (60 °F)		1990 System: at Maximum Sag	
Classification	Rigid Parts	Wires, Conductors, & Cables	Rigid Parts	Wires, Conductors, & Cables
Structure clearance	6.7/22.0	NA	6.7/22.0	NA
Effectively grounded equipment cases	NS	NA	NS	NA
Communication cables; guys; messengers; 230E1 neutrals	NA	7.6/25.0	NA	7.2/23.5
Surge-protection wires	NA	8.2/27.0	NA	7.2/23.5
Open communication conductors	NA	8.2/27.0	NA	7.3/24.0
230C1 supply cables	NA	7.6/25.0	NA	7.2/23.5
230C2 & C3 supply cables				
0–750 V	NA	8.2/27.0*	NA	7.3/24.0
> 750 V	NA	8.2/27.0*	NA	7.5/24.5
Open supply				
0–750 V	NS	8.2/27.0*	NS	7.5/24.5
751 V–22 kV	NS	8.5/28.0*	NS	8.1/26.5
22–50 kV	NS	8.8/29.0*	NS	8.1/26.5+†
Contact conductors				
0–750 V	NA	6.7/22.0*	NA	6.7/22.0*
> 750 V	NA	6.7/22.0*	NA	6.7/22.0*

NA: not applicable.
NS: not specified.
* EXCEPTIONS are provided.
† Requires voltage adder.

Table 232-1, Category 2. Clearances over roads, streets, and other areas subject to truck traffic

(Alleys, nonresidential driveways, and parking lots were moved to Category 3 in the 1993 Edition.)

The intention of this rule to cover general areas traversed by trucks and other vehicles, whether paved or unpaved (see Category 4) was clarified in the 1977 and succeeding editions. The clearance values have not changed for conductors above 22 kV except for the reduction allowed in the 1984 Edition and the 1990 uniform coordination. For purposes of these rules, trucks are defined as any vehicle exceeding 2.45 m (8 ft) in height. Areas not subject to truck traffic are areas where truck traffic is not normally encountered or not reasonably anticipated.

Footnote 6 (applying to guys and communication service drops) was revised in the 1977 Edition to limit application to those streets and roads that are residential in nature; it did not apply to arterial highways with high traffic volumes and a considerable number of trucks. The wording was also clarified to indicate that reduced clearances apply only at the side of the traveled way and that the basic clearance 5.5 m (18 ft) should still be maintained at the center of the traveled way. This footnote was deleted in the 1990 Edition because of the change in requirements for communication lines.

It should especially be noted that, like all clearances in the Code, these clearances are based upon the dimensions of the potential conflicting activity (i.e., a 4.3 m [14 ft] high truck) plus appropriate adders for basic sag, etc. No allowance is included in the clearance requirements for resurfacing of the roadway in later years. It is left to the discretion of the designer to decide if (1) extra clearance should be added at the time of construction or (2) the line should be reconstructed when the roadway is resurfaced. It is common for designers to add an extra foot or so of clearance at the time of construction to allow for several roadway resurfacings. It also is common to add a foot or more, depending upon span length, to allow for change in the straightness and plumbness of poles *after* installation and for errors in stringing tensions and sags. In snow areas, it is common to allow additional room for seasonal changes in the road level. Many utilities use a basic value or a percentage of the span length, whichever is greater, for an extra construction clearance so that they have limited concern for such future changes.

This category covers areas traversed by forklifts, such as lumber yards and some loading docks. Other loading docks are covered by Rule 234C and Table 234-1 with matching requirements. As shown in NESC Appendix A, the clearances over these areas are based upon a 4.3 m (14 ft) high truck. If such forklifts are reasonably expected to extend higher than 4.3 m (14 ft) (including the load) when they are located underneath the line, an appropriate additional clearance should be provided to account for the expected height above the 4.3 m (14 ft) included in the table value.

This rule was applied in the 1977 Edition to "parking lots subject to truck traffic;" in the 1981 Edition, "nonresidential driveways and other areas subject to truck traffic" was added. Previous editions did not specify clearances for these areas; Rule 200C, Rule 210, and Rule 211 of the prior editions applied. These items, along with alleys, were moved to Category 3 in the 1993 Edition.

Table 232-1—Category 2.
Clearances Over Roads, Streets, and Other Areas Subject to Truck Traffic

| Potential Conflict | passage of truck |
| Dimension of Conflict: | truck height of 4.6 m (14 ft) |

Basic Clearances (m/ft)

	1987 System: at 15 °C (60 °F)		1990 System: at Maximum Sag	
Classification	Rigid Parts	Wires, Conductors, & Cables	Rigid Parts	Wires, Conductors, & Cables
Structure clearance	4.6/15.0	NA	4.6/15.0	NA
Effectively grounded equipment cases	4.9/16.0 §	NA	4.6/15.0	NA
Communication cables; guys; messengers; 230E1 neutrals	NA	5.5/18.0*	NA	4.8/15.5*
Surge-protection wires	NA	5.5/18.0*	NA	4.8/15.5*
Open communication conductors	NA	5.5/18.0*	NA	4.9/16.0*
230C1 supply cables	NA	5.5/18.0*	NA	4.8/15.5*
230C2 & C3 supply cables 0–750 V	NA	5.5/18.0*	NA	4.9/16.0*
> 750 V	NA	5.5/18.0	NA	5.0/16.5
Open supply 0–750 V	4.9/16.0	5.5/18.0	4.9/16.0	5.0/16.5
751 V–22 kV	5.5/18.0	6.1/20.0	5.5/18.0	5.6/18.5
22–50 kV	5.8/19.0	6.4/21.0	5.5/18.0+†	5.6/18.5+†
Contact conductors 0–750 V	NA	5.5/18.0*	NA	5.5/18.0*
> 750 V	NA	6.1/20.0*	NA	6.1/20.0*

NA: not applicable.
NS: not specified.
* EXCEPTIONS are provided.
† Requires voltage adder.
§ 1987 Edition, Rule 286E1.

Table 232-1, Category 3. Clearances over driveways, parking lots, and alleys

Since driveways and parking lot facilities may be subject to truck traffic, additional provision for clearances has been made where trucks, campers, and similar vehicles are involved, similar to Rule 230-24(b) of the National Electrical Code (NEC).

Residential driveways are considered to be an entity unto themselves; they are not considered to be subject to truck traffic in that the full 4.3 m (14 ft) of clearance are not required. Residential driveways are driveway locations near or adjacent to a residence where the occupier will routinely park or store automobile-sized vehicles, such as family cars or pickup trucks. No garage is required. However, *not all driveways leading to residences are classified as residential driveways.* Some such driveways, or portions thereof, may be considered as general use driveways where such area is expected to be used by delivery trucks, emergency life support vehicles, moving vans, and the like. Such driveways or portions of driveways would be considered as *nonresidential driveways.*

The Code recognizes that camper vans and similar vehicles may cross under the lines. Because of the large number of accidents with moving vans, campers, and similar vehicles whose body height or citizens band (CB) antenna height nears that of a full-size truck, there was a strong desire to require full roadway clearances over all driveways and parking lots in the 1977 revision. These clearances can be provided in many cases. However, in a great number of cases, it is not practical to provide those clearances because of the lay of the land and house attachment limitations. In a large number of cases, there is no expectation of the use of any vehicle taller than a normal automobile or pickup truck in the driveway; in such cases, it would be neither necessary nor cost-effective to provide clearances for full-size trucks. Usually in the latter cases there is other, more appropriate access for moving vans available.

Other types of activities in residential driveways include normal pedestrian movement as well as pedestrian movement with ladders and other maintenance tools. The rules recognize the relative hazard potential from various conductors and cables to all of these uses of the land underneath the wires; cables that represent primarily a mechanical hazard are allowed lesser clearances. Since the activities in commercial areas that are not subject to truck traffic present much the same relative conflict, such areas are allowed the same clearances as residential driveways. In the 1977 Edition, the clearance for communication conductors, etc., was allowed to remain at 3 m (10 ft) above grade. However, the clearance required for open-supply conductors of 0–750 V, and supply cables of all voltages meeting Rule 230C2 or Rule 230C3 was increased to 4.6 m (15 ft) to be the same as that required for Spaces or Ways Accessible to Pedestrians Only. This increase was required because of a recent history of accidents in which careless homeowners contacted such conductors with campers, CB antennas, etc., and a general history of conflicts during movement of ladders around houses.

This category was applied to commercial areas not subject to truck traffic in the 1977 Edition. Prior editions did not specify clearances for those areas; Rule 200C, Rule 210, and Rule 211 of those editions applied.

When supply cables of 0–750 V meeting Rule 230C2 or Rule 230C3 were moved in the 1984 Edition to join communication conductors, etc., for most purposes, those cables were also required to have a clearance of 3.6 m (12 ft) above commercial areas not subject to truck traffic and residential driveways. The clearance values were revised in the 1990 Edition to match the coordinated uniform clearance system and specifically matched the mechanical and electrical components with a special set of reference dimensions at the same time that it dramatically changed the footnoted requirements.

In the 1987 Edition, service drops could automatically be allowed to meet reduced clearance values—*this is no longer the case!* Footnote 7 to Table 232-1 of the 1990 Edition allows the reduced clearances for low-voltage service drops *only* when the *building* being served is not high enough to allow the table clearance values with normal construction. This implied a *residential* building (even though it also used the term *or other installation)*, but this limitation to residential buildings was not explicitly stated until Footnote 7 and Footnote 8 were changed in the 2007 Edition to apply only to residential buildings. The 1990 change created a conflict with electricians following the NEC, which was not changed at the same time. The NEC still allowed (as the NESC had done) a 120/240V service drop to be as low as 3.6 m (12 ft) above a residential driveway, but the NEC also allows the drip loop to be at 3.0 m (10 ft) above grade, regardless of building height. The 1997 NESC change emphasizes that the controlling dimension is the height of the building. In essence, Footnote 7 does not apply to two-story buildings or to many one-story buildings with gable roofs. It normally only applies (or, at least, only applies in full) to low, flat-roofed or hip-roofed buildings.

Note that the reduced clearances of Footnote 7 apply only to clearances over a *residential* driveway. They do not apply over *commercial* or *general-use* driveways. Many residences do not have driveways restricted to residential use, part or all of their driveways for general use, including delivery trucks, moving vans, and emergency vehicles. In the case of Figure H232-1FN7a, there is room to park trucks and emergency vehicles on the street in front of the house and the sidewalk would be expected to be used for ambulance gurneys, hand trucks of movers, etc. In that case, Footnote 7 of Table 232-1 applies to the driveway.

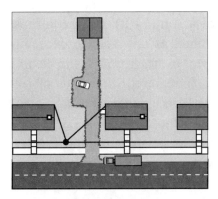

Figure H232-1FN7a
Residential driveway beside house

However, in Figure H232-1FN7b, all of the truck-sized vehicles would be expected beside and in front of the house, so that portion of the driveway would be for general use and a lift pole would be required (Footnote 7 would not apply). In many farmhouse areas, all driveways are subject to trucks. Note also that some recreational vehicles are now 3.6 m (12 ft) in height, thus limiting application of Footnote 7 in some installations.

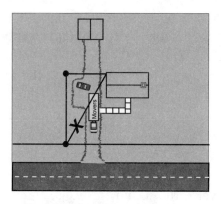

Figure H232-1FN7b
General use driveway beside house

The new clearance values for all construction that is not a service drop with a *residential building* height constraint provide clearance for a full truck height of 4.3 m (14 ft). For the purposes of the clearances rules, trucks are defined as any vehicle exceeding 2.45 m (8 ft) in height. In areas subject to truck traffic, the clearance for a maximum truck of 4.3 m (14 ft) is required. Areas not subject to truck traffic are areas where truck traffic is neither *normally encountered* nor *reasonably anticipated* (see Footnote 9 to Table 232-1).

As in other areas, it is incumbent upon the designer to proactively consider the activity that is expected under the line during its lifetime. If careful consideration is given to anticipating land uses, the intention of the Code is met. It is recognized that no person can be omniscient;

even if a land use occurs at a later time that careful, reasoned consideration did not anticipate, the intention of the NESC has been met. It is not enough that an action be *possible*; it must be *expected* under practical conditions.

In the 2002 Edition, Footnote 13 was revised to limit application of the reduced clearances to driveways, parking lots, and alleys not subject to truck traffic.

Table 232-1—Category 3.
Clearances Over Driveways, Parking Lots, and Alleys

| Potential Conflict | passage of autos, workers |
| Dimension of Conflict: | 4.3 m (14 ft)* |

Basic Clearances (m/ft)

	1987 System: at 15 ˚C (60 ˚F)		1990 System: at Maximum Sag	
Classification	Rigid Parts	Wires, Conductors, & Cables	Rigid Parts	Wires, Conductors, & Cables
Structure clearance	NS	NA	NS	NA
Effectively grounded equipment cases	4.6/15.0	NA	4.6/15.0	NA
Communication cables; guys; messengers; 230E1	NA	3.0/10.0	NA	4.8/15.5*
neutrals	NA	3.6/12.0	NA	4.8/15.5*
Surge-protection wires	NA	3.6/12.0	NA	4.8/15.5*
Open communication conductors	NA	3.0/10.0	NA	4.9/16.0*
230C1 supply cables	NA	3.0/10.0	NA	4.8/15.5*
230C2 & C3 supply cables				
0–750 V	NA	3.6/12.0*	NA	4.9/16.0*
> 750 V	NA	4.6/15.0	NA	16.5
Open supply	3.6/12.0*			
0–300 V		3.6/12.0*	4.9/16.0*	5.016.5*
301–750 V	14.0/3.0	4.6/15.0	4.9/16.0	5.0/16.5
751 V–22 kV	5.5/18.0	6.1/20.0	5.5/18.0	5.6/18.5
22–50 kV	5.8/19.0	6.4/21.0	5.5/18.0+†	5.6/18.5+†
Contact conductors				
0–750 V	NA	5.5/18.0*	NA	5.5/18.0*
>750 V	NA	6.1/20.0*	NA	6.1/20.0*

NA: not applicable.
NS: not specified.
* EXCEPTIONS are provided.
† Requires voltage adder.

Table 232-1, Category 4. Clearances over other land traversed by vehicles, such as cultivated, grazing, forest, orchard, etc.

(This category was added in the 1977 Edition.)

The Sixth and prior editions of the Code specified safe clearances for railroads, roads, driveways, and "spaces or ways accessible to pedestrians only." This latter category was interpreted by some to cover clearances over cultivated or grazing land, *but this was never the intention of those who wrote the Code*; such areas plainly were subject to farm trucks, tractors, wagons and/or similar vehicular traffic. This situation was clarified in the 1977 Edition of the Code by the addition of a separate category for "cultivated or grazing land" and retention of a category for lands which are so inaccessible that only pedestrians would be involved.

It should be noted that equestrians are not considered as pedestrians; Footnote 9 was revised to make it clear that, if horseback riding is anticipated in the area, the area does not qualify for the "spaces or ways accessible to pedestrians only" category; the reduced clearances would endanger riders on horseback, if not the horses themselves. See the discussion of Footnote 9 under the next section on Table 232-1, Category 5—*Clearances over spaces or ways subject to pedestrians or restricted traffic only.*

Prior to the 1977 Edition, the Code relied on Rules 210 and 211 to apply to these areas, rather than specify clearance requirements that might be unduly restrictive in some areas. This new category reflects the changed nature and frequency of conflicts between human activity and overhead lines away from roadways. As overhead lines were more widely used in farm and wooded areas, as specialized biomass harvesting equipment increased in use, and as more areas were accessed by riders on horseback, conflicts began to occur and minimum requirements were required to be specified.

It is distinctly impossible for a national code to reflect the possible use of every conceivable type of equipment in every area. To do so would overly penalize construction in areas that do not utilize such equipment. For example, the small farms in the rolling hills of the Piedmont areas of southeastern states do not use equipment as large as that used in the new, large farms being started in the southeastern coastal plains. Neither type of farm uses either the large equipment commonly found in the midwestern grainlands or the tall, specialized harvesters and trimmers used in the orchards of the deep South and West.

The clearance requirements specified in the NESC are for equipment with a maximum operating height of 4.3 m (14 ft). This will allow ordinary road vehicles and equestrians to move freely; the clearances required are the same as those for roadways. Where vehicles with greater operating heights are expected, it is appropriate to increase the clearances shown by the increase in operating height above 4.3 m (14 ft). In recognition of the hazard to overhead lines presented by oversized farm equipment, at least one major manufacturer is attempting to limit operating heights to 4.3 m (14 ft).

The presence of trees does not affect required clearances to ground; see Rule 218A (Rule 281A of the 1987 and prior editions).

The Code does not specifically address required ground clearances where snow accumulation may be significant. In some areas of the country, accumulated snow would prohibit the use of normal vehicles and would not bring extraordinary vehicles into use. In other areas, snow tractors may be expected to be in general use; increased clearances that consider expected snow accumulation may be appropriate in these latter areas.

Since the vertical clearances are based upon a maximum vehicle height of 4.3 m (14 ft), it should be obvious that the minimum clearances specified in the Code are not adequate for oversize haulage trucks such as may be found around some mining sites, especially if the line is located above an area where the truck body is expected to be raised for dumping. In such areas, it is reasonable to increase the clearances by the difference between the oversized haulage truck operating height and the 4.3 m (14 ft) included in the Code.

The 2002 Edition added Footnote 26 to specify how to determine appropriate clearance *if* a line is designed to clear an oversized vehicle of known height. In essence, the known height of the oversized vehicle would be substituted for the 4.3 m (14 ft) truck height now included (see NESC Appendix A and the Handbook discussion at the beginning of Rule 232).

Table 232-1—Category 4.
Clearances Over Other Lands Traversed by Vehicles

Potential Conflict	passage of trucks			
Dimension of Conflict:	truck height of 4.3 m (14 ft)			
	Basic Clearances (m/ft)			
	1987 System: at 15 °C (60 °F)		1990 System: at Maximum Sag	
Classification	Rigid Parts	Wires, Conductors, & Cables	Rigid Parts	Wires, Conductors, & Cables
Structure clearance	NS	NA	NS	NA
Effectively grounded equipment cases	NS	NA	4.6/15.0	NA
Communication cables; guys; messengers; 230E1 neutrals	NA	5.5/18.0	NA	4.8/15.5
Surge-protection wires	NA	5.5/18.0	NA	4.8/15.5
Open communication conductors	NA	5.5/18.0	NA	4.9/16.0
230C1 supply cables	NA	5.5/18.0	NA	4.8/15.5
230C2 & C3 supply cables				
0–750 V	NA	5.5/18.0	NA	4.9/16.0
> 750 V	NA	5.5/18.0	NA	5.0/16.5
Open supply				
0–750 V	16.0	5.5/18.0	4.9/16.0	5.0/16.5
751 V–22 kV	5.5/18.0	6.1/20.0	5.5/18.0	5.6/18.5
22–50 kV	19.0	6.4/21.0	5.5/18.0+†	5.6/18.5+†
Contact conductors				
0–750 V	NA	NS	NA	NS
> 750 V	NA	NS	NA	NS

NA: not applicable.

NS: not specified.

† Requires voltage adder.

Table 232-1, Category 5. Clearances over spaces and ways subject to pedestrians or restricted traffic only

Footnote 9 was added in the 1977 Edition as shown below to clarify the intention of the Clearances Subcommittee with respect to the limits of this area:

"Spaces and ways accessible to pedestrians only are areas where vehicular traffic is not normally encountered or not reasonably anticipated. Land subject to (but not limited to) highway right-of-way maintenance equipment, logging equipment, all-terrain vehicles, etc., shall not be considered as accessible to pedestrians only."

The second sentence was dropped in the 1981 Edition. The principal concern was that inclusion of that sentence caused confusion and might result in unnecessary exclusion of areas intended to be included under this definition. Obviously, highway right-of-way maintenance equipment is reasonably anticipated along highways and is redundant. Logging equipment may or may not be reasonably anticipated. Most "all-terrain" vehicles of the "dune-buggy" or "Baja California" type are, by design, built low enough that, even with a CB radio antenna, they will not generally be conflictive with the clearances required under this category. The designer is not relieved of the responsibility to think ahead, but, as in the other sections of the Code, it is not intended that every *possible* occurrence be provided for if such occurrence is not also *normally encountered or reasonably anticipated*. However, the designer should be cautioned not to use this provision as a crutch; the requirement is to be proactive about (1) using reason to anticipate the activities that can be expected under the line during its life and (2) placing the area in the appropriate category.

Among the areas intended to be included in this category are swamps, steep hillsides, and other areas where vehicles are not normally encountered nor reasonably anticipated. It should be noted that the open conductor clearances of this category would be generally appropriate for swamps subject to most airboats, the reduced clearances allowed for cables would not be appropriate in these areas. Equestrians (riders on horseback) are not pedestrians; this category does *not* include areas where horseback riding is anticipated. The reduced clearances allowed under this category can endanger riders, if not horses themselves. It is not inappropriate to think of full-size horses as "trucks," since the horse alone is generally in excess of 2.45 m (8 ft) high at the head; with a rider on top, the equestrian assembly stands higher than the special reduced clearances for cables and guys; if carrying work tools, the rider may be endangered by the normal clearances required for open conductors.

These clearances may very well be appropriate, however, if the area is limited to the use of animals of lesser height, such as pony riding.

Because some people were found not to understand what equestrians were, the 1993 Edition changed the term to *riders on horseback*. This immediately prompted a call from a camel farm as to whether the clearances applied to areas where *camels* are raised or ridden. This caused the 2002 change to add riders on large animals other than horses into Category 5 of Table 232-1.

The intention of Footnote 9 should be obvious; if the expected height of *vehicles or other mobile units* (i.e., nonpedestrians and nonvehicle, but mobile)—or *riders on horses or other large animals*—exceeds 2.45 m (8 ft), the area is considered as Category 4—*other lands*. If the height of the expected activity under the line is restricted to 2.45 m (8 ft), the area is considered as a space or way accessible to pedestrians or restricted traffic only. The term *other mobile units* was added in Footnote 9 in the 1987 Edition to differentiate between cows and other moving things (including camels), so that the height that was normally encountered or reasonably anticipated could be used to determine if Category 5 (cows) or Category 4 (horses and camels) applied. As of the 2002 Edition, Footnote 9 now specifically recognizes the limited areas where large animals other than horses are ridden and where the animal is large enough for the rider to extend more than 2.45 m (8 ft) above ground. An example would be camel or elephant riding areas at fairgrounds, zoos, or breeding farms (see Figure T232-1_Cat5).

Figure T232-1_Cat5
Clearances for riders on large animals approach that of a full-size truck

The increasing safety problems caused by low service conductors near buildings have been recognized in the 1977 and later editions. The lowest point allowed in the drip loop of the service conductors at the entrance of the building (and meeting the limits prescribed) was retained at 3 m (10 ft) above grade. This was to allow entrance to flat-roofed buildings where greater heights cannot be attained without inappropriate mast heights. The assumption has been that the drip loop, which does not increase in sag, would be the lowest point; all other points would be higher as the service conductors extended toward the pole-attachment height. The wording of early editions, however, allowed those service conductors to be 3 m (10 ft) above grade at *any point* in the span (which effectively allowed even lower clearances under high-temperature operation or ice load). The 1981 Edition was revised to limit application of the 3 m (10 ft) clearance to drip loops only. The 1984 Edition recognized that portions of the service span within 4.6 m (15 ft) of the building are (1) less subject to sag changes and (2) generally less prone to inadvertent contact by people carrying ladders or driving camper vehicles through a yard area not ordinarily expected to have vehicular traffic under the line.

This area was exhaustively reviewed again for the 1990 Edition. Just as with Footnote 7 and Category 3—*Driveways, Parking Lots, and Alleys*, Footnote 8, applicable to Category 5, was revised to limit its application to buildings that were not tall enough to allow an attachment height using normal mast heights that would allow the table value clearances. The 2007 Edition further limited application of the footnote to *residential* buildings, which was the originally intended application. See the discussion of Footnote 7 in the discussion of Category 3 of Table 232-1. Like Footnote 7, Footnote 8 only applies where the residential building height does not allow meeting Table values (see Figure H232-1FN7/8). A special set of Reference Heights was used to create the required clearances. A Reference Height of 2.45 m (8 ft) was used with the applicable electrical and mechanical components to determine the clearance required of items that essentially produce a mechanical interference problem. A Reference Height of 3 m (10 ft) was used with the applicable mechanical and electrical components for open-supply conductors. This recognizes the expected relative differences in safety issues presented by someone carrying a ladder across a yard.

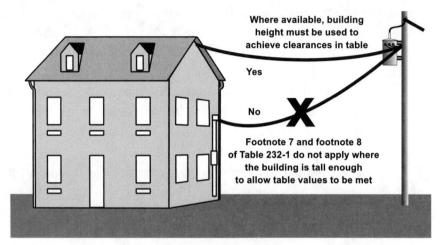

Figure H232-1FN7/8
Applicability of Footnote 7 and Footnote 8 of Table 232-1

Table 232-1—Category 5.
Clearances Over Space and Ways Subject to
Pedestrians and Restricted Traffic Only

Potential Conflict	human activity
Dimension of Conflict:	2.45 m (8ft) or 3 m (10 ft)

Basic Clearances (m/ft)

Classification	1987 System: at 15 °C (60 °F)		1990 System: at Maximum Sag	
	Rigid Parts	Wires, Conductors, & Cables	Rigid Parts	Wires, Conductors, & Cables
Structure clearance	NS	NA	NS	NA
Effectively grounded equipment cases	3.0/10.0§	NA	3.4/11.0*	NA
Communication cables; guys; messengers; 230E1 neutrals	NA	2.45/8.0	NA	2.9/9.5
Surge-protection wires	NA	4.6/15.0	NA	2.9/9.5
Open communication conductors	NA	3.0/10.0	NA	3.6/12.0
230C1 supply cables	NA	3.0/10.0	NA	3.6/12.0*
230C2 & C3 supply cables				
0–750 V	NA	3.6/12.0*	NA	3.6/12.0*
> 750 V	NA	4.6/15.0	NA	3.8/12.5
Open supply				
0–300 V	3.6/12.0*	3.6/12.0	3.6/12.0*	3.8/12.5
301–750 V	4.0/13.0	4.6/15.0	3.6/12.0	3.8/12.5
751 V–22 kV	4.0/13.0	4.6/15.0	4.3/14.0	4.4/14.5
22–50 kV	4.3/14.0	4.9/16.0	4.3/14.0+†	4.4/14.5+†
Contact conductors				
0–750 V	NA	4.9/16.0	NA	4.9/16.0
> 750 V	NA	5.5/18.0	NA	5.5/18.0

NA: not applicable.
NS: not specified.
* EXCEPTIONS are provided.
† Requires voltage adder.
§ 1987 Edition, Rule 286E1.

Table 232-1, Category 6. Clearances over water areas not suitable for sailboating or where sailboating is prohibited

(This category was added in the 1977 Edition.)

These clearances are intended to provide safe clearances for a fisherman holding a fishing rod to traverse under the line and to engage in the sport.

The NESC does not include criteria for determination of whether a water area is suitable for sailboating or for determination of the appropriate water height from which to measure the clearance. Areas that are not suitable for sailboating are so diverse in nature that these determinations are left to the judgment and experience of the designer with respect to the conditions encountered.

Many of these areas are so tortuous, narrow, or rocky, or have such swift currents that they are unsuitable for maneuvering a sailboat. However, even though sailboats may not reasonably be anticipated in these areas, these areas may still be entirely suitable for a canoe, raft, or small boat during periods of appropriate water flow. While the use of some nonsailboating water areas may increase for canoeing, etc., during periods of high water, the use of others may be reduced because currents become too swift or turbulent. The appropriate level for measuring clearances will depend on the local conditions at the site.

These clearances were developed carefully during the 1977 revision; they have not changed except for the adjustments required in the 1990 Edition to make them match the coordinated clearance system introduced in that edition (see Appendix A of the NESC).

Table 232-1—Category 6.
Clearances Over Water Areas Not Suitable for Sailboating

| Potential Conflict | human with fishing poles |
| Dimension of Conflict: | 3.8 m (12.5 ft) |

Basic Clearances (m/ft)

	1987 System: at 15 °C (60 °F)		1990 System: at Maximum Sag	
Classification	Rigid Parts	Wires, Conductors, & Cables	Rigid Parts	Wires, Conductors, & Cables
Structure clearance	NS	NA	NS	NA
Effectively grounded equipment cases	NS	NA	NS	NA
Communication cables; guys; messengers; 230E1 neutrals	NA	4.6/15.0	NA	4.3/14.0
Surge-protection wires	NA	4.6/15.0	NA	4.3/14.0
Open communication conductors	NA	4.6/15.0	NA	4.4/14.5
230C1 supply cables	NA	4.6/15.0	NA	4.3/14.0
230C2 & C3 supply cables				
0–750 V	NA	4.6/15.0	NA	4.4/14.5
> 750 V	NA	4.6/15.0	NA	4.6/15.0
Open supply				
0–750 V	NS	4.6/15.0	NS	4.6/15.0
751 V–22 kV	NS	5.2/17.0	NS	5.2/17.0
22–50 kV	NS	5.2/17.0	NS	5.2/17.0+†
Contact conductors				
0–750 V	NA	NS	NA	NS
> 750 V	NA	NS	NA	NS

NA: not applicable.
NS: not specified.
† Requires voltage adder.

Table 232-1, Category 7. Clearances over water areas suitable for sailboating including lakes, ponds, reservoirs, tidal waters, rivers, streams, and canals

(This category was added in the 1977 Edition.)

Prior to the 1977 Edition, the Code did not specify clearances over waterways or lakes, since navigable streams and waterways come under the jurisdiction of the U.S. Army Corps of Engineers or other governmental agencies. Because trailers or transporters had extended the use of pleasure boats to areas other than those under the jurisdiction of the governmental agencies, and because the nature of the boats themselves had changed, a new category was added in the 1977 Edition to provide for the safety of the boating public.

For any practical purpose, these clearances were not necessary until the late 1960s when (1) the catamaran was popularized and (2) the CB radios and other tall-antennaed radios came into general use on motorboats. Before that time, water-borne vessels were not high enough to provide conflict, except in areas that were already governed by the requirements of a U.S. Army Corps of Engineers crossing permit.

The first recorded case of an outside request to the NESC Committee for clearance requirements over water areas appears to be a letter of 29 November 1973, resulting from the annual meeting of the Atlantic Fisheries Biologists. This was fully two years after the Clearances Subcommittee started considering this problem on its own and in detail. Because of the interest in this rule, a full account of its codification is presented here.

The first public draft of Clearances Subcommittee recommendations for clearance requirements over water areas was released on 15 August 1973. Although the requirements of that draft were applicable to all waters, Footnote 18 would have required increases where vessels whose maximum height above water exceeded 7.6 m (25 ft). The burden was on the utility to guess what height of vessel would be under the lines in the future. This was felt by the utilities and the state utilities commissions alike to be an unreasonable burden upon the general ratepayers.

The commissions suggested that clearance requirements should be based on the size of the particular body of water and the boats in use on that water; once the clearance requirements were set, they would not be expected to have to be changed. It would then be up to the manufacturers and users of the sailing equipment to do their part.

The next draft, released on 1 March 1 1975, included essentially the same proposed requirements, even though the work of the Clearances Subcommittee had already caused it to conclude that a more flexible and definitive set of requirements would have to be developed. Since that development was not complete, the 1975 draft was issued without those changes. Coincident with the printing and publication of that draft, a number of contacts were being made with state utilities commissions, sailboat manufacturers, and sailboat users throughout the country. A preliminary draft of further, unapproved revision proposals was provided to all state utilities commissions on 2 December 1974, in a letter requesting available accident data. Those proposals, patterned after the earlier Michigan and Wisconsin work, formed the basis of the eventual requirements.

The nongovernment and nonutility groups that provided information on motorboat and sailboat use and dimensions included, but were not limited to: the National Boating Federation, the North American Yacht Racing Association, the South Atlantic Yacht Racing Association, the Lake Michigan Yachting Association, the Florida Sailing Association, the Carolina Sailing Club, the High Rock Yacht Club, and the Boating Industry Associations (representing the Boat Manufacturers Association, Outboard Motor Manufacturers Association, Trailer Manufacturers Association, and Marine Accessories and Services Association). A number of other groups in the West and Midwest provided information, but the names of these groups are no longer available.

In addition, at least a dozen state departments of wildlife, natural resources, or the equivalent provided information. Each of the state utilities commissions was contacted in 1974, 1975, and 1976. The U.S. Army Corps of Engineers provided information on activities within its areas of responsibility (the Corps adopted the water clearances of the 1977 Edition of the NESC).

The entire records of the hearings in the states of Wisconsin and Michigan, in which water clearances based upon the size of the body of water were considered, were studied by the NESC Clearances Subcommittee. A number of data from outside sources were forwarded to the Clearances Subcommittee as a result of the two known articles in national sailing magazines about the pending revisions.

The results of the 1974 poll of state utilities commissions included responses from 20 states: Arizona, California, Colorado, Florida, Hawaii, Indiana, Iowa, Kansas, Maine, Maryland, Michigan, Missouri, New Hampshire, New York, North Carolina, Ohio, Oregon, South Dakota, Vermont, and Wisconsin. Many of these states did not have accident records or complete records. For those states that did supply data, there were 30 recorded sailboat accidents. Ten of the sailboats were catamaran types; eight more appeared to be catamarans from the mast height and other information in the responses. Three sailboats were definitely not catamarans, and four appeared not to be, leaving five that could not be identified in any fashion. The mast heights above water were as follows: 26 ft or less (7), 27–28 ft (6), 29–30 ft (2), 32 ft (1), 36 ft (1), and unknown (10). Three boats were on trailers with mast heights above ground of 24 ft, 31 ft, and 33 ft. The records of this information remain available, but the later responses of other states and groups are no longer available. Although one of the previous accidents is the only recorded sailboat accident that had occurred in California through 1974, none of the other information gained about prevalent sailboat use in California remains.

A significant effort was made by Clearances Subcommittee members to gather information concerning prevalent sailboat use in western and midwestern areas, as well as along the major rivers in the United States, in order to augment that which had been obtained on sailing in the Atlantic and Great Lakes states. In addition, members of the Edison Electric Institute and the Southeastern Electric Exchange also contributed to this effort. The major outside assistance came from the Special

Safety Committee of the South Atlantic Yacht Racing Association; this committee was appointed specifically to provide input to the process of developing the 1977 NESC clearance requirements over water areas and was instrumental in developing the information used in relating expected mast height to the size and type of the body of water.

During the spring and summer of 1975, the Clearances Subcommittee received information from the above sources concerning the heights of sailboat masts above water and above ground. A representative sample of these dimensions was shown in the 1 April 1976 public draft. The sample shown in the 1976 draft are only those data that had been provided in an official response by the National Association of Regulatory Utilities Commissioners (NARUC) to the 1975 draft and are only a small portion of the data that were available at the time. Unfortunately, the information in the 1976 draft and the testimony of the National Boating Federation and others in the Wisconsin hearings appear to be the only parts of that type of information that still exist.

The final requirements recognized that the size of boats found in use varied with the size of the body of water and were based upon surveys of boats actually in use throughout the country.

The development of the clearance requirements themselves was not the only problem facing the Clearances Subcommittee; one of the most difficult problems was the definition of the limits of the area over which the clearances would be required. The limit for controlled impoundments was relatively easy to define—the surface existing at the design high-water level. It should be noted that the design high-water level may be significantly above the spillway level where the flow rate over the spillway is controlled to limit flooding downstream.

The surface area for noncontrolled impoundments was more difficult to define. The Clearances Subcommittee recognized the accidents that had occurred as a result of boaters sailing during floods into areas normally not available for sailing, usually further up river than normally possible. It also recognized that many areas covered during flood only are still unsuitable for sailing due to trees, rocks, etc., that inhibit sailing; to require increased clearances over those areas is not appropriate.

At the time of that development, the Clearances Subcommittee had been informed that new federal government flood-control programs either had already defined, or soon would define, the annual 5-year, 10-year, 50-year, and 100-year flood levels for all water areas of NESC interest in the country. The annual and the 10-year flood levels were used in the 1977 Edition to define the area over which increased clearances would be required and the level from which they would be measured, respectively. This is a practical requirement and is consistent with available data. However, since then, most areas have available data on 10-year flood levels. As of 2002, Footnote 18 allows the 10-year flood level to be used as "normal flood level" for uncontrolled water flow areas, if the 10-year data is available.

Unfortunately, the government programs have still not defined flood levels in all needed areas. As a result, the height measurement requirement was generalized in the 1984 Edition; it is the responsibility of the designer to determine the "normal" flood level appropriate for use in measuring required clearances.

During the time of the a Clearances Subcommittee's deliberations, there were several definitions of water levels issued in the *Federal Register* (Friday, July 25, 1975, vol. 40, number 144, Part IV, Permits for Activities in Navigable Waters or Ocean Waters), which were found to be useful. On page 31321, first column, paragraph (d)(2)(ii), the term *ordinary high water mark* was used as a measuring point. It was defined as being inundated 25% of the time. This was used to refer to inland waters. For tidal waters, *mean high water*—defined as the average of all high tides—was used on page 31325.

The requirement to use the largest surface area of any one-mile-long segment of rivers, etc., resulted from subcommittee study of actual river use. This requirement recognizes the area needed to turn the larger boats and generally fits actual use. Because of the increasing use of marinas and marina-type housing developments and because of accidents involving sailors leaving the main body of water to picnic or relax in a small cove or tributary, these bodies of water require the same clearances as the main body of water. It should be noted that, especially in marina areas and canals off major bodies of water, the larger sail-boats may travel the narrower channels under power (with mast up); thus such an area may be expected to be traveled by sailboats for ingress and egress, even though the area is too narrow for sailing per se.

During the last two years of the 1977 Edition revision effort, Clear-ances Subcommittee members and those members of the sailing com-munity described earlier made every attempt to document sailboat use throughout the nation. The available information indicated that the new requirements would provide safe clearances over sailing areas in the United States for the types of sailboats then in use.

The unobstructed *water* areas available at the controlling water level are to be used to calculate the acreages required to determine the cor-rect part of Table 232-1. The acreage of land areas included in rigging and launching areas are not added to the water areas for this purpose.

These clearance requirements remained unchanged until the minor adjustments of the 1990 Edition that were required to make these clear-ances meet the coordinated uniform clearance system.

The 2002 Edition substituted values in squared kilometers for the former hectares. The rules recognize that access of sailboats to certain areas may be restricted to a bridge or other overwater obstruction. When this occurs, two clearances must be determined over the area with restricted access: the clearance over the smaller body of water as if it were standing alone, and the amount that the overwater obstruction decreases the clearance required for the larger body of water. The vessel heights to be used in this determination are contained in Table 232-3. The greater of the two clearances is required.

Example 232-1: The clearance required for a 7200 volt-to-ground conductor above a lake is 6.2 m (20.5 ft), 8.7 m (28.5 ft), 10.5 m (34.5 ft), or 12.3 m (40.5 ft), depending upon the size of the body of water. Assume that a 8.5 km^2 (850 ha; 2100 acres) lake is split by a 8.2 m (27 ft) high bridge that restricts the access to a 0.73 km^2 (73 ha; 180 acres) portion of the lake (over which the line was to be built) from the remaining 7.77 km^2 (777 ha; 1920 acres). The clearance required above the 0.73 km^2 (73 ha; 180 acres) side for vessels entering under the bridge from the 7.77 km^2 (777 ha; 1920 acres) side is 9.6 m (31.5 ft) (10.5 m – 0.9 m [34.5 ft – 3 ft]). Because this is greater than 8.7 m (28.5 ft),

9.6 m (31.5 ft) is the controlling clearance. The clearance required over the 0.73 km^2 (73 ha; 180 acres) side is the greater of that required for 0.73 km^2 (73 ha; 180 acres) or that required for 7.77 km^2 (777 ha; 1920 acres) as reduced by the bridge obstruction. The clearance required above a 0.73 km^2 (73 ha; 180 acres) lake is 8.7 m (28.5 ft) (from Table 232-1, 0.08 – 0.8 km^2 [8 – 80 ha; 20 – 200 acres]). The clearance required above a 7.77 km^2 (777 ha; 1920 acres) lake is 10.5 m (34.5 ft) (from Table 232-1, 0.08–0.8 km^2 [80 – 800 ha; 200 – 2000 acres]), based upon a Reference Height of 9.1 m (34.5 ft) (from Table 232-3).

In areas of stable water level, it would make no difference in safety if the bridge clearance at high water was used in this determination, since the clearance is required to exist above the high-water level. In areas of rapidly fluctuating water levels, such as some tidal areas and some areas downstream of hydroelectric or flood-control dams, it may be appropriate to consider the bridge clearance at low water if the water level could be expected to rise significantly after a boat sailed under the bridge.

After the 1977 Edition of the NESC was codified, each of the state utilities commissions or other agencies that were responsible for administration of the new edition was informed of the new require-ments for clearances over water. Each agency was requested to do whatever was in its own power or to work with other agencies to post water areas as to the clearances of utility facilities that could be expected in the area. This has been a slow process, but it is one that is being carried out in sailing areas across the nation as a result of joint cooperation of all concerned.

Where a regulatory body constrains the activity over water and issues a specific crossing permit, the constraint of the permit applies.

Table 232-1—Category 7. (1987 System)
Clearances Over Water Areas Suitable for Sailboating

Potential Conflict Dimension of Conflict:	passage of sailboat mast masts of 16, 24, 30, and 36 feet

Basic Clearances @ 60 °F (feet)

	Wires, Conductors, and Cables				
Classification	Rigid Live Parts	Less Than 20 acres	20–200 acres	200–2000 acres	Over 2000 acres
Structure clearance	NS	NS	NS	NS	NS
Effectively grounded equipment cases	NS	NS	NS	NS	NS
Communication cables; guys; messengers; 230E1 neutrals	NA	18.0	26.0	32.0	38.0
Surge-protection wires	NA	18.0	26.0	32.0	38.0
Open communication conductors	NA	18.0	26.0	32.0	38.0
230C1 supply cables	NA	18.0	26.0	32.0	38.0
230C2 & C3 supply cables					
0–750 V	NA	18.0	26.0	32.0	38.0
> 750 V	NA	18.0	26.0	32.0	38.0
Open supply					
0–750 V	NS	18.0	26.0	32.0	38.0
751 V–22 kV	NS	20.0	28.0	34.0	40.0
22–50 kV	NS	21.0	29.0	35.0	41.0
Contact conductors					
0–750 V	NA	NS	NS	NS	NS
> 750 V	NA	NS	NS	NS	NS

NA: not applicable.
NS: not specified.

Table 232-1—Category 7. (1990 System)
Clearances Over Water Areas Suitable for Sailboating

Potential Conflict: passage of sailboat mast
Dimension of Conflict: masts of 4.9, 7.3, 9.0 and 11.0 m (16, 24, 30, and 36 ft)

Basic Clearances @ **Maximum Sag** (m/ft)

Classification	Wires, Conductors, and Cables				
	Rigid Live Parts	Less Than 0.08^2 (8 ha) (20 acres)	Over 0.08 –0.8 km² (8–80 ha) (20–200 acres)	Over 0.8–8 km² (80–800 ha 200 ha) (200–2000 acres)	Over 8 km² (800 ha) (2000 acres)
Structure clearance	NS	NS	NS	NS	NS
Effectively grounded equipment cases	NS	NS	NS	NS	NS
Communication cables, guys, messengers, 230E1 neutrals	NA	5.3/17.5	7.8/25.5	9.6/31.5	11.4/37.5
Surge-protection wires	NA	5.3/17.5	7.8/25.5	9.6/31.5	11.4/37.5
Open communication conductors	NA	5.5/18.0	7.9/26.0	9.8/32.0	11.6/38.0
230C1 supply cables	NA	5.3/17.5	7.8/25.5	9.6/31.5	11.4/37.5
230C2 & C3 supply cables					
0–750 V	NA	5.5/18.0	7.9/26.0	9.8/32.0	11.6/38.0
> 750 V	NA	5.6/18.5	8.1/26.5	9.9/32.5	11.7/38.5
Open supply					
0–750 V	NS	5.6/18.5	8.1/26.5	9.9/32.5	11.7/38.5
751 V–22 kV	NS	6.2/20.5	8.7/28.5	10.5/34.5	12.3/40.5
22–50 kV	NS	6.2/20.5+†	8.7/28.5+†	10.5/34.5+†	12.3/40.5+†
Contact conductors					
0–750 V	NA	NS	NS	NS	NS
> 750 V	NA	NS	NS	NS	NS

NA: not applicable.
NS: not specified.
† Requires voltage adder.

Table 232-1, Category 8. Clearances over public or private land and water areas posted for rigging or launching sailboats

During the investigations for setting the proper clearances over water detailed in the previous Category 7, the increase in height required to step (raise) a mast while the boat is on a trailer or in the water was considered. Boats that are generally trailerable constitute the bulk of the problem; it is easier to step the mast while the boat is still on the trailer, rather than wait until the boat is in the water. Prior to the 1977 Edition, an increasing number of accidents began occurring as a result of careless sailers either (1) stepping masts while the boat was parked under a supply line, (2) driving trailers underneath supply lines after the mast was stepped, or (3) carrying catamarans out of the water back to a trailer with the mast still up. The NESC Subcommittee investigations identified the fact that an additional 1.5 m (5 ft) of clearance would provide adequate clearance for stepping masts on trailered boats.

For the medium-sized boats, which are not generally transported but are moored in a marina area, the 1.5 m (5 ft) of additional clearance will also allow stepping of the mast. The mast on large sailboats is not generally removed except during major maintenance in a specialized maintenance facility, if even then.

It should be noted that there was never any intention of requiring all lines near water areas to meet the clearances required by this rule; to do so would present an unnecessary and inappropriate burden on the general body of ratepayers and generally would not result in an increase in safety. The NESC Clearances Subcommittee intentionally specified that these requirements apply only to posted areas and, also intentionally, did not specify the form of posting—recognizing that it could be by signs, launching ramps, or other special facilities or land improvements or use which indicate that the area is intended for such use. Because of questions about which areas were required to have clearances for sailboats, the title of Category 8 was changed in 2002 to specifically include established boat ramps and associated rigging areas, *as well as* any other areas that are posted for rigging and launching sailboats. Posting of such other areas must be done with one or more signs. It is recognized that many areas adjacent to launching sites are posted to prohibit raising masts in the area. Under various state, federal, and local regulations, it is encumbent on those responsible for sailboat sales establishments, whether near or far from water, (1) to limit the raising and lowering of masts to areas without electric supply lines or (2) to arrange with the operator of the electric supply lines to relocate or raise existing lines to appropriate heights.

In general, if a body of water is suitable for sailboating, then areas on its shoreline that are likely to be used to beach sailboats while the sailors rest, picnic, change parties, etc., should have the clearances specified in this rule, because such activities are normally encountered or reasonably anticipated. However, the adjacent land areas are not required to meet these clearances unless they are obviously intended for transportation of the sailboat overland to or on a trailer, since such activity would otherwise not be normally encountered and rarely would be reasonably anticipated. Sailors who, for whatever reason, transport a sailboat overland in areas not so intended, whether or not subject to a charge of trespass, are responsible for the effects of those actions. It is not enough that it is physically *possible* to transport a sailboat over an area for these requirements to apply; there must also be a reasonable expectation based upon the character of the site that the area will serve as a launching or rigging site.

The obviousness of public access through a site should be considered; where public access is obviously available, these clearances for rigging and launching areas may in some cases be appropriate. However, where public access has not been specifically and obviously provided, these clearances are not required. For example, a picnic area located between the water and the parking lot would require the clearances for a rigging and launching area *if it was expected* that people would carry sailboats through the area to the trailer in the parking lot. Most such areas have enough trees, tables, or other obstructions that the reasonable expectation would be for someone choosing not to wait until the ramp was clear to walk around the area, instead of carrying a sail-boat through it. In the latter case, the picnic area could be considered as a Category 4 area if garbage trucks were expected, or as a Category 5 area if the area were restricted to personnel or restricted-height vehicles.

The required differential of 1.5 m (5 ft) above the clearance required for the associated body of water has not changed.

Table 232-1—Category 8. (1987 System)
Clearances Over Areas for Rigging or Launching Sailboat

Potential Conflict:	beached sailboat or sailboat on a trailer
Dimension of Conflict:	mast levels of 21, 29, 35, and 41 ft

Basic Clearances @ **60 °F (feet)**

	Wires, Conductors, and Cables				
Classification	Rigid Live Parts	Less Than 20 acres	20–200 acres	200–2000 acres	Over 2000 acres
Structure clearance	NS	NA	NA	NA	NA
Effectively grounded equipment cases	NS	NA	NA	NA	NA
Communication cables; guys; messengers; 230E1 neutrals	NA	23.0	31.0	37.0	43.0
Surge-protection wires	NA	23.0	31.0	37.0	43.0.
Open communication conductors	NA	23.0	31.0	37.0	43.0
230C1 supply cables	NA	22.0	31.0	37.0	43.0
230C2 & C3 supply cables					
0–750 V	NA	23.0	31.0	37.0	43.0
> 750 V	NA	23.0	31.0	37.0	43.0
Open supply					
0–750 V	NS	23.0	31.0	37.0	43.0
751 V–22 kV	NS	25.0	33.0	39.0	45.0
22–50 kV	NS	26.0	34.0	40.0	46.0
Contact conductors					
0–750 V	NA	NS	NS	NS	NS
> 750 V	NA	NS	NS	NS	NS

NA: not applicable.
NS: not specified.

Table 232-1—Category 8. (1990 System)
Clearances Over Areas for Rigging or Launching Sailboat

Potential Conflict: beached sailboat or sailboat on a trailer
Dimension of Conflict: mast levels of 6.4, 8.8, 10.5, and 12.5 m (21, 29, 35, and 41 ft)

Basic Clearances @ **Maximum Sag** (m/ft)

Classification	Rigid Live Parts	Wires, Conductors, and Cables			
		Less Than 0.08 km^2 (8 ha) (20 acres)	Over 0.08–0.8 km^2 (8–80 ha) (20–200 acres)	Over 0.8 km^2–8 km^2 (80–800 ha) (200–2000 acres)	Over 8 km^2 (800 ha) (2000 acres)
Structure clearance	NS	NA	NA	NA	NA
Effectively grounded equipment cases	NS	NA	NA	NA	NA
Communication cables; guys; messengers; 230E1 neutrals	NA	6.8/22.5	9.3/30.5	11.1/36.5	12.7/42.5
Surge-protection wires	NA	6.8/22.5	9.3/30.5	11.1/36.5	12.7/42.5
Open communication conductors	NA	7.0/23.0	9.4/31.0	11.3/37.0	13.1/43.0
230C1 supply cables	NA	6.8/22.5	9.3/30.5	11.1/36.5	12.9/42.5
230C2 & C3 supply cables					
0–750 V	NA	7.0/23.0	9.4/31.0	11.3/37.0	13.1/43.0
> 750 V	NA	7.1/23.5	9.6/31.5	11.4/37.5	13.2/43.5
Open supply					
0–750 V	NS	7.1/23.5	9.6/31.5	11.4/37.5	13.2/43.5
751 V–22 kV	NS	7.7/25.5	10.2/33.5	12.0/39.5	13.8/45.5
22–50 kV	NS	7.7/25.5+†	10.2/33.5+†	12.0/39.5+†	13.8/45.5+†
Contact conductors					
0–750 V	NA	NS	NS	NS	NS
> 750 V	NA	NS	NS	NS	NS

NA: not applicable.
NS: not specified.
† Requires voltage adder.

Table 232-1, Category 9. Clearances along (but not overhanging) roads, streets, or alleys

This category may, at first, appear unchanged in recent editions, but careful attention to the changes in the NOTES and the title will show significant changes. The 1977 Edition retained the requirements of earlier editions. The reductions of Footnote 10 in urban districts were no longer allowed in the 1981 Edition. Experience has shown that, through misapplication of the rule and especially because of changes in the character of land use (there now seem to be few places in urban areas where trucks will not pull off the road under a line), this reduction in clearances is no longer appropriate. Similarly, the continual change of once-rural areas to urban uses, without adjustment by the operating utilities to maintain the clearances required over the now-urban area, led to the reclassification of this category in the 1984 Edition. All roads, streets, and alleys (except those portions covered by Category 10), whether urban or rural, are now included in Category 9. This change clears up the uncertainty caused by the former lack of specific address to rural roads where vehicles *were* expected under the line.

Minor adjustments in clearance requirements occurred with the uniform clearance coordination of the 1990 Edition.

Application of Footnote 13 was not intended to apply to truck areas and was no longer reference in Category 9 in 2002.

Table 232-1—Category 9.
Clearances Along and Within Rights-of-Way but Not
Overhanging the Roadway of Roads, Streets, and Alleys

| Potential Conflict: | truck pulling or parking under line |
| Dimension of Conflict: | 4.3 m (14 ft) |

Basic Clearances (ft)

	1987 System: at 15 °C (60 °F)		1990 System: at Maximum Sag	
Classification	Rigid Parts	Wires, Conductors, & Cables	Rigid Parts	Wires, Conductors, & Cables
Structure clearance	NS	NA	NS	NA
Effectively grounded equipment cases	4.6/15.0§	NA	4.6/15.0	NA
Communication cables; guys; messengers; 230E1 neutrals	NA	5.5/18.0*	NA	4.8/15.5*
Surge-protection wires	NA	5.5/18.0*	NA	4.8/15.5*
Open communication conductors	NA	5.5/18.0*	NA	4.9/16.0*
230C1 supply cables	NA	5.5/18.0*	NA	4.8/15.5*
230C2 & C3 supply cables				
0–750 V	NA	5.5/18.0*	NA	4.9/16.0*
> 750 V	NA	5.5/18.0*	NA	5.0/16.5*
Open supply				
0–750 V	4.9/16.0*	5.5/18.0	4.9/16.0	5.0/16.5
751 V–22 kV	5.5/18.0	6.1/20.0	5.5/ 18.0	5.6/18.5
22–50 kV	5.8/19.0	6.4/21.0	5.5/18.0+†	5.6/18.5+†
Contact conductors				
0–750 V	NA	5.5/18.0*	NA	5.5/18.0*
> 750 V	NA	6.1/20.0*	NA	6.120.0*

NA: not applicable.
NS: not specified.
*EXCEPTIONS are provided.
† Requires voltage adder.
§1987 Edition, Rule 286E1.

Table 232-1, Category 10. Clearances along roads where it is unlikely that vehicles will be crossing under the line

This category was revised in the 1977 Edition so that it applied only in areas along rural roads where a vehicle is unlikely to travel under the line. This change was made as a result of increasing conflicts with farm trucks, logging trucks, and other vehicles that turned off a rural road to go into a field or forest. Except in the limited areas meeting the requirements of Category 10, clearances along a roadway are now generally required to be the same as those over the roads themselves. The 2007 Edition removed the 1977 restriction on applying this rule to rural areas only and allowed the rule to apply in urban areas where vehicles under the line are unlikely.

It is the responsibility of the designer to consider the *expected use of the area during the lifetime of the installation*. The mere existence of trees or a ditch along the line does not mean that vehicles are unlikely. It is common for loggers to fell a tree or two into a ditch and drive over them for access. It is also common for farmers to add additional entrances to fields when planting multiple crops where one might be damaged by transit of farm machinery to work on the other at the wrong time of year. In areas changing from urban to rural, driveways are often installed to prepare a home construction site long before or concurrent with notification of the utility of the need for service. In many areas, the ditches are shallow enough, particularly swale-type ditches, that they can easily be traversed by wreckers maneuvering to hook up to or remove a disabled vehicle or a farm, logging, or construction vehicle. Many lines have been drug down by being snagged by trucks or equipment crossing under the line. Thus it behooves the utility with overhead lines to consider the character of the land, as well as the changes that are reasonably expected during the life of the line, when installing overhead lines.

There should be some permanent terrain feature (such as a cut or fill) or land-use regulation that would be expected to prohibit vehicles from crossing or parking under the line for this category to apply. Figure H232-5 illustrates this point. If truck access to the area is available from off the right-of-way, such as an adjacent field, the area back of the ditch on the right-of-way is considered to be Category 9, not Category 10.

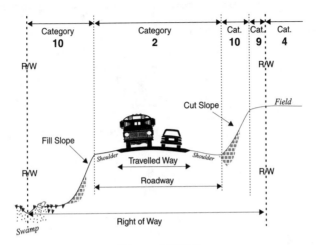

Figure H232-5
Clearance categories applying to road right-of-way

Table 232-1—Category 10.
Clearances Along and Within Rights-of-Way but Not Overhanging Rural Roads—Where Vehicles Under the Line Are Unlikely

| Potential Conflict: | highway maintenance equipment |
| Dimension of Conflict: | 3.6 m (12 ft) |

Basic Clearances (m/ft)

Classification	1987 System: at 15 °C (60 °F)		1990 System: at Maximum Sag	
	Rigid Parts	Wires, Conductors, & Cables	Rigid Parts	Wires, Conductors, & Cables
Structure clearance	NS	NA	NS	NA
Effectively grounded equipment cases	NS	NA	4.0/13.0*	NA
Communication cables; guys; messengers; 230E1 neutrals	NA	4.0/13.0*	NA	4.1/13.5*
Surge-protection wires	NA	4.3/14.0	NA	4.1/13.5*
Open communication conductors	NA	4.0/13.0*	NA	4.3/14.0*
230C1 supply cables	NA	4.3/14.0*	NA	4.113.5*
230C2 & C3 supply cables				
0–750 V	NA	4.3/14.0*	NA	4.3/14.0*
> 750 V	NA	4.6/15.0	NA	4.4/14.5
Open supply				
0–750 V	4.0/13.0*	4.6/15.0	4.3/14.0*	4.4/14.5
751 V–22 kV	4.9/16.0	5.5/18.0	4.9/16.0	5.0/16.5
22–50 kV	5.2/17.0	5.8/19.0	4.9/16.0+†	5.0/16.5+†
Contact conductors				
0–750 V	NA	5.5/18.0*	NA	5.5/18.0*
> 750 V	NA	6.1/20.0*	NA	6.1/20.0*

NA: not applicable.
NS: not specified.
*EXCEPTIONS are provided.
† Requires voltage adder.

232B2. Clearance to Unguarded Rigid Live Parts of Equipment (1990 and later editions)

(The requirement to use Table 232-2 for these clearances was moved to this rule number from Rule 232C in the 1990 Edition.)

The previous discussions are generally applicable except that sag is not a consideration. Drip loops of service drops do not move significantly and are generally, like vertical and lateral conductors, considered to be rigid. Rule 232 keeps the service drop and its drip loop together under Table 232-1; the drip loop is not under Table 232-2. For low buildings only drip loops are allowed to have lesser clearances above a residential driveway than the service drop (see Footnote 7 of NESC Table 232-1).

232B2. Sag Increase (1987 and prior editions *only*)

(Prior to the 1990 Edition, the table clearance values were required at 15 °C [60 °F] and limited to spans of the Basic Span Length of Rule 232A and 50 °C [120 °F] maximum operating temperature. Thus additional clearances were required at 15 °C [60 °F] for longer spans or operation above 50 °C [120 °F] to recognize the additional expected change in sag. With the respecification of clearance requirements at maximum sag in the 1990 Edition, this rule was deleted.)

In the Sixth Edition, clearances are based upon 60 °F, no wind, with allowances for greater sag. This allowance is limited by Rule 232B1a(3) to 75% or 85% of the "maximum sag increase." A typical 69 kV wood H-frame line might be required under the Sixth Edition to have 25 ft of clearance (at 60 °F) over a road at midspan and 22 ft of clearance if the road is adjacent to the H-frame structure. At 100 °C (212 °F) or greater conductor temperatures, which some electric utilities commonly designate as a thermal-loading limit, the sag can increase as much as 13 ft. Without a corrective adder, the clearance over the road could then be as little as 12 ft. Since many trucks approach 14 ft in height, the conductor could contact these trucks when the conductor is at these high conductor temperatures. For such lines, additional clearances are required at the 60 °F condition to recognize the additional sag change at higher temperatures and the effects of long spans.

Electric lines supported on suspension insulators that are free to swing in the direction parallel with the conductors are subject to a magnifying effect on the sag increase. If the crossing at the midpoint in a span is longer than the ruling span, the increase in sag with temperature varies with the square of the ratio of the actual span to the ruling span. For instance, the sag in a 600 ft ruling span might be 10 ft at 60 °F and 20 ft at 200 °F. If the actual span is 700 ft, the sag would be 13.6 ft at 60 °F and 27.2 ft at 200 °F. The increase in sag would be 10 ft with a 600 ft span and 13.6 ft in a 700 ft span in a section of line having a 600 ft ruling span.

The basic clearances in Table 232-1 (1977–1987 Editions; Table 1 of the Sixth and prior editions) for the basic span lengths are specifically stated in terms of 15 °C (60 °F) and final sag conditions to aid in measurement. The values shown are large enough to allow for increased sag beyond the measurement conditions due to ice loading or 50 °C (120 °F) conductor temperature operation. Whether the ice-loading condition or the 50 °C (120 °F) conductor temperature condition determines the maximum sag will depend upon the individual case. Where spans are longer than the basic length or maximum conductor operating temperature is higher than 50 °C (120 °F), other rules require *additions* to the basic clearances *measured* at the 50 °C (60 °F), final sag condition.

Under the 1977 revisions, clearances for most short-span electric distribution lines are computed in the same manner as in the Sixth Edition. Clearances for spans longer than the basic spans must have an allowance for *all* of the increase in sag.

It is intended in Rule 232B2 (1977–1987 Editions) that the owner of the supply wires establish the maximum conductor operating temperature. Some lines are limited by distance and load to operating temperatures barely over summer ambients. Others may operate as high as 150 °C (300 °F) in emergencies. Many electric utilities have established different maximum conductor operating temperatures for each of their lines. Because practice varies widely, and appropriately so, the NESC does not specify the conductor operating temperature or otherwise limit the operation of the line; the Code specifies the clearance that must be met during all conditions of operation. These conditions include short-term emergency overloads if these overloads produce more severe sags than long-term overloads or normal operation.

The method of adjusting the clearance for points of crossing not at midspan were made uniform throughout Section 23 in the 1977 Edition. Formerly, the multipliers found in Rule 232B2e (Rule 232B1b in the Sixth and prior editions) applied to the *sum* of the basic clearance and the additional clearance for longer span length. This was revised in the 1977 Edition so the multipliers apply to the additional clearances only. The multipliers are based on the following parabolic sag formula:

$$S_x = 4 S_m (X - X^2)$$

where

S_x = Increase in sag at any point **x** in the span

S_M = Increase in sag at the midpoint in the span

X = Distance to nearest support in fraction of the span length

The reductions in additional clearance requirements allowed by Rule 232B2e for crossings that are not at midspan apply only to the additional clearances required by Rule 232B2c (for long spans) and Rule 232B2d (for high conductor temperatures); they do not apply to the additional clearances required by either Rule 232B1a, Rule 232B1b, or Rule 232B1c.

Example: Large supply conductors having a maximum sag increase of 1.2 m (4 ft) are to cross a highway in the *heavy loading district* with a 300 m (1000 ft) span. The point of crossing is 22.8 m (75 ft) from the nearest support. No increase is assumed in this example to be required to meet Rule 232B1c. The maximum conductor operating temperature is 50 °C (120 °F).

The pre-1990 NESC basic span length for the heavy loading district was 53 m (175 ft). The excess span length over the basic span is $300 - 53 = 247$ m ($1000 - 175 = 825$ ft). the additional clearance required at 15 °C (60 °F) is 30 mm (0.1 ft) for each full 3 m (10 ft) of span length greater than the basic span.

In this example, there are 247/3 = 82.3 (825/10 = 82.5) increments of extra span length. since there are only 82 full increments of extra span length, the additional clearance required by the basic formula for the long span at 15 °C (60 °F) is 82 times the 30 mm (0.1 ft) value, or 2.5 m 98.2 ft). However, since the maximum sag increase for this conductor is only 1.2 m (4 ft), the additional clearance is limited to 1.2 m (4 ft) at midspan.

Because the crossing is at 75% of the span, the *additional* clearances required for *midspan* crossings may be reduced. Interpolating from the table under Rule 232B2e, the additional clearance required is therefore reduced to 0.275 × 1.2 m = 3300 m (0.275 × 4 ft = 1.1 ft).

The extra clearance requirements of Rule 232B do not apply to certain facilities in general and, in particular, to communications facilities that run along and within the limits of public highways. It is recognized that, in general, trolley conductors are frequently supported and will not be subject to large increases in sag. Also, trolley conductors must be restricted to certain heights in order for a trolley pole to contact them. Guys, of course, are normally fairly short and of such a high strength that they are not subject to large increases in sag.

Cables supported on messengers are subject to some increase in sag, however. In general, because of the weight of lead sheath cables, the addition of an ice load produces comparatively small increases in sag. Also, heavy cables are rarely strung to long spans. Logically, extra clearance for long spans is not necessary in these cases. With the advent of very lightweight cables (e.g., cables weighing about as much as their supporting strand), long spans are quite feasible, however, and sag increase under ice loading can be significant. The EXCEPTION to Rule 232B for communication cables on messengers should be applied with discretion, particularly where long-span construction is involved in either the medium- or heavy-loading areas.

Communication conductors, because of their small size and generally low strength, can be subject to considerable increase in sag under ice loading. However, it is not reasonable to require extra clearance to compensate for this increase in sag where low hanging conductors will not create a hazardous condition. Thus, wires that are run along, but not crossing, public roads do not require extra clearance for longer spans, unless they would interfere with normal traffic under storm-loaded or high-temperature conditions. Hence, this EXCEPTION is not to be applied to situations where conductors overhang the traveled way.

232B3. Clearance to Support Arms, Switch Handles, and Equipment Cases

(This rule was created in the 1977 Edition and includes the provisions of previous Rule 286F—Transformers. It moved from 286E to 232B3 in the 1990 Edition.)

The previous rule was expanded to include other equipment cases. Since an ungrounded case can have the same voltages as the equipment inside if the equipment fails, *ungrounded* equipment cases that contain equipment connected to circuits of more than 150 V are required to have the same clearances as specified for rigid live parts in Rule 232C.

The rule does not cover the installation of such equipment in areas not specified. As in other areas of the Code, the intended definitions of terms not specifically defined in the Code, such as *walkway*, are those found in a conventional dictionary.

Clearances to equipment cases were added to Rule 232B3 and Table 232-2 in the 1990 Edition. Switch handles were added in 2007. Also in 2007, application of the previous Footnote 1 was simplified by (1) renumbering Footnote 1 (applying to Row 1b *driveways, parking lots, and alleys* to Footnote 6 and to the center column *unguarded rigid live parts of 0 to 750 V*) and (2) restricting its application to such *areas not subject to truck traffic*. Whereas the previous editions used only Footnote 1b in Row 1d *spaces and ways subject to pedestrians or restricted traffic only*, the former Footnote 1a was eliminated, so that the remaining portion of Footnote 1 (old Footnote 1b) could apply as it did to this cell. In substance, the only change was to limit the application of the former Footnote 1 to areas where trucks were not expected. Riders on large animals other than horses were recognized in Footnote 5.

Footnote 7 to Table 232-1 allows effectively grounded switch handles and equipment cases to be mounted at a lower level than specified in the body of the table for accessibility if such items do not unduly obstruct a walkway.

232B4. Street and Area Lighting

(Previous Rule 286E2, Rule 286E3, and Rule 286E7 and Rule 286F were deleted in the 1977 Edition; see 1977–1987 Editions, Rule 286E, Rule 286F, Rule 215C and Rule 286E, respectively. The Section 28 rules were moved in the 1990 Edition to 232B3 and 234J2 in the 1990 Edition.)

Rule 232B4—*Street and Area Lighting* was moved from Rule 286G1 and Rule 286G4 in 1990 (Rule 286G2 and Rule 286G4 moved to Rule 420F). This appears to be an awkward placement because Rule 232B4 is a climbing space issue and Rule 232B4b is similar to the insulator placement requirements of Rule 279A2b.

Because a worker climbing a structure will not always be aware of a luminaire located above, the luminaire should be so placed as to give adequate climbing and working space. In cities where the height of the luminaire above ground is prescribed by ordinance, the location of the nearest support arm, span wire, or equipment should be chosen so as to give ample clearance from the luminaire. Luminaire leads should be carefully located and lamp brackets should be effectively insulated from the current-carrying parts.

In the case of externally wired luminaire fixtures, the construction should be such as to avoid the possibility of the wires coming into contact with the metal parts of the fixture or its supports. This can often be accomplished by extending the vertical run on the pole to a point below the boom of the luminaire fixture. Where the brackets are internally wired, care should be taken in protecting the insulation on the lead-in wires from abrasion at the point where they enter the bracket. Other insulation, in addition to a weatherproof covering, is recommended both for these wires and for vertical wires on the pole.

232C. Additional Clearances for Wires, Conductors, Cables, and Unguarded Rigid Live Parts of Equipment (1990 and later editions)

(Prior to the 1990 Edition, Rule 232A included a Basic Span Length for each loading area of Rule 250, and Table 232-1 included expected sag change from the required 15 °C (60 °F) measurement condition to the position under either maximum ice loading or 50 °C (120 °F) thermal operation. For longer spans or higher temperature operation, Rule 232B required additional clearances at the 15 °C (60 °F) position so that the truly intended clearances would not be violated under ice loading or maximum thermal loading conditions. The sag adders of Rule 232B2 were deleted with the 1990 specification of clearance requirements at the maximum sag position, rather than the 15 °C (60 °F) position. The clearance requirements were moved in the 1990 Edition for rigid live parts from 232C to 232B and the requirements for additional clearances for higher voltages in 1987 Rule 232B1 and Rule 232C2 were combined into a revised Rule 232C.)

Increased clearances are required for conditions that exceed the basic conditions of Rule 232A, including spans exceeding specified limits (1987 and prior editions only), voltages exceeding the table heading limits (50 kV for 1987 and prior editions; 22 kV thereafter), conductor temperatures above 50 °C (120 °F) (1987 and prior editions only), and high altitudes. The rule states that these increases are cumulative where more than one applies.

232C. Clearance to Live Parts of Equipment Mounted on Structures (1977–1987 Editions *only*)

This rule was added in the 1977 Edition. Generally, the requirements correspond to Table 232-1, except that the values in Table 232-2 are 0.6 m (2 ft) less, since sag is not a factor for rigid parts. The discussion for Rule 232B also applies to this rule.

Rule 232C—*Supply Pole Wiring at Underground Risers* from previous editions was essentially included within this rule in the 1977 Edition.

232C1. Voltages Exceeding 22 kV

(The requirements of this rule were contained in Rule 232B1 and Rule 232C2 of the 1987 and prior editions. They were combined here in the 1990 Edition.)

Rule 232B1 of the 1977 Edition provided the continuity between the Sixth and 1977 Editions for the additional clearances for voltages exceeding 50 000 V. The additional clearances of the Sixth Edition were retained, except that they are limited to supply lines operated up to 470 kV (814 kV phase-to-phase). Above that level, additional clearances based on the 10 m (0.4 in)-per-kV adder above 50 kV are questionable, because of the saturation effects of large air gaps. There is one very important difference between the voltage adder calculations in the 1977 and later editions and those required earlier—as of 1977, circuits operating above 50 kV to ground were required to use the **maximum operating voltage**, rather than the nominal voltage. The voltage adder was started at 22 kV in the 1990 Edition and the 22–50 kV column of values used in the 1987 and prior editions of Table 232-1 and Table 232-2 was eliminated.

For lines operated above 98 kV (170 kV phase-to-phase), alternate clearances permit a reduction of clearances where the line-switching surge performance is known (see Rule 232D).

Rule 232B1b and Rule 232C2b of the 1977 Edition (Rule 232C1b of the 1990 Edition) recognized for the first time the reduction of air-dielectric strength with increasing elevation. The *additional* clearances above 50 kV are required to be increased at a rate of 3% per 300 m (1000 ft) in excess of 1000 m (3300 ft) above mean sea level. The 3% rate is the ANSI C57.12.00 (1973) standard rate applied to substation equipment for elevation correction. The reference elevation of 1000 m (3300 ft) was selected on the basis that most of the supply and communication lines in the U.S. were at elevations below 1000 m (3300 ft).

Rule 232C1c (Rule 232B1c and Rule 232C2c of the 1977–1987 Editions) recognizes the potential safety hazard associated with electrostatically induced currents that may occur when insulated equipment and objects are in close proximity to transmission lines.

Recent investigations reveal that such potential safety hazards can exist where conductor clearances above ground, roads, etc., are determined solely from flashover considerations. The induced current magnitude of 5 ma rms is based on the recommendation by the IEEE Working Group on Electromagnetic and Electrostatic Effects of Transmission Lines. In the 1993 Edition, the application of Rule 232C1c was limited to ac systems; the 5-ma limitation applies to ac systems, not HVDC systems.

One of the research items available to the Clearances Subcommittee was Research Report No. 1, 3 September 1969, Project MR005.08-0030B, of the Naval Medical Research Institute, titled "Minimum Thresholds for Physiological Responses to Flow of Alternating Electric Current Through the Human Body at Power-Transmission Frequencies." Report No. 1 included the results of a survey of all available information about electric shock to humans, including children, at power-transmission frequencies of 50 and 60 Hz. Reliable quantitative data at these frequencies was available for three measurable physiological responses to electrical stimulation: (1) the perception of electric current flow, (2) uncontrollable muscular contraction, and (3) death. Relevant threshold conditions for response to minimum currents include the size and resistance of the body and the duration and pathway of current flow. 1% of the general populace can perceive from 0.1 to 0.5 ma of 50–60 Hz current, depending upon the type of hand contact made with an electrically energized circuit. A safety threshold of 5 mA, recommended for the general population including children, is based upon the conclusion that any 50–60 Hz current in excess of the release threshold of an individual should be regarded as hazardous and potentially lethal. 99% of adult male workers should be able to release 9 mA of 50–60 Hz current.

Rule 232B1c was modified in the 1981 Edition to recognize that there are a variety of methods that may be suitable for reducing electrostatic effects, e.g., using choke coils on an electric fence or grounding it. The size of the vehicle for which the calculation is required is intentionally not specified; to do so would penalize areas where only smaller equipment is expected.

232D. Alternate Clearances for Voltages Exceeding 98 kV Alternating Current to Ground or 139 kV Direct Current to Ground

The alternate clearance method added in Rule 232D in the 1977 Edition provides a direct method of determining the required conductor clearances above ground, rail, etc. It permits the clearance under specified conductor sag conditions to be determined from (1) the reference height of vehicles or other conflicting activities and (2) the electrical clearance thereto based on switching-surge flashover considerations. This rule recognizes that the 10 mm (0.4 in)-per-kV adder of Rule 232C1 (Rules 232B1 and 232C2 in the 1977–1987 Editions) *may* not be required for many higher-voltage installations with low switching-surge factors.

Through the use of the alternate clearance method, transmission line clearances may be reduced and optimized according to individual utility needs. As in the case for the basic clearances derived by Rule 232A, Rule 232B, and Rule 232C, the computed clearances derived by the alternate method of Rule 232D may need to be increased to satisfy the electrostatic effects. The clearances derived from Rule 232D are *allowed* as alternates to those required by other rules for voltages between 98 and 470 kV rms. The clearances of Rule 232D are *required* for voltages in excess of 470 kV. Table 232-3 includes reference heights to which electrical components of clearances calculated by Rule 232D3 must be added. The reference heights are based upon generally expected conflicting activity under the line. The reference of 67 m (22 ft) for rail tracks resulted from an agreement with railroad representatives and is the height required of bridges over railroads. A truck height of 4.3 m (14 ft) is the reference for vehicle areas. The height of a worker with hand tools was used for pedestrian areas.

Rule 232D3 is based on the equation derived by L. Paris and R. Cortina in *IEEE Transactions on Power Apparatus Systems*, vol. PAS-87, no. 4, pp. 947–957, Apr. 1968. This equation is used because of its simplicity and its adaptability to the various clearances considered throughout Section 23. Furthermore, comparisons with flashover-test data from other sources show favorable agreement with those obtained by the Paris-Cortina equation. The electrical component of clearance derived from Rule 232D3a represents a three-sigma (three standard deviation) voltage withstand against a designed switching-surge level for a conductor-to-plane configuration, corrected to elevation up to 450 m (1500 ft). A 20% margin of safety is included for intangibles. Rule 232D3c was revised in the 1993 Edition to apply the 5 mA limitation only to ac systems; it was not appropriate for dc systems.

The application of Rule 232D4 with very low switching-surge magnitudes may result in inadequate clearances for lightning exposure of lines rated above 230 kV. Rule 232D4 establishes a floor for the alternate method.

Figure H232-6 illustrates the range of voltage and switching-surge factors where Rule 232D permits reductions in the required clearance over roads. The sloping straight line labeled *Rule 232* is the required clearance according to Table 232-1 adjusted for voltage according to Rule 232C1. The horizontal straight line labeled *Rule 232D4* is the lower limit of the required clearance. Figure H232-1 shows that the basic method in Rule 232B and Rule 232C do not apply above 814 kV line-to-line (470 kV line-to-ground).

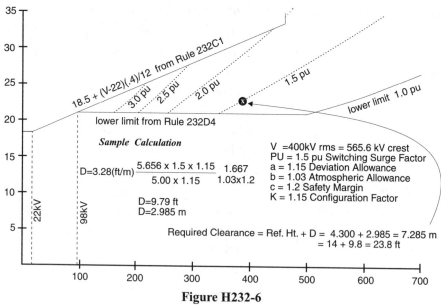

Figure H232-6
Alternate clearance for conductors over roads, streets, etc.

233. Clearances Between Wires, Conductors, and Cables Carried on Different Supporting Structures

(The Sixth and prior editions mixed both conductor movement and electrical clearance requirements in the same numerical values. These requirements were revised in the 1977 Edition but were left in generally the same format. This rule was revised in the 1981 and 1984 Editions to change the format and to separate conductor movement considerations from clearance considerations. Rule 233A was revised in the 1990 Edition for clarity and the 22–50 kV column was deleted from Table 233-1, thus starting some voltage adders at 22 kV. Table 233-1 was rearranged in the 1993 Edition for clarity and the 22–50 kV row was deleted, thus starting all voltage adders at 22 kV.)

The clearances specified in this rule are intended to ensure safe operation under either fair weather conditions or adverse weather conditions. The clearances recognize that the sag of conductors increases under both ice loading and high-temperature operation. The required clearances have been developed and revised over a long history of continual study of the interaction of crossing lines. A general history of the addition and modification of these requirements is discussed here before undertaking a discussion of the present requirements.

It will be noted that the wire-crossing clearances specified in some cases are greater when the crossing occurs within 1.8 m (6 ft) of one of the crossing structures (Footnote 3 and Footnote 7 associated with Table 3 of the Sixth and prior editions). This is to provide additional clearance for the protection of workers on such structures (see Rule 234B).

The recommendation that wire crossings be made on a common crossing pole recognizes that this kind of crossing generally represents the safest kind of construction but, obviously, is not always practical. Clearances specified in this rule do not apply to common crossing poles.

Effectively grounded neutrals of circuits not over 22 000 V phase to ground have the same clearances as guys and messengers (see the discussion of Rule 230D). Note that wire-crossing clearances are based upon the phase-to-ground voltage in the case of effectively grounded systems (see Rule 232A).

The requirement that clearances are to be measured with the upper crossing conductor at its final unloaded sag and the lower conductor at its initial unloaded sag was added in the Fifth Edition. The intermediate limit on open supply wires in Table 3 (now Table 233-1) was also changed in the Fifth Edition from 7500 V to 8700 V "between wires." As a part of the latter change, a Footnote 10 was added to allow conductors up to 8700 V phase to ground to continue to have a clearance above communication conductors of 4 ft. Messengers were generally treated the same as the conductors with which they were associated. Additional clarifying NOTES were added as detailed in the following paragraphs.

In the Fifth Edition, the same clearance was required for a 7200 V single-phase, delta *or* wye-connected system, regardless of the voltage to ground of any one of these systems. Specific EXCEPTION to this was provided in Footnote 10 of Table 3, allowing the same clearance for certain multigrounded supply circuits (not exceeding 15 000 V between wires, or 8700 V to ground), as for a supply circuit not exceeding 8700 V between wires (5000 V to neutral or ground). This footnote was added to retain a practice that had developed in prior years as a result of interpretations of the previous conflicting Code requirements. This practice had not been found to increase hazards unduly, so long as the communication-line span crossed over is relatively short. However, where this span is long and the communication-conductor sags are large, conductors can whip up into the supply line in the event of sudden release of ice or wind-induced "dancing" of the conductors. To obviate this, the provision was inserted that, at 60 °F, with no wind, the supply conductor at the upper level must not sag below the line of sight between the points of support of the communication conductors in the crossing span. It was concluded that the communication conductors will rarely pass above this level, even with large dancing amplitudes.

Provision of adequate clearances for conductors over guys, span wires, and messenger wires can be of as much importance as the clearance between two systems of conductors. In the case of messenger wires supporting communication cables, it is necessary that safe clearances be provided from supply conductors to allow workers safe access to all parts of the span.

The specified clearance of 2 ft was intended to be the minimum clearance provided where fire-alarm wires or private communication wires were involved. In cases where communication circuits for public use crossed, conflicted, or were on joint structures with each other, the clearance of 2 ft was allowed by Footnote 2 of Table 3 to be reduced where desired.

Footnote 7 of Table 3 of the Fifth Edition applies where a crossing or colinear wire is within 6 ft of, but not attached to, a support of the wires crossing beneath it. This rule was added because the sag in the upper wires may offer a hazard to line workers if they are on the lower line structure when the upper wires are sagged excessively under load. To alleviate this danger, added clearance between the facilities is required.

A recommendation was added in the Sixth Edition that crossings should be made on a common structure where possible. The Sixth Edition also generally changed tables and rules to refer to phase-to-ground voltages, if the circuit was wye-configured (or phase-to-phase voltages, if delta-configured). Table 3 remained essentially the same as that of the Fifth Edition except that old Footnote 10 was no longer needed. The new Footnote 10 is an unrelated NOTE.

The extensive use of conductors of new types and combinations of materials, as well as the use of relatively small conductors in long spans, was recognized in the Fifth Edition by the development of a new method for determining increased clearances for longer spans. The basis of the new method is outlined in the following paragraphs.

The various sags that are of interest in considering clearances are (1) the initial unloaded sag, (2) the final unloaded sag, and (3) the maximum total sag. Sag increase is the arithmetic difference between final unloaded sag at 60 °F, and total sag or 120 °F sag, whichever is the greater. The sag increase of particular interest is the "maximum sag increase" (msi), which is defined in Rule 233B1(a) of the Sixth and prior editions.

Conductors have greater sags when loaded with ice or operated at high temperatures than they do under normal conditions. The amount of the increase in sag is a controlling factor in providing safe clearances. On the basis of data obtained from the conductor manufacturers, curves of the sag increase were plotted according to span length for all of the commonly used conductors in each of the three loading districts. The results were that, as span length increased, the *sag increase* also increased. However, as spans were increased, the sag increase became larger at a decreasing rate. A maximum sag increase was eventually reached at some span length, beyond which the sag increase is less than this maximum. The maximum-sag-increase values were determined for most of the more commonly used conductors. (Such values are obtainable from conductor manufacturers.) They were based on the assumption that the conductors would be strung with the minimum sags, and therefore the maximum tensions, permitted by Rule 261F4 of the Fifth

Edition. If conductors are strung with less than these maximum tensions (which was and is common practice, especially in areas with heavy ice), the maximum sag increase will be less and will result in providing greater crossing clearances than required by the rules.

Study of the sag-increase curves for all conductors indicated that, although they differed widely, the sag increase was greater for the smaller conductors than for the larger conductors. In order to avoid unduly penalizing the larger conductors, the clearance requirements specified different clearance increments for "small" and "large" conductors.

In order to ensure a margin of safety, it was decided that there should be at least a 18-in clearance at wire crossings with total sag in the upper conductors and initial unloaded 60 °F sag in the conductor at the lower level. Additional curves were therefore drawn of the sag increase plus 18 in plotted as a function of span length. It was apparent that, in any span where the sag increase is not over 2.5 ft (that is, where the sag increase plus 18 in is not over 4 ft), no increase in clearance because of span length is necessary at crossings where the basic clearance required by Table 3 is 4 ft. This provided the basis for determining the basic span lengths specified in Rule 233A2, beyond which increased clearances were required.

The maximum-sag-increase figures were used in setting limits on the amount of additional clearance that would otherwise continue to increase indefinitely with each 10 ft of increase in span length. The maximum sag increases of certain of the smaller conductors could not be determined from existing data; most of these indeterminate cases occurred in the light-loading district. Since these values were not available, clearance-increase limits for these conductors could not be established and it was, therefore, necessary to add the applicable clearance increment as computed for the length of span involved in each such case. Limiting the additional clearance in the more usual cases to 75% to 85% of the maximum sag increase, depending upon the loading district, was empirical but was regarded as ensuring adequate clearances. It should be noted that the span length at which the maximum sag increase for a given conductor occurs bears no relation, unless by coincidence, to the length of span for which a required clearance is being determined.

The point of maximum sag in a conductor, even where its supports are at different elevations, is approximately at midspan. Sag increases are less at other points in the span than they are at midspan, and smaller clearance increases are therefore permitted for crossings at such points. This was accomplished by means of reduction factors that were determined from the catenary curve shown in Figure H233-1 and Table H233-1, given in the rule for different points of crossing, and expressed in percentage of crossing-span length. Catenary curve values were developed for level spans but apply well for differences in elevation of attachment points as much as 10% of the span length.

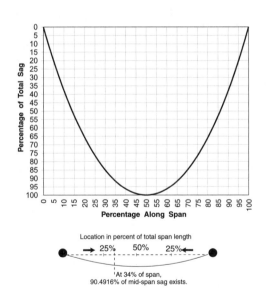

Percent of Span	Percent of Total Sag	Percent of Span	Percent of Total Sag	Percent of Span	Percent of Total Sag
0	0.000000	17	58.418360	34	90.491578
1	4.271367	18	60.984311	35	91.651605
2	8.430784	19	63.460999	36	92.734637
3	12.479916	20	65.849414	37	93.741107
4	16.420383	21	68.150513	38	94.671416
5	20.253760	22	70.365214	39	95.525938
6	23.981582	23	72.494405	40	96.305014
7	27.605338	24	74.538938	41	97.008955
8	31.126480	25	76.499629	42	97.638043
9	34.546415	26	78.377263	43	98.192530
10	37.866511	27	80.172592	44	98.672638
11	41.088097	28	81.886333	45	99.078559
12	44.212460	29	83.519172	46	99.410455
13	47.240852	30	85.071762	47	99.668458
14	50.174483	31	86.544724	48	99.852673
15	53.014527	32	87.938647	49	99.963172
16	55.762119	33	89.254090	50	100.000000

Figure H233-1
Catenary curve and percentage total sag

Table H233-1
Changes in Requirements for Crossings Not at Midspan:
Assuming 3 Ft of Additional Sag at Midspan

Sixth Edition 1977 Edition

% of Crossings Span	Reduction Factor (RF)	Required Clearance* = RF(7 ft)	Reduction Factor (RF)	Required Clearance = RF(3 ft) = 4 ft
For a Basic Clearance of 4 ft and Total Clearance of 7 ft at Midspan				
5	0.35	4.00 ft[a]	0.19	4.57 ft
10	0.47	4.00 ft*	0.36	5.08 ft
15	0.60	4.20 ft	0.51	5.53 ft
20	0.71	4.97 ft	0.64	5.92 ft
25	0.82	5.74 ft	0.75	6.25 ft
30	0.90	6.30 ft	0.84	6.52 ft
35	0.96	6.72 ft	0.91	6.73 ft
40	1.00	7.00 ft	0.96	6.88 ft
45	1.00	7.00 ft	0.99	6.97 ft
50	1.00	7.00 ft	1.00	7.00 ft
For a Basic Clearance of 6 ft and Total Clearance of 9 ft at Midspan:				
5	0.47	6.00 ft*	0.19	6.57 ft
10	0.58	6.00 ft*	0.36	7.08 ft
15	0.68	6.12 ft	0.51	7.53 ft
20	0.78	7.02 ft	0.64	7.92 ft
25	0.85	7.65 ft	0.75	8.25 ft
30	0.92	8.28 ft	0.84	8.52 ft
35	0.98	8.82 ft	0.91	8.73 ft
40	1.00	9.00 ft	0.96	8.88 ft
45	1.00	9.00 ft	0.99	8.97 ft
50	1.00	9.00 ft	1.00	9.00 ft

a The minimum clearance is the value required by Table 3 of Rule 233A.

When a new crossing is constructed, and until the upper conductor has been subjected to load, the upper conductor will have initial unloaded sags. The clearance value appropriate for measurement is the minimum clearance given in the rules plus the difference between the initial and final unloaded sags for the upper conductor. The final unloaded sag must include the effects of both storm loading and long-term creep.

It should be noted that, although Rule 232B specifically did not require extra clearance for long spans for messenger-supported cable, no such EXCEPTION appeared in Rule 233B. Although the sag increase of messenger-supported cables may be small enough to be ignored in considering ground clearances in some spans and loading conditions, the conductor crossing clearances is small enough that the sag increase of messenger-supported cables is relatively significant and cannot be ignored.

It is obvious that there should be some increase in clearance as the voltage of the conductors increases. Table 3 (now Table 233-1) gives definite steps of clearance increase for voltages up to 22 000 or 50 000, depending upon the column or row. Above this voltage, a uniform increment per 1000 V is applied.

A few inches of displacement of the free end of a suspension insulator toward a crossing span that it supports, such as when a conductor breaks, can reduce the clearance of such a span by as many feet. If, however, there are suspension insulators at both supports, only the differential displacement is involved, and this will be relatively less than with suspension insulators at one support only. The general rule ignores the resulting change in sag in this situation although it is possible that in some cases the change will be material. Rules 232B3 and 232B4 of the Sixth and prior editions were so worded as to modify the clearances to provide for these conditions.

These rules were deleted in the general 1977 revision. Application of Rule 233 of the prior editions is illustrated in Example H233X-1.

Example H233X-1 (Sixth Edition): Supply Line Crossing Communication Line

Conditions: A 600-ft span of large supply conductors carrying 110 000 V between conductors crosses a communication line at a point 200 ft from the nearest support. The supply line is not effectively grounded. Maximum sag increase (msi) for the conductors is 3.5 ft; the loading district is heavy. Find the required clearance.

(1)   *The basic clearance* from Table 3, Rule 233A, is 6 ft.

(2)   *The increased clearance required for span length exceeding the limits of Rule 233A.* From Rule 233B1(a), for heavy loading, 0.15 ft extra clearance is required for each 10 ft in excess of 175. Since there are 42 full 10 ft increments of additional span length [(600 − 175 = 425)/10 = 42.5], the required long span adder = 0.15(42) = 6.3 ft. However, the increase need not exceed 75% of the msi in the heavy loading area, which is 0.75 x 3.5 or 2.625 ft. The required clearance would be 6 + 2.625 = 8.625 ft. Generally, this would be rounded up to 8.7 ft.

(3)   *From Rule 233B1(b),* for crossings not at midspan, this may be reduced by the factor for a crossing at one-third (33.3%) of span length. For a 6-ft basic clearance, the factor would be approximately 0.96. Thus 8.625 x 0.96 = 8.28 ft, or 8.3 ft.

(4)   *The extra clearance required for voltages over 50 000.* From Rule 233B2, 0.4 in per 1000 V of excess is required. This is 0.4(110 − 50) = 24 in or 2 ft.

(5)   *The total clearance* is the sum of (3) and (4) above, or 8.3 + 2 = 10.3 ft or 10 ft-4 in. This is the clearance at 60 °F with the upper conductor at final unloaded sag with no wind or ice and the lower conductor at initial sag (see Rule 233A).

The contrast between this calculation and that required by the current edition can be seen by examination of the same example calculated by those methods, as shown later in Example 233-2. The current rules require an 11 ft-1 in clearance; the change is primarily as a result of the 1977 revisions deleting the percentage reduction of maximum sag increase and applying nonmidspan corrections to apply to the increased clearances only.

In the 1977 Edition, supply cables and messengers meeting Rule 230C1 were separated from open-supply conductors, etc. The rule was also extensively revised in the 1977 Edition. The previous rule requirements were combined under 233A—*At Crossings* (for vertical clearances at a crossing), and a new Rule 233B—*Other Than at Crossings* was added to specify horizontal and vertical clearances for adjacent wires, conductors, and cables *not* carried on the same supporting structures.

The 1977 revision of Rule 233 recognized most of the factors that influenced the revision of Rule 232. Both rules, for example, are changed to make it clear that temperature means *conductor* temperature, not ambient air temperature. Vertical clearances required at crossings were contained in Rule 233A in the 1977 Edition and moved to Rule 233C in the 1981 Edition.

Under the Sixth Edition, vertical clearances at crossings are based on 60 °F, without wind displacement, with an allowance for greater sag. This allowance is limited by Rule 233B1a to 75% or 85% of the "maximum sag increase." A typical 69 kV wood H-frame line might be required under the Sixth Edition to have 7 ft of clearance over another electric line at midspan, and 4 ft of clearance if the crossing is adjacent to the H-frame structure. At 200 °F conductor temperature, the sag can increase by as much as 13 ft. It is apparent that these supply conductors could contact each other, causing an electrical fault and interruption to electric service. A major blackout in the Pennsylvania, New Jersey, and Maryland area in the late 1960s was caused by an overloaded 230 kV line sagging into an electric distribution line. In recognition of such events, the formerly allowed percentage reductions in sag increases were deleted in the 1977 Edition.

In addition, the method of adjusting for nonmidspan crossings was changed in the 1977 Edition. Formerly, a reduction factor was applied to the complete clearance, thus incorrectly recognizing sag changes. The reduction factors were adjusted in the 1977 Edition and were applied only to the *increase* in sag required at midspan. Currently, where additional clearance is required by reason of sag increase in the longer spans, this added clearance is adjusted for location of the crossing in the span by the use of parabolic curve factors, just as in Rule 232. This is a more straightforward and logical method than that used in the Sixth and prior editions. It should be noted that the nonmidspan crossing reduction factor does *not* apply to voltage adders.

For supply voltages exceeding 50 kV, many of the same considerations that resulted in the 1977 revision of Rule 232 apply also to Rule 233. The switching-surge-factor approach (Paris–Cortina Equation) is presented as an alternative means of calculation in both rules. The configuration factor for conductor-to-conductor orientation is different from the factor for conductor-to-plane orientation used in Rule 232.

Both methods of calculating clearances between supply conductors crossing over other supply conductors appropriately assume that the two closest crossing conductors are completely out of phase and their respective phase-to-ground potentials are additive, although they could be exactly in phase in some cases. In effect, this can increase the required clearances, however slightly. Because communication cables are assumed to be at ground potential, no such increase exists. In the past, there has been some concern expressed over the fact that communication workers doing work on cables crossing directly under supply lines would have less clearance to the supply conductor than would the cable itself if they climb above the cable. But, since additional clearances are required for supply conductors crossing communication cables, and the experience has been satisfactory, no further increase appears warranted.

Both methods of calculating clearances above 50 kV specify use of the *maximum* operating voltage to ground. The ratio of nominal voltage to maximum operating voltage is not uniform, and the difference can be significant in the case of EHV conductors.

The correction factor for elevations of more than 450 m (1500 ft) above sea level is just as valid for conductor-to-conductor clearances as for conductor-to-ground covered in Rule 232.

The word "minimum" was deleted from the title of Table 233-1 to avoid the possible misinterpretation that table values apply under either storm-loaded or high-temperature conditions. It should be noted that "triplex" and "quadruplex" service and secondary cable facilities are covered in this table because they are considered to be cables meeting Rule 230C3.

Footnote 4 was expanded in the interest of consistency to cover both electrified railroad feeders and trolley feeders.

The percentage reductions in required sag increase formerly allowed by Rule 233B1a were deleted in the 1977 Edition; the limit is now the full maximum sag increase.

Rule 233B—*Other Than at Crossings*—of the 1977 Edition (Rule 233B—*Horizontal Clearance*—of the 1981 and later editions) was similar to Rule 234A of the Sixth Edition. It was revised and transferred to Rule 233 in the 1977 Edition to group all clearances between conductors supported on separate structures. This rule is to be applied under conditions similar to those for crossings, except that the effect of wind on the conductors is to be considered; both conductors are to be displaced in the same direction. The 1981 Edition revision clarified the original intention of the 1977 Edition Clearances Subcommittee; the conductors were to be blown in the same direction, not in opposing directions, and the clearance was required at the closest proximity of the conductors under the same ambient conditions.

The horizontal clearances required by Rule 233B must exist when the wind is blowing in the direction that would yield the closest approach between the conductors where both are displaced by a 290 Pa (6 lb/ft^2) wind. The code recognizes that, if *both* of the parallel lines are so close to large buildings that reflected wind off the buildings would decrease the effective wind on the conductors all along the span, a value of 190 Pa (4 lb/ft^2) may be used. It is rare that both lines would be so sheltered. The 2002 Edition made it clear in both Rule 233A1a(2) and figure 233-2 that trees cannot be depended upon for such shelter (they are temporary and variable).

The 1977 Edition of Rule 233B increased the minimum horizontal clearance from four feet to five feet, the same clearance as required from a conductor to a structure of another line. The 5-ft horizontal clearance of Rule 233B is 3 ft greater than the basic 2-ft vertical clearance in Table 233-1. This extra 3 ft will take care of most nonsynchronous movement under wind load (see Example H233BX). Unfortunately, the 1977 language did not properly specify the additional clearance required for transmission voltages. This was corrected with the 1981 reorganization of Rule 233, but the manner of doing so effectively removed the extra 3 ft of clearance to allow for nonsynchronous movement in the wind on a gradual basis up to 129 kV. This was corrected in the 2007 Edition. As of 2007, the voltage adder starts at

22 kV for horizontal clearances as it has for vertical clearances and the extra 3 ft horizontal clearance is properly maintained at all voltages.

Example H233BX: **Horizontal clearances Between Conductors of Parallel Lines Supported on Different Structures**

If the nearest conductor of Line A would have a horizontal displacement of 2.6 m (8.5 feet) at 15 °C (60 °F) under a 6 lb/ft^2 wind pressure, while the nearest conductor of Line B had a horizontal displacement of 3.8 m (12.5 ft), the closest approach mismatch would occur when the wind displace Line B toward Line A. At that time, the conductors would be 2.6 − 3.8 = 1.2 m (8.5 − 12.5 = 4.0 ft) closer together than at rest. Thus the minimum clearance between the conductors at rest would be 1.20 m (4.0 ft) plus the 1.50 m (5.0 ft) required by Rule 233B = 2.70 m (9.0 ft) to which must be added any applicable voltage adder.

(The rule was restructured in the 1981 Edition to separate conductor movement considerations from clearance considerations. The wording of that change was improved in the 1984 Edition. In the restructured Rule 233, the application requirements were placed in Rule 233A, horizontal clearance requirements remained in Rule 233B, and vertical clearance requirements were moved to Rule 233C.)

In order to ascertain whether a particular installation meets the clearance requirements of the NESC, it is necessary to do two things:

(1) *Determine the clearance that must be met.* This will depend upon the possible voltage potential between the items in potential conflict and upon the safety factors to be used, among other things.

(2) Determine the conditions under which the *point of maximum potential conflict between the items exists.* This will depend upon both (a) the ability of the items to change location, such as with wind, ice, or thermal loading and (b) their positions when at rest.

Determination of the point of maximum potential for conflict is generally easy when one of the items in potential conflict does not move, such as when measuring the clearance from a conductor to a building. In that case, any adders for extra movement of the conductor due to loading or for voltages greater than those applicable to the basic clearance, combine to make one easily measured and easily calculated *total clearance* requirement.

However, the problem is complicated somewhat when both of the potentially conflicting items can move. To aid in the determination of whether a conductor meets the clearance required from another conductor, the required clearance is calculated separately and is not mixed in with adders resulting from conductor movement due to loading. It makes no difference whether the required clearance is calculated first or the locations of the potentially conflicting conductors are calculated first.

It must be recognized that while conductors of different lines can have different thermal or ice loadings, they can be expected to have the same wind loading. If one conductor is under a wind loading, then the other conductor is assumed to be under the same loading in the same direction. The amount of displacement under the load would, of course, depend upon the conductor characteristics. However, one line can have a very light thermal loading while the other has a heavy thermal loading. Likewise, one conductor can have ice loading while the other, if its thermal loading is great enough to prevent ice, can essentially be at its 15 °C (60 °F) location. At crossings, it is not unusual for greater levels of ice to build on one line than the other when one line runs across the wind and the other runs with the wind.

It is important to note that the position of maximum potential conflict may not be when the conductors are under some wind loading; the conductors may be at their closest proximate position with thermal or ice loading but without wind displacement. Wherever the closest proximate position is located, the distance between the conductors must be at least that of the required clearance.

In recognition of these conditions, Rule 233 was restructured into three sections in the 1981 Edition. Rule 233A—*General* included the requirement that the clearances between conductors required by Rule 233B—*Horizontal Clearance* and Rule 233C—*Vertical Clearance* must be met under any *simultaneously occurring* conductor positions.

Rule 233A defined two new terms, *clearance envelopes* and *conductor movement envelopes*. Clearance envelopes are defined by the required horizontal and vertical clearances. Conductor movement envelopes delineate the limits of possible conductor positions.

The conductor movement envelopes are used to determine the point at which the two conductors will be in closest proximity while both are experiencing the same ambient air temperature and wind loading. The clearance envelopes are used to determine which clearance requirement, horizontal or vertical, is applicable. It should be stressed that the wind direction and pressure must be the same but, obviously, the blowout (displacement from rest) of the conductors will differ according to conductor characteristics and the loading on the conductors. It must also be recognized that a lower conductor may be unloaded while an upper conductor is loaded, either from ice buildup or high-temperature operation. It should be noted that appropriate horizontal wind displacement is to be considered in all cases. While this may not apply in most crossings, it definitely could be a factor when the lower crossing line is on an incline and the wind displaces the upper conductors in that direction.

Line DE of Figure 233-2 is shown as straight. It is recognized that this would not normally be the exact path. When a conductor is at Position E due to high-temperature operation, a relatively small amount of wind will serve to rapidly cool the conductor; as a result, the conductor tends to at first approach Position C more rapidly than Position D. The actual path will, therefore, tend to be more concave than shown. On the other hand, a conductor that has sagged to Position E as a result of ice loading will first tend to displace horizontally on an arc until enough ice breaks off to lighten the conductor. As a result, the actual path of an ice-loaded conductor would be expected to be first convex and then concave as the wind increased going from E to D. Although not exact in its path delineation, Line DE is recognized as appropriate from a safety standpoint. It is required by Footnote 4 of Figure 233-2 to be considered as straight unless the actual *concavity* characteristics are known.

The title of Table 233-1 was changed in the 1981 Edition to show that these requirements are for vertical clearances only. The horizontal clearance required is 5 ft for voltages of 129 kV or less, plus 0.4 in per kilovolt in excess. Alternate clearances are allowed above 98 kV alternating current to ground or 139 kV direct current to ground.

The revision of this rule in the 1981 Edition also clarified the application of clearance requirements to situations where conductors cross over or near anchor guys from other structures. Previous to the 1981 Edition, there were vertical clearance requirements in Table 233-1, but no horizontal clearance requirements at all, if the case was considered as a crossing. Under the wording of those editions, the reasonable conclusion was to treat such anchor guys as part of the "other structure" and apply Rule 234B. This was one of the situations that prompted the complete revision of Rule 233 in the 1981 Edition. The current Rule 233 specifically applies to such situations.

The revision in the 1984 Edition clarified the intention of the Clearances Subcommittee that the point of closest proximity may be with one or both conductors *within* the limits of maximum conductor movement defined by the conductor movement envelopes.

The relevant positions of the wires, conductors, or cables *on* or *within* their respective conductor movement envelopes are those that can occur when (1) both are *simultaneously* subjected to the same ambient air temperature and wind-loading conditions and (2) each is subjected *individually* to the full range of its icing conditions and applicable design electrical loading. The revised Figure 233-1 is intended to clearly show the intention of measuring the required clearance with both conductors under the same ambient conditions. In fact, the range of possible locations for the conductors under any one ambient condition is limited to a single vertical slot for each; the slot represents its possible range of vertical movement under the full range of changes in loading that can occur during that ambient condition.

Although a graphical method of using conductor movement envelopes and clearance envelopes to ensure required clearance under the "worst case" are illustrated in the 1981 and 1984 Editions, it is made clear in the 1984 Edition it that other methods of ensuring required clearance are acceptable.

Table 233-1 was revised in the 1984 Edition to recognize the change
to the 22 kV breakpoint in open-supply-wire clearance requirements.
Example H233X-2 illustrates the required method of calculation.

EXAMPLE H233X-2 (1984 Edition)

Conditions: A 600 ft span of large supply conductors carrying 110 000 V
between conductors (115.5 kV maximum operating voltage) crosses a communication line at a point 200 ft from the nearest support. The supply line is not
effectively grounded, but it is promptly de-energized upon a ground fault.
Maximum sag increase for the conductors is 3.5 feet; the loading district is
heavy. Find the required clearance.

(1) *The basic vertical clearance* from Table 233-1, Rule 233C, is 6 ft.

(2) *The additional clearance for voltages above 22 kV* required by Rule 233C2
     is 0.4 inches per excess kilovolt. The required additional clearance is
     $0.4(115.5/1.732 - 22) = 1.5$ ft.

(3) *The total vertical clearance* required under *all conditions* is
     $(1) + (2) = 6 + 1.5 = 7$ ft-6 in.

(4) The additional sag at midspan for long spans required by Rule
     233A1b(3)(a) is 0.15 ft per 10 ft of span length greater than the basic span.
     Since there are 42 full 10 ft increments of additional sag beyond the 175 ft
     basic span $[(600 - 175 = 425)/10 = 42.5]$, the calculated additional sag con-
     sideration is $0.15(42) = 6.3$ ft, as in Example 233-1. The limit on addition-
     al sag requirements contained in Rule 233A1b(3)(b) reduces the additional
     sag required to 3.5 ft.

(5) No additional sag is required for high-temperature operation in this exam-
     ple.

(6) The reduction factor for crossings not at midspan allowed by Rule
     233A1b(5) applies only to the additional clearance required. The reduction
     factor of a crossing at one-third span is approximately 0.88. The span
     adders required in (4) and (5) are reduced to $0.88(3.5 + 0) = 3.08 = 3$ ft-1
     in.

(7) The total consideration for nonbasic conditions in this example is $(3) + (6)$
     $= 7$ ft-6 in $+ 3$ ft-1 in $= 10$ ft-7 in. This is almost the same as 10 ft-3 in
     calculated under the Sixth Edition.

Comparison with the previous calculation in Example 233-1 shows
the slight effective increase in clearances required in the 1977 and later

editions of the Code by the requirement to use the full sag increase and the change in the method of adjustment for nonmidspan crossings.

Part 1 of the NESC gives required clearances from live parts to ground, but it does not give clearances between conductors crossing over other conductors or buses. Although the vertical clearance requirements of Rule 233 are not specifically *required* for clearances of incoming high-side conductors that cross over a rigid bus in a substation, unless such conductors are considered by the operator to be outside of the substation, Rule 233 provides a reasonable guide for such clearances.

Footnote 1 to Table 233-1 was revised in the 2002 Edition to specify no clearance between guys or span wires that are electrically interconnected, since there is no issue of a hazardous electrical potential between. Thus, in practice, the only remaining concern is mechanical damage if they were so close as to rub or pull upon one another during certain conditions. Obviously, such potentially damaging mechanical contact must be avoided.

It should be noted that, in the case of parallel lines, Rule 234B (which governs clearances to other supporting structures) may be the controlling rule. Rule 234B also controls when non-joint-use communications structures are added at midspan in joint-use lines. In so-called "skip-span" construction, this provides adequate working room for communications workers on those structures.

The 1990 Edition further revised the rule to require maintenance of the required clearances under the full range of thermal and icing conditions applicable to each conductor when both are subjected to the same ambient air-temperature and wind-loading conditions. This includes checking the "worst-case" winter conditions, with the lower conductor without ice and the upper conductor still covered with ice (at the various applicable temperatures); and checking the worst-case summer conditions, where the upper conductor is thermally loaded and the lower conductor is without electrical load at initial sag. This allows room for the lower line to be re-installed at initial sag while the upper is energized.

These requirements consider the lower conductor to be at the same temperature as the ambient air considered in calculating the position of the upper conductor. Thus, it is the responsibility of the designer to consider where the upper conductor will be located at various combinations of ambient air temperature and maximum ice or thermal loading, while considering that the lower conductor is completely unloaded and at the same ambient temperature.

For example, the greatest winter mismatch in an icing area would generally be with the upper conductor at 0 °C (32 °F), with full ice covering and the ambient air as cold as it can be and still have the line losses heat the conductor up to 0 °C (32 °F), just before melting the ice off the conductor. If the upper conductor has heavy electrical loads, it may be able to heat up an ice-covered conductor to 0 °C (32 °F) on an −18 °C (0 °F) night. In that case, the lower conductor would be considered to be at −18 °C (0 °F), initial sag while the upper conductor was at 0 °C (32 °F) with full ice covering a temperature mismatch of 18 °C (32 °F). However if the upper conductor had only a small electrical load and the coldest ambient temperature at which its line losses could heat an ice-covered conductor up to 0 °C (32 °F) was at −1 °C (30 °F), then the expected temperature mismatch would only be 1 °C (2 °F) (see Figure H233-2 and Example H233CX).

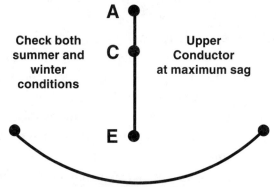

A

Check both
summer and **C**
winter
conditions

Upper
Conductor
at maximum sag

E

Lower conductor at coldest ambient
temperature that produces point E
on upper conductor
Figure H233-2
Crossing conditions

Example H233CX: Crossing Clearances

At a line crossing in the Heavy Ice Loading District, the upper line is a 198 m (650 ft) span of 4/0 Penguin ACSR conductor that is strung with the sags shown in Table H233CX-1. On a –18 °C (0 °F) night, the line losses can heat a conductor covered with 12.5 mm (0.50 in) radial ice to 0 °C (32 °F). In the summer, the line can reach 100 °C (212 °F).

The lower line is a 122 m (400 ft) span of #2 AWG Sparrow that is strung with the sags shown in Table H233CX-2.

For this example, it is assumed that the upper line was de-energized when the lower conductors were strung at 15 °C (60 °F).

Table H233CX-3 shows that the conductors will move closer together by 1.466 m (4.81 ft) from the installation conditions to the winter mismatch conditions. Table H233CX-4 shows that the conductors will move closer together by 1.009 m (3.31 ft) from the installation conditions to the summer mismatch conditions. Thus winter conditions govern in this case.

In this example, the installation clearance at 15 °C (60 °F) could not be less than 1.466 m (4.81 ft) plus the required clearance from Table 233-1 (600 m [2.0 ft]) plus any voltage adders. As a practical matter, an allowance would also be made for later pole movement, errors in stringing tensions/sags, etc., to ensure continued code compliance during the life of the installation.

Table H233CX-1
Upper Conductor: 4/0 Penguin 650 Ft Span, Heavy Loading

Temperature		Radial Ice		Final Sag	
(°C)	(°F)	mm	(in)	m	(ft)
0	32	12.5	0.50	5.096	16.72
15	60	0	0	4.194	13.76
100	212	0	0	5.581	18.31

Table H233CX-2
Lower Conductor: #2 Sparrow 400 Ft Span, Heavy Loading

Temperature	Temperature	Radial Ice (In)		Final Sag (Ft)	
(°C)	(°F)	mm	(in)	m	(ft)
−18	0	0	0	2.606	8.55
15	60	0	0	3.170	10.40
41	105	0	0	3.548	11.64
49	120	0	0	3.667	12.03
69	156	0	0	3.914	12.84

Table H233CX-3
Sags and Movement (in Ft) From 60 °F to Winter Mismatch Conditions

Winter Mismatch Conditions				
Upper conductor with 12.5 mm (0.5 in) radial ice @ 0 °C (32 °F), final sag Lower conductor with no ice @ −17.8 °C (0 °F), initial sag				
Upper Conductor (Final Sag)		Lower Conductor (Initial Sag)		Change in Clearance: (lower movement −upper movement)
0 °C (32 °F) w/ice	5.096 m (16.72 ft)	−17.8 °C (0 °F) no ice	2.606 m (8.55 ft)	
15.6 °C (60 °F)	4.194 m (13.76 ft)	15.6 °C (60 °F)	3.170 m (10.40 ft)	
Sag Change:	0.902 m (+2.96 ft)		−0.564 m (+1.85 ft)	−1.466 m (−4.81 ft)

Table H233CX-4
Sags and Movement (in Ft)
From 60 °F to Winter Mismatch Conditions

Summer Mismatch Conditions				
Upper conductor @ 212 °F, final sag **Lower conductor @ 105 °F, initial sag**				
Upper Conductor **(Final Sag)**		**Lower Conductor** **(Initial Sag)**		**Change in Clearance (lower movement –upper movement)**
100 °C (212 °F) w/ice	5.581 m (18.31 ft)	40.6 °C (105 °F)	3.548 m (11.64 ft)	
15.6 °C (60 °F)	4.194 m (13.76 ft)	15.6 °C (60 °F)	3.170 m (10.40 ft)	
Sag Change:	+1.387 m (+4.55 ft)		+0.378 m (+1.24 ft)	–1.009 m (–3.31 ft)

Since these requirements take into account actual sag conditions for the manner in which the conductors are installed, the long-span adder system was deleted.

As part of the general movement in the Code to have the basic clearances tables apply to distribution class voltages, with voltages for transmission level voltages, the 22–50 kV column was deleted in Table 233-1 in the 1990 Edition, thus starting the normal voltage adder of 10 mm (0.4 in) per kilovolt at 22 kV for the *upper* conductor. At the time, the "lumpiness" of the requirements was under study for the 1993 Edition.

That study work was not completed in time for the 1993 Edition, but two changes were made to Table 233-1. The table was rearranged to have both columns and rows read in a low-to-high voltage order. The 22–50 kV row was deleted, thus eliminating the 0.6 m (2 ft) jump in some clearance requirements and starting the voltage adder at 22 kV for the lower conductor. The voltage-adder issue is based upon what voltage could be impressed upon what circuit, without regard to which conductor was in the upper position and which conductor was in the lower position. The 1997 Edition separated the requirements for supply and communication guys and made slight adjustments to the clearances in Table 233-1.

The 2002 Edition limited the lesser clearances in Table 233-1 to effectively grounded guys, span wires, and messengers, not ungrounded ones. Also consent of regulatory authority is no longer required when communication parties agree to lesser clearances between their conductors, guys, and messengers. Clearances may not be reduced if electric supply lines are on joint-use structures.

IR 535 issued 23 May 2005 addressed the classification of traffic signal messengers for purposes of using Table 233-1. The lines that feed traffic signal bulbs are classified as electric supply lines, not communication lines. Typically when two messengers are used for larger intersections, the upper one is attached in the supply space; the lower one may be attached in the supply space or in the communication worker safety zone in accordance with Rule 238, but the messengers are still considered to be supply messengers for purposes of Rule 233 crossing clearances.

Table 233-2 presents values for alternate clearances between transmission voltage lines computed under the procedures required by Rule 233C3. Almost every value changed in the 2007 Edition as a result of the coordinated effort to respecify significant digits on values and assure that appropriate rounding up to the correct decimal had been accomplished. Some of the changes were more than 0.1 ft, but no significant change in safety occurred as a result of these respecifications.

234. Clearance of Wires, Conductors, Cables, and Equipment From Buildings, Bridges, Rail Cars, Swimming Pools, and Other Installations

(Rule 234 was broadened to include clearances to swimming pools, rail cars, signs, chimneys, antennas, and tanks containing nonflammables in the 1977 Edition. Earlier editions of the Code did not cover some commonly encountered structures. Additionally, the method of specifying horizontal clearances changed in the 1977 Edition and the method of specifying both horizontal and vertical clearances was modified in the 1990 Edition to use the coordinated uniform clearance system.)

234A. Application

(Prior to the 1977 Edition, Rule 234A included clearances from another line. That rule was incorporated into Rule 233 in the 1977 Edition.)

Starting with the 1977 Edition, Rule 234A covers the conditions for basic vertical and horizontal clearances. Basic horizontal clearances apply with the conductor at 15 °C (60 °F) and final unloaded sag. The effects of both storm loading and long-term creep are to be considered in development of the final unloaded sag. Vertical clearances are to be measured without wind displacement of the conductor. Horizontal clearances apply with the conductor displaced by a 690 Pa (6 lb/ft^2) wind in the 1977–1987 Editions; the 1990 Edition returned to specifying "at rest" horizontal clearances in the tables and required consideration of wind effects in a separate calculation. The definition of *final sag* includes the effects of the loading specified; the final sag used for determining conductor blowout is intended to include the vector effect of the wind force on the conductor. Horizontal clearance was specified in earlier editions, but no reference was made to wind displacement of conductors, supporting insulators, or structures (see Rule 234C).

The effect of wind upon structure sway and suspension insulator swing is also to be considered. The rule allows the use of a lower wind pressure on poles, conductors, cables and other components in protected areas. This rule recognizes that when large buildings, such as tall warehouses, are beside a complete span, wind blowing toward the building will be less on a conductor or cable due to back pressure from wind bouncing off the building. The temporary nature of some buildings should lead to careful consideration before such reductions are used. Trees are not considered as shelter. Horizontal displacement due to wind is a function of the diameter, density, and stretch coefficients of conductors, cables, and cable messengers. For a conductor of the same type, doubling the diameter will double the wind load but quadruple the weight per lineal foot. Thus, larger conductors have greater sags but lesser wind deflection angles (see Figure H234A-1). Copper conductors are heavier and more conductive than aluminum conductors and, thus, have greater sags, smaller diameters, and smaller wind deflection angles than aluminum conductors (see Figure H234A-2, Figure H234A-3, and Figure H234A-4). All-aluminum conductors are lighter and do not stretch as much as ACSR conductors. Aluminum conductors with steel reinforcement (ACSR) are strong but stretches, both because of the steel core. For a given wire size, more steel increases both strength and stretch. For the same current-carrying capacity, aluminum conductors have a larger diameter, less density, and greater wind deflection angles than copper. For example, compare the relative movement of #6 AWG HD cw, #4 AWG ACSR, and #4 AWG AAC in Figures H234A-2, Figure 234-3, and Figure H234A-4. Thus required clearances can sometimes be obtained by using heavier conductor for spans where pole locations are limited.

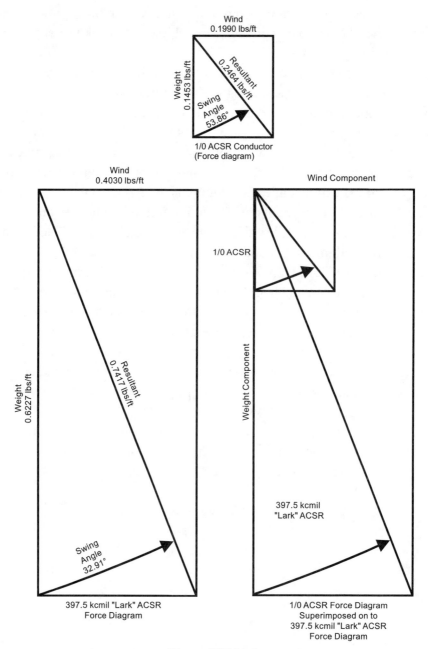

Figure H234A-1
Relative wind and weight forces and wind deflection angles per linear foot of two conductors

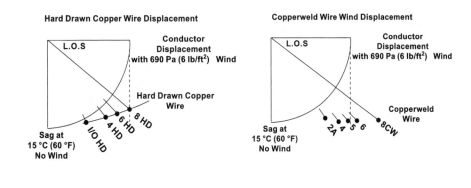

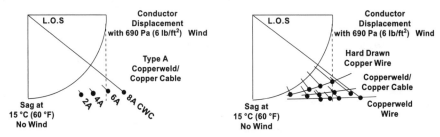

Figure H234A-2
Wind displacement of copper conductors and
cable relative to sag at 15 °C (60 °F)

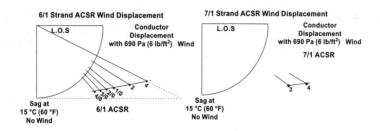

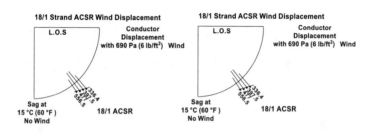

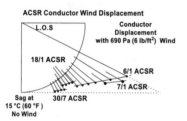

Figure H234A-3
Wind displacement of ACSR conductors relative to sag at 15 °C (60 °F)

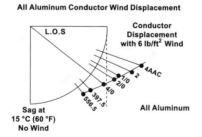

Figure H234A-4
Wind displacement of all-aluminum conductors relative to sag at 15 °C (60 °F)

An EXCEPTION was added in Rule 234A1 in the 1984 Edition that effectively reduced the horizontal clearances required by deleting consideration of wind displacement for communication conductors and cables, guys, messengers, surge-protection wires, neutral conductors meeting Rule 230E1, and various supply cables; these items normally only present mechanical, rather than electrical, safety considerations. The clearances required without wind displacement were revised in the 1990 Edition to meet the coordinated uniform clearance system. In the process, the EXCEPTION moved from the application rule, 234A, to the horizontal clearance rule, 234B1, by including a table that identified the specific types of items not required to be considered as displaced by wind. This resulted in confusion; as a result, the 1997 Edition also added back an EXCEPTION to the horizontal clearance rule, Rule 234B1. Even though the term *messengers* was in the EXCEPTION, the 2007 Edition clarified that this also applied to insulated communication conductors and cables, as it typically does in other areas of the Code where these items are grouped together.

It should be noted that, as of the Sixth Edition, clearance zones are no longer squared off in the corners. The transition zone clearance effectively creates a diagonal connecting the horizontal and vertical clearances. This increases the area available to conductors without a reduction in practical safety. As shown in NESC Figure 234-1, the vertical clearance can be considered as a stiff rod that is always in the vertical position when over or under a roof or projection; only when it gets to the edge and would otherwise "fall off" does the building end (of this hypothetical stiff rod) stop moving laterally and allow the opposite end to lean out (as a diagonal clearance) until it intersects and stops the vertical projection of the horizontal clearance limit. Over the roof itself, the vertical measurement is always vertical, not diagonal, regardless of roof slope. This is also shown in Figure H234-1, Figure 234-2, and Figure 234-3, discussed later.

See Rule 234C for additional discussion and for illustrative diagrams.

Horizontal displacement of conductors under wind loading is required to consider deflection of suspension insulators if they are used. Deflection of the structure itself is not required to be considered unless the structure or supported wire, conductor, or cable exceeds 18 m (60 ft) in height above the grade below. Most structures shorter than 18 m (60 ft) above grade do not move enough under wind loading to appreciably affect the total horizontal displacement.

CAUTION: Clearances are specified under a 290 Pa (6 lb/ft^2) (i.e., 22 m/s [50 mph]) wind; storm wind loadings will move conductors to greater horizontal displacements than used to determine the required clearance, but not so great as to contact the structure, so long as structure deflections are controlled. Small conductors (particularly #ACSR) stretch enough in storm winds that care should be taken to ensure that the structure is not unduly flexible, especially in the high wind areas and special wind regions of the ASCE 7 wind map (ANSI/ASCE, Minimum Design Loads for Buildings and Other Structures) (see Figure 250-2).

234B. Clearances of Wires, Conductors, and Cables From Other Supporting Structures

Rule 234B covers clearances of wires, conductors, and cables from other supporting structures. Rule 234B also applies to traffic signal, roadway lighting, and area lighting structures, regardless of whether the electric equipment thereon is fed from an overhead or underground circuit or is physically located on a public right-of-way or on private land. Rule 234B applies to communication lines as well as supply lines.

The *horizontal clearance* required with the conductor at rest was increased in the 1977 Edition to 5 ft for voltages up to 50 kV with the conductor blown toward the structure by a 6 lb/ft^2 wind. In the 1990 Edition, the value of 1.5 m (5 ft) was retained for the "at rest" requirement and a table of requirements (ranging from 1.07–1.4 m [3.5–4.5 ft]) was added for requirements under the 290 Pa (6 lb/ft^2) wind-displacement conditions. Unless the two line structures are close and horizontal wire movement is 150 mm (6 in) or less, the wind displacement plus the required clearance will determine the required at rest location.

As of 1990, the rule also requires a *vertical clearance* of 1.4 m (4.5 ft) at maximum sag (1.8 m [6 ft] at 15 °C [60 °F] plus adders for the 1977–1987 Editions) for voltages below 22 kV (15 kV in the 1977–1984 Editions and 8.7 kV previously) plus an additional 300 mm (1 ft) of clearance for voltages up to 50 kV. Earlier editions allowed a clearance of 900 mm (3 ft) without differentiating between horizontal or vertical requirements. The clearances were increased to allow room for maintenance of the other structures and to allow the other structures to be replaced while the line in question remains in service. This rule applies both to separate lines and to situations where conductors carried at the higher level are attached only to alternate poles (i.e., so-called "skip-span" construction) (see Figure H234B).

Figure H234B
Skip-span pole not having vertical clearance required by Rule 234B2 of horizontal clearance at the structure required by Rule 235B

Conductors that are attached to structures are usually made easily obvious to workers by crossarms, insulators, etc., that support them. Workers who are ascending, descending, or performing a task on a structure of one line, and are concentrating on those facilities, may not be readily aware of the proximity of a conductor from a second line that approaches, but is not attached to, that structure. This problem is exacerbated when the conductor closely approaches the climbing-space side or top of the structure.

In order to limit accidental contact, a greater clearance from structures is required for unattached conductors than for attached conductors.

It should be recognized that, if conductors of one line are not kept well away from structures of a second line, they may move into dangerous proximity if the structures of either line settle, or are otherwise pulled out of line by, the addition of service drops, changes in conductor tension with loading, or other lateral forces. This is especially likely to create a problem when the conductors of one line straddle the poles of the second line. This rule is intended to inhibit the latter construction unless the structures of the two lines are close together and the span lengths are reasonably equal.

Where structures of one line and conductors of a different line are in close proximity, it is generally preferable to attach the conductors of one line to the structures of the other by means of clearance arms or pole extensions. This will eliminate the possibility of (1) accidental contact between the conductors and structures or (2) reduction in the climbing space of one line or the other. Otherwise, the greater clearances of this rule are necessary. A new NOTE was added to Rule 234B2 in the 1981 Edition to clarify that, as of the 1977 Edition, the clearances of wires, conductors, and cables from the guys of an adjacent structure are specifically covered in Rule 233.

Rule 230E determines whether a neutral conductor must meet the full clearance requirements of Rule 234B. If the neutral conductor meets the requirements of Rule 230E1, it is considered to be equivalent to a messenger-neutral meeting the requirements of Rule 230C and, thus, EXCEPTION 1 of Rule 234B applies. If the neutral conductor meets the requirements of Rule 230E2, the full clearance is required as specified in Rule 234B.

> *NOTE:* EXCEPTION 2 of Rule 234B, added in the 1981 Edition, applies only to the vertical clearance requirement and may only be used *in lieu* of EXCEPTION 1. The clearance reductions of the two EXCEPTIONS are *not* cumulative.

Although Rule 234B does not specifically mention the clearances of energized live parts of equipment on one structure to another structure, the same clearances required for conductors applies, since conductors attach to the bushings on the equipment.

Rule 234B was reorganized into 234B1 (horizontal clearances) and 234B2 (vertical clearances) for clarity in the 2002 Edition. The EXCEPTION in Rule 234B1a now applies for effectively grounded guys, messengers, and neutrals. As a result, wind deflection of ungrounded span guys must be considered. As a companion to the change to the EXCEPTION under Rule 234B1, the 2007 Edition also clarified EXCEPTION 1 to the vertical clearance requirements of Rule 234B2 by adding e*ffectively grounded messengers* and *insulated conductors and cables* to the EXCEPTION to recognize all of the group of relatively benign items normally grouped together.

234C. Clearances of Wires, Conductors, Cables, and Rigid Live Parts From Buildings, Signs, Billboards, Chimneys, Radio and Television Antennas, Tanks, and Other Installations Except Bridges

The title of this rule was changed in 1977 to reflect the application of the rule to installations other than buildings. Originally, tanks containing nonflammables only were included; tanks containing flammables were not covered. The application was changed to be to all tanks in the 1984 Edition. Unfortunately, an apparent editorial error left the reference to tanks containing nonflammables in Rule 234F1c. The intention is for that rule to apply to all tanks.

The clearances of this rule apply from the nearest conductor surface to the nearest surface of a building or its projections or attachments.

Table 234-1 of the 1977 Edition is an expansion of Table 4 of the Sixth Edition. The required clearances for certain voltages were increased in some instances.

The required horizontal clearance to buildings was increased to provide more working space between the wires, conductors, or cables and the building surface. As discussed in the following paragraphs, it is recognized that it is not possible, much less practical, to completely eliminate accidents that occur as a result of inattentive people pushing, carrying, or dropping long metal objects, such as antennas, ladders, pipes, and poles into contact with electric lines. This rule provides adequate space for workers with small hand tools to maintain a building or other structure. It also provides adequate clearance for trained workers to use specialized maintenance tools.

The required vertical clearances were increased in the 1977 Edition for all voltages to reflect changes made in the measurement conditions and for coordination. Clearances over "roofs accessible to pedestrians" are the same as required in Table 232-1 for "spaces or ways accessible to pedestrians only." Clearances over roofs accessible to vehicular traffic, such as parking garages, are basically the same as required in Table 232-1 for clearances over roads, streets, and alleys.

It is recognized that "roofs not accessible to pedestrians" may have workers on such roofs occasionally. The vertical clearances over such roofs were increased because the average man is taller than he was in 1920 when this rule was last revised. Many men and some women can reach the 8 ft level specified in the Sixth Edition for 300 to 15 000 V, especially with small hand tools such as a hammer. In addition, the 1977 Edition required the vertical clearance to exist at 60 °F. Previously, the rule was not clearly specified and the values appeared to be "closest-approach" values to exist at maximum sag.

Footnote 3 to Table 234-1 defines attributes of a roof, balcony, or other area that is *readily accessible to pedestrians*. Roofs accessible by a ladder that comes down to or near the ground is accessible but, if the ladder stops more than 2.45 m (8 ft) above grade, the roof is not accessible. If a carport roof is accessible through a window, it is accessible. Basically, if a person wishing to access the roof or balcony must get a ladder, use a tool (such as a key), or use extraordinary (not casual) effort, the area is not accessible. Prior to the 2007 Edition, this footnote used the term *special* tool; the modifying term *special* was eliminated in 2007, because it was undefined and not necessary. If a roof is accessible by keys available to tenants, it is accessible; if the roof key is only available to maintenance personnel, the roof is not considered to be accessible.

The vertical clearances over signs, billboards, chimneys, antennas, and tanks are reduced from those required over building roofs. It is recognized that these structures will generally be accessible only to skilled workers and that, while these facilities may require a person to be *at* the top of the structure for maintenance, they *do not* require walking on top of the structure. Any structure that is expected to have personnel walking erect thereupon should be treated the same as a building. Table 234-1 was revised in 1997 to add Row 2b(1) in Table 234-1 to require the same clearances used for roofs or balconies accessible to pedestrians for catwalks and other surfaces on nonbuilding installations upon which personnel walk.

Fence walls are generally treated as "other installations" in Table 234-1, unless they are so wide and the access is such that they are considered as a "projection" from a building that is accessible to the public.

Example: The backyards of two houses are dug out of a hillside in the area where they join. The hillside is held back by a retaining wall that is 1.5 m (5 ft) high. The wall extends between the two yards to separate them. If the fence/wall is wide enough to provide steady footing, it **is** considered the same as a portion of a building or other installation that is accessible to the public.

EXCEPTIONS 1 and 2 of Rule 234C3d (Rule 234C5c in 1977, Rule 234C4c in the 1981–1987 Editions, Rule 234C3c in 1990) were added in the 1977 Edition to improve coordination with the National Electrical Code (NEC). Likewise, Rule 234C3d(2) was added in the 1984 Edition to be consistent with Rule 230-24(c) of the NEC.

The 2002 Edition of Rule 234C3a specified required construction of service drops for 0–750V (any 230C cables or 230D covered wire) and >750V (230C1 cables). Rule 234C3c was revised in 2002 to specify a clearance of 75 mm (3 in) for cables allowed to be attached to buildings, rather than referring the user to Table 235-6.

Since the earliest edition of the Code, there has been a requirement to leave a space to raise ladders for fire fighting either adjacent to, or starting no more than 2.45 m (8 ft) away from buildings of three stories of 15 m (50 ft) or greater in height. This space must be a minimum of 1.8 m (6 ft) wide. This requirement started at a time when fire-fighting equipment was mostly manual and was less sophisticated than it generally is today. This clearance allowed firefighters to place ladders on buildings and move hoses to the fire without being impeded by the utility lines. Although the methods of fighting fires have improved dramatically in recent years, and this requirement may not always be necessary from a fighting standpoint, it is still necessary from the standpoint of providing adequate space for (1) placement of ladders for use in removal of persons who may be trapped in a building and (2) penetration of walls with water spray equipment. It is also useful, of course, during building maintenance. This provision is not a requirement where the fire department has an *unvarying* rule *not* to use ladders where conductors are present.

Frequent inquiry was made several decades ago about the possibility of receiving a shock from a high-voltage conductor through a hose stream. For short distances between nozzle and conductor, such shocks are possible if the stream of water is "solid." At some distance from the nozzle, even a solid stream of water breaks up into discrete particles that do not form a continuous conducting path. Tests run in the 1930s showed that, when this distance is reached, no shock can be received through the hose stream. So-called "fog nozzles" in common use today propel a spray of such discontinuity as to present little problem, even at close ranges.

Editor's Note: *These tests were run with water from city water systems and may not be valid for water containing special chemicals that increase the ion content.*

In general, conductors approaching buildings should be isolated by clearance or be guarded. This rule provides adequate clearances for unattached conductors that pass by or over buildings, etc. It requires attached conductors to be either guarded or made inaccessible. Specifically addressed are clearances from roofs, balconies, and windows.

Significant changes have been made in this rule in recent editions. Table H234-1 shows the changes in horizontal clearance requirements since the first codification.

Table H234-1
Changes in Required Horizontal Clearances of
Open Conductors to Buildings

Edition/Rule No.		Voltage Classification and Clearances			
2nd 247	Clearance–R	300 g– 7500 p 3 ft (a)	7500 p– 15 000 p 8 ft (b)	15 000 p– 50 000 p 10 ft (b)	Above 50 000 p 10 ft (b)
3rd 228	Clearance–R	300 g– 7500 p 3 ft (a)	7500 p– 15 000 p 8 ft (b)	15 000 p 50 000 p 10 ft (b)	Above 50 000 p 10 ft (b)
4th 234C	Clearance–R(c)	300 g– 7500 p 3 ft	7500 g– 15 000 p 8 ft	15 000 p– 50 000 p 10 ft	Above 50 000 p 10 ft + 0.5 in/kV (d)
5th 234C	Clearance–R(c)	300 p– 8700 p 3 ft	8700 p– 15 000 p 8 ft	15 000 p– 50 000 p 10 ft	Above 50 000 p 10 ft + 0.5 in/kV (d)
6th 234C	Clearance–R(e)	300 g– 8700 g 3 ft	8700 g– 15 000 g 8 ft	1500 g– 50 000 g 10 ft	Above 50 000 g 10 ft + 0.4 in/kV (d)

Table H234-1
Changes in Required Horizontal Clearances of
Open Conductors to Buildings *(Continued)*

Edition/Rule No.	Voltage Classification and Clearances			
1977 234A C,&F		(g)0g– 8700 g	8700 g– 15 000 g	15 000 g– 50 000	Above 50 000 g(h) 10 ft + 0.4 in/kV(d)(f)
	Clearance–W(e)	**5 ft (f)**	**8 ft (f)**	**10 ft (f)**	
1981	Same as 1977 except (i)				
1984 234A(1) C&F		0 g– 8700 g	8700 g– 22 000 g	22 000 g– 50 000 g	Above 50 000(i) 7ft + 0.4 in/kV(d)(f)
	Clearance–W(e)	**5 ft (f)**	**6 ft (f)**	**7 ft**	
1987	Same as 1984 except (j)				
1990 234A, C&G		0 g– 750 g	751 g– 22 000 g	Above 22 000 g(i)	
	Clearance–R	**5.5 ft**	**7.5 ft**	**7.5ft + 0.4 in/kV(k)**	
	Clearance–W	**3.5 ft(f)**	**4.5 ft(f)**	**4.5ft + 0.4 in/kV(f)(k)**	

g = Voltage phase to ground

p = Voltage phase to phase

R = At rest

W = Displaced by wind

Notes on Code Requirements:

(a) "Conductors—shall be so arranged that they do not come nearer than":

(b) "Conductors—shall be so arranged that they clear the surfaces—by":

(c) "Conductors—shall not come closer...than":

(d) The added clearance required per kilovolt in excess of 50

(e) "The—clearance—shall be as listed"

(f) With conductor "displaced from rest by a 6 lb/ft^2 6-pound-per-square-ft wind at final sag at 60 °F." After 1990, 290 Pa wind @ 15 °C could also be used.

(g) All clearances are based on the maximum operating voltage.

(h) An alternate method is allowed above 140 kV to ground.

(i)   An alternate method is allowed above 98 kV to ground.

(j)   Add 610 mm (2 ft) to obtain "at rest" clearance if sag is 900 mm (3 ft) or less at 15 °C (60 °F).

(k)   The added clearance required per kilovolt in excess of 22 kV.

The first codified edition included horizontal clearance requirements for conductors of 300 V to ground and higher. Although other rules of the NESC gave conditions under which the clearances must be met, this rule did not. The rule required that "conductors—shall be so arranged that they" either "do not come nearer than" or "clear the surfaces—by" the clearances shown. It is apparent that the language of the Code gave latitude of placement to the designer as long as those clearance requirements were never violated.

The original clearances were retained in the Third Edition. The same clearance values were retained in the Fourth Edition but 0.5 in per kilovolt over 50 kV was added as well as a requirement that "conductors—shall not come closer—than" those values. The rule title was also changed to be "Minimum Clearances" in the Fourth Edition. The wording and requirements of the Fourth Edition were generally retained in the Fifth Edition. Interpretations issued during this time indicated that it was unclear whether conductor displacement was required.

The basic clearance values of the Fifth Edition were retained in the Sixth Edition, but three significant changes were made: (1) the voltage ranges were changed from phase to phase to phase to ground, thus reducing clearances required for some voltages; (2) the voltage adder was reduced to 0.4 in/kV in excess of 50 kV; and (3) the former clearance language was changed to be "the—clearance—shall be as listed." The heading of "Minimum Clearances" was retained. Unfortunately, the third change did not help to determine whether wind displacement of conductors was required. It was not uncommon for utilities to ignore wind displacement of conductors passing buildings except for long-span construction or where suspension insulators were used.

The voltage classification change had a dramatic effect on required clearances for several common distribution voltages, as shown in Table H234-2.

Table H234-2
Horizontal Clearance Required From Building for
Phase Conductors of Three Voltage Classifications

	120/ 240 V Triplex Conductor	277/ 480 V Open Conductor	2.4/12.5 kV Three-phase Open Conductor	7.2/12.5 kV Three-phase Open Conductor	14.4/24.9 kV Three-phase Open Conductor	19.9/34.5 kV Three-phase Open Conductor
2nd (1916)R	—	—	3'	8'	10'	10'
3rd (1920)R	—	—	3'	8'	10'	10'
4th (1927)R	—	—	3'	8'	10'	10'
5th (1941)R	—	3'	3'	8'	10'	10'
6th (1961)R	—	—	3'	3'	8'	10'
1977W	3'5'(a)	5'(a)	5'(a)	5'(a)	8'(a)	10' (a)
1981W	3'	5'(a)	5'(a)	5'(a)	8'(a)	10' (a)
1984, 87W	3'	5'(a)	5'(a)	5'(a)	6'(a)	6' (a)
1990R(b)	5.0'	5.5'	7.5'	7.5'	7.5'	7.5'
W(b)	—	3.5'(a)	4.5'(a)	4.5'(a)	4.5'(a)	4.5'(a)

(a) Horizontal clearance required after displacement of conductor from a 6 lb/ft^2 wind at 60 °F and final sag. After 1987, the SI units of a 290 Pa wind pressure at 15 °C were added.

(b) Use whichever produces the greater clearance when the conductor is at rest.

After conductors began to be installed at the reduced Sixth Edition clearances, accidents were recorded of such a nature as to require an increase in horizontal and vertical clearance requirements in the 1977 Edition. The difficulty of balancing the need to achieve the most practical level of safety without creating an unmanageable burden in areas with close rights-of-way, such as in alleys or along small residential lots, caused that revision to become one of the most heated and protracted of the revision process. It was recognized that there is no *possible* way to effectively eliminate accidents resulting from human carelessness around electric lines. Beyond a point, further increases in clearances from buildings only resulted in increased expenditures of ratepayer money without an effective increase in safety for the ratepayer.

As a part of the review effort, the representatives of the National Association of Regulatory Utilities Commissioners representatives queried state utilities commissions for available accident data. This was included with the previously discussed query about sailboating accidents. Respondents with information were New Hampshire, Maine, Ohio, Iowa, New York, Colorado, Hawaii, Florida, Oregon, Missouri, California, Wyoming, and North Carolina. Arizona responded but had no available information. The answers to these inquiries are summarized in Tables H234-3 and H234-4. Various Clearances Subcommittee members were able to obtain from other states additional accident records that were invaluable in considering the complete nature of the problem. None of those records are available today.

TableH234-3
Conflicting Fatal Activity and Actual Horizontal
Clearances of Conductors From Building

| Fatal Activity | Clearances | |
	Known Distances	# Unknown
1—Installing antennas	20 in, 5 ft-6 in, 14 ft-0 in, 15 ft, 18 ft-9 in, 19 ft-0 in, 19 ft-6 in, 20 ft, 22 ft	9
2—Carrying ladders: at service entrance	4 ft-2 in, 6 ft, 7 ft-2 in, 8 ft, 8 ft-6 in, 8 ft-11 in, 9 ft-6 in, 10 ft, 11 ft, 20 ft, 25 ft, 28 ft	1
3—Painting with brush or long roller handle	2 ft, 3 ft-0 in, 3 ft-1 in, 4 ft-6 in, 6 ft-6 in, 8 ft, 8 ft-6 in, 8 ft-9 in, 9 ft, 14 ft	7
4—Loading tanks or silos	5 ft-6 in, 7 ft-2 in, 9 ft	5
5—Carrying construction materials	6-1/2 in, 2 ft-5 in, 4 ft-4 in, 5 ft, 5 ft-4 in, 6 ft, 6 ft-3in, 6 ft-5 in, 9 ft, 11 ft, 12 ft-5 in, 25 ft, 41 ft	6
6—Installing flashing or trim	4 ft-11 in, 7 ft-7 in, 8 ft, 10 ft	2
7—Person "touched wire"		5
8—Unknown		34

TableH234-4
Conflicting Fatal Activity and Actual Vertical Clearance

Fatal Activity	Clearances	
	Known Distances	**# Unknown**
1—Installing antenna		1 above roof
2—Carrying construction materials	15 ft-6 in above roof	
3—Person "touched wire"	4 ft-6 in & 4 ft-7 in above roof	
4—Person or auto under low service drop	7 ft	2

It should be noted that covered conductor was involved in two ladder-related, one painting-related, and two low-service-drop-related accidents.

The known horizontal clearances had the following frequency of occurrence in these accident data: (1) at service entrance, (2) 1–2 ft, (4) 2–3 ft, (1) 3–4 ft, (5) 4–5 ft, (5) 5–6 ft, (4) 6–7 ft, (6) 7–8 ft, (7) 8–9 ft, (3) 9–10 ft, (2) 10–11 ft, (2) 13–14 ft, (1) 14–15 ft, (2) 18–19 ft, (3) 19–20 ft, (1) 21–22 ft, (2) 24–25 ft, (1) 27–28 ft, (1) 41 ft.

These data and the other available data indicated two things: (1) the clearances should be increased, and (2) there was a limit beyond which there was no discernible increase in safety for an increase in clearances. It was quite obvious in some cases that, because of the type of careless human activity involved, an additional clearance of even several more feet would not have prevented or lessened the severity of the accident. In one such case, a person carried a ladder *across* the street, so that he would not step on it, and contacted a conductor on the other side of the street. In another such case, an antenna was dropped from a roof onto a line, with tens of feet to spare. On the other hand, it was quite obvious that additional clearance could have been useful in some cases. In one such case, a painter on a ladder contacted a conductor running by the gable end of a house. Later, a second painter came to finish the job.

Even though he knew about the first accident, he also contacted the same conductor while painting. The clearance was just under 5 ft.

There were some Clearances Subcommittee members who initially recommended leaving the requirements as they were in the Sixth Edition; others recommended increasing the requirements even beyond those required by the Fifth Edition. The data indicated that the horizontal clearances of the Fifth Edition were reasonable. After careful scrutiny of the data, the Subcommittee reached a consensus that the minimum clearance of open-supply conductors from buildings should be greater than 5 ft. On the other hand, it was not clear that there would be a discernible increase in safety for short-span construction if the clearances for the lower-voltage conductors were increased to or beyond the original 8-ft requirement.

The requirements of the 1977 Edition both (1) recognize the capabilities and requirements of utility construction and (2) provide adequate clearances to allow maintenance of buildings and other structures near electric supply lines. The minimum horizontal clearance for open-supply conductors is 5 ft *plus* the horizontal displacement of the conductor from a 6 lb/ft^2 wind at final sag with a conductor temperature of 60 °F.

Prior to the 1977 Edition, the clearance was measured with the conductor at rest. There were two reasons for adding the blowout requirement. The first was to recognize that several accidents occurred while emergency repairs were being made during storms. The second was that this was an effective way to move the horizontal clearance requirement beyond the critical distance under which increased clearances appeared to be capable of positively affecting the outcome of careless human activity.

Similar analysis of the data available for accidents above roofs indicated that it was necessary to increase the clearance to 10 ft to allow workers to repair roof areas with normal small hand tools. There was no discernible increase in safety beyond 10 ft. This change was made to reflect the change in Rule 234A in the specification of the conditions under which the clearances are to exist. The clearances above buildings in 1977 were clearly specified to exist at 60 °F, final sag, and included the usual 18-in allowance for additional sag for ice loading or thermal loading up to 120 °F. These clearances apply to roofs that are not accessible to pedestrians (See Footnote 3 to Table 234-1), regardless of building type. They provide adequate clearances for common maintenance activity with small hand tools; they are not appropriate for and do not apply to areas that are accessible to pedestrians. Such areas may be used for sun decks or other purposes; greater clearances are required for areas that do not meet the requirements of Footnote 3. The term *readily accessible* was further refined in the 1993 Edition based upon previous Clearances Subcommittee discussions and Official Interpretations. The 1993 revision was not a change in requirements but, rather, a clarification of previous intent.

NESC Figure 234-1 was added in the 1977 Edition to show precisely where vertical and horizontal clearances are to be applied to buildings and other installations and projections therefrom. Figure 234-1 shows that the vertical clearance above a roof remains a vertical, not diagonal, measurement regardless of roof slope. Only in the transition zone between the horizontal and vertical clearances is the vertical clearance used as a diagonal. (As of the 1997 Edition, the horizontal clearance is used as the diagonal, if the required horizontal clearance is greater than the vertical clearance; see NESC Figure 234-1(c).)

Figure H234-1 shows the clearance requirements of the Fifth and prior editions. Figure H234-1 also shows the effect of the use of the diagonal clearance beginning with the Sixth Edition.

An additional clearance for long spans to be applied to both vertical and horizontal clearances was required in the Sixth Edition. Those requirements continued in the 1977 Edition. Continuance of the adder for vertical clearances was appropriate; however, neither the long-span sag adder nor the high-temperature sag adder should have been applied to the 1977 Edition horizontal clearances, because in the 1977 Edition, horizontal wind displacement of the conductor was required. Since actual displacement was considered, the adders were not necessary for horizontal clearances; they were deleted in the 1981 Edition. Figure

H234-2 shows the intention of the 1977, 1981, and 1984 Editions (see Rule 234F).

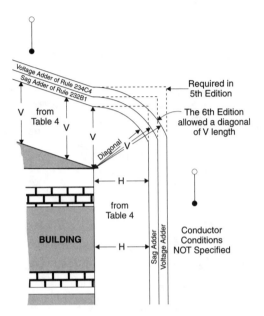

Figure H234-1
Clearances to buildings—Fifth and Sixth Editions

With the complete revision of the 1990 Edition came a respecification of the clearance requirements and adjustment to meet the coordinated uniform clearance system (see Appendix A of the NESC). The values shown in Table 234-1 were changed back to apply at rest (some careless users did not read the rules that went with the tables and used the former values for at-rest conditions without adding conductor displacement under wind), and a separate requirement was detailed in both the rules and Table 234-1. Another 10 years of accident data was reviewed and the appropriateness of the 1977 changes was confirmed. With the 1990 requirement of vertical clearances above structures to be met at maximum sag conditions, the former sag adder requirements were deleted (see Figure H234-3).

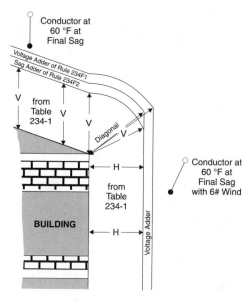

Figure H234-2
Clearances to buildings—1977–1987 Editions

The 1997 Edition recognized that some spans may be so long or so slack as to have enough sag (and thus corresponding horizontal deflection under wind loading) that the horizontal clearance required with the conductor at rest is greater than the vertical clearance with the conductor at maximum sag. Thus, the transition zone between the horizontal and vertical clearances should recognize this condition. The value of the diagonal clearance in the transition zone is now the greater of the required vertical and horizontal clearances (see Figure H234-3 and Figure H234-4).

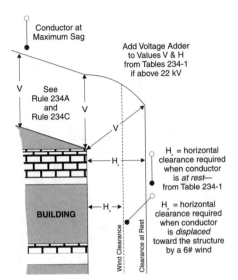

Figure H234-3
Clearance to buildings—1990 Edition and later Editions

Although minor shifting of some clearances occurred in the 1990 Edition to match the coordinated uniform clearance system, the vertical clearances above buildings increased significantly. There was no accident data to suggest that the change was necessary, but rather, there was no reason found for the vertical clearances to vary from the uniform system.

Rule 286F of the 1987 and prior editions required rigid live parts of equipment and ungrounded equipment cases to have the same clearances as required for their associated conductors. In the 1990 Edition Rule 286F moved to Rule 234J and rigid live part clearances were added to Table 234-1 and Table 234-2. The transformers shown in Figure H-T234-1 were added after construction of the building without changing to a taller pole; as a result the clearances to the transformer bushings and the jumpers do not meet the requirements of Table 234-1 or Rule 234J.

Figure H_T234-1_1
Transformer bushing and jumper clearance less than
required by Table 234-1 and Rule 234J

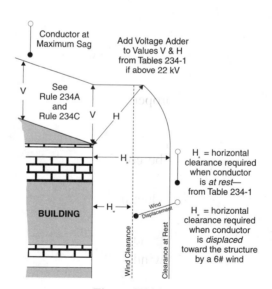

Figure H234-4
Clearance to buildings—1997 Edition; if
horizontal clearance is greater than vertical clearance

Rule 234C does not cover buildings in transit. Such situations as movement of buildings are special cases that are governed by Rule 012. Note that the OSHA "10-ft rule" of 29 CFR 1926 Subparts O and N effectively prohibits workers from riding atop a home during a move under most energized power lines, because they cannot maintain the required 10 ft clearance. Even if the power lines are covered with insulating sleeves by the utility, OSHA prevents ordinary workers from touching the insulated lines, so there is little need for workers not qualified as line workers to be on a house during a move. If, as required by OSHA, the house mover makes appropriate arrangements with the affected utilities prior to a move, there will be little need for a person to be on the roof at any time during the move. In addition, OSHA fall-protection requirements generally make having someone on a roof during a house move impractical.

NOTE: The OSHA transit clearance for the building itself moving under a line is only 4 ft.

Rule 234C does not specify clearances to the edge of a utility right-of-way. However, the required clearances to buildings should be considered, since buildings may be expected to be constructed at the right-of-way line in many cases. Increasingly, however zoning set-backs limit construction of building near property lines.

When applying Rule 234C to motorized signs, flagpoles, and other structures that have movable portions, such movement is required by the NOTES to Table 234-1 to be considered. Wind-deflected movable attachments, such as flags and wind vanes, are assumed to be deflected in the same direction as the conductor under the same wind conditions, similar to the requirements of Rule 233. Whether additional clearance is required for wind loading depends upon the deflection of conductors toward the support, when the wind is in one direction, relative to the deflection of the movable portion toward the conductors, when the wind is in the opposite direction.

In the 1981 Edition, previous Rule 234C1—*General* was deleted; it contained gratuitous information and was not a specific requirement. Requirements for providing for firefighters are detailed elsewhere in the

Code. Notice, however, that the changes to Rule 234C in the 1990 Edition essentially provide firefighter space next to the building. The remaining rules under 234C were renumbered.

If a portion of a structure is not to be maintained in place, then reduced clearances are allowed to that portion in some cases by Footnote 1 to Table 234-1 (see the following paragraphs). For example, where a flagpole itself requires no maintenance and the flag is removed for maintenance, the clearance may be able to be reduced. Table 234-1 includes a Footnote 1, which allows reduced clearances when maintenance activity is not expected between the conductors and the affected installation. Such installations as unpainted masonry structures (without windows, etc.) stainless steel tanks etc., were intended to be covered by this NOTE; antennas, because of their special problems, were not.

Footnote 1 to Table 234-1 was modified in the 1997 Edition to include the other items covered by the rule along with the covered conductors that were listed previously. This reduction in clearances only applies when people are not expected to pass between the covered facilities and the structure and, thus, has little application. However, it may be useful in certain alley situations or other close quarters near structures that are not expected to be painted or otherwise maintained from that side. It should be noted that energized facilities that are merely covered in accordance with Rule 230D, and do not meet the requirements to be considered as insulated, are not considered as safe for either line workers or the public to touch without deenergization or auxiliary insulating protection (see NESC Rule 443A2 and OSHA regulations).

NOTE: Footnote 1 to Table 234-1 only applies to certain horizontal clearances; it does not apply to any vertical clearances (see IR 527 issued 28 May 2002).

CAUTION: Users of NESC tables should note exactly to which rows, columns, or specific cells in the table that particular footnotes apply.

The 1997 Edition added Footnote 5 and Footnote 13 to Table 234-1 to clarify the treatment of guys near buildings and other installations. In addition, Footnote 14 was added to clarify that the vertical clearance applicable to roofs *not* accessible to pedestrians applies above railings, walls and parapets around roofs and balconies, regardless of the accessibility of the roof or balcony itself. Footnote 14 was revised in 2002 to allow lesser clearance over a wall or railing if access via outside stairway is available, since no one is expected to stand erect on the rail or wall to access the area by ladder.

CAUTION: This change recognizes that people are not expected to stand erect on such items, because they are normally too thin to provide appropriate balance. Some walls may be so wide as to effectively be a walkway—if so, they should be treated as to their accessibility. However, since OSHA generally requires all protection around such places, it should be indeed rare to find a wall that itself is considered to be accessible to pedestrians.

Table 234-1, Row 2a, was split into two rows to recognize the differences between areas accessible versus not accessible to pedestrian, and match similar treatment in Part 1 of Table 234-1.

Rule 234C2, which requires supply conductors and rigid energized parts not meeting the clearances of Table 234-1 to be guarded, was modified in 1997 to respecify guarding requirements for rigid live parts. Prior to 1997, a NOTE indicated that metal-clad cables were considered as guarded. In 1997, the term *metal-clad cables* was replaced at various places in the Code with the term *cables meeting Rule 230C1.* In effect, all cables meeting Rule 230C1 were then considered as guarded under this rule. However, metal-clad cables meet Rule 230C1. Upon further consideration, this language was corrected in the 2007 Edition (1) to return to the long used requirements and limit application so that only cables meeting Rule 230C1a are considered to be guarded and (2) to properly state the EXCEPTION language as part of the rule. NOTES are not allowed to contain requirements or permissions— NOTES are not part of the Code (see Rule 015F).

Rule 234C does not specifically cover clearances of overhead conductors from stored materials, areas where cranes or other special loaders may be used, and well sites. Specification of these clearances was intentionally omitted for several reasons, among which were the following: (1) any clearance requirement that provided for the "worst case" would, by definition, penalize the general case; (2) it is not the intention of the NESC to limit uses of rights-of-way when adequate safety precautions can be provided, by use of appropriate tools and equipment, by agreement or otherwise, under Rule 012; (3) OSHA covers approach distances for cranes from supply facilities; (4) the need to install or remove well casings occurs so seldom as to be generally within reason for the utility, upon reasonable notification, to sleeve (insulate) or otherwise protect supply facilities from accidental contact; and (5) it is not possible to predict every action of an unthinking landowner. See the previous discussion of IR 159 under Rule 232. Clearances to special installations should be determined by agreement and in accordance with Rule 012.

Rule 234C3 applies where attachment to a building is necessary for an entrance. It is not intended as an exemption to Rule 217B. Rule 234C3d (Rule 234C3c before 1993) specifies clearances of service drops to buildings. The basic vertical clearance required for service drops by Rule 234C3d(1) through the 2002 Edition was 2.45 m (8 ft) from the highest point of roofs or balconies over which they pass. This matched NEC requirements. The 2007 Edition increased the basic vertical clearance in Rule 234C3d(1) to 3.0 m (10 ft). Porches and attached decks were also added in 2007. The increase in basic clearance recognized that the activities expected in the areas covered by the rule are not unlike those expected in a yard or on a roof. The new values were increased to better match requirements for similar expected activities in Row 1b(2) of Table 234-1 and in Footnote 8 to Table 232-1. This can be difficult to achieve with some building configurations, particularly when the service drop must pass over a garage or carport (which are often installed after the original house). There are two EXCEPTIONS to the rule that apply when access to the roof or balcony is limited. The 2007 Edition also revised the additional EXCEPTIONS for clarity, with the former EXCEPTION 2 being rolled into EXCEPTION 1.

Note that the reduced vertical clearance requirement of 900 mm (3 ft) in EXCEPTION 1 to Rule 234C3d(1) (Rule 234C3c(1) in 1990 and earlier editions) changed from 900 mm (3 ft) to 600 mm (2 ft) in the 1990 Edition and then changed back to 900 mm (3 ft) in a Tentative Interim Amendment and in the 1993 Edition. The 900 mm (3ft) clearance matches the NEC and is not low enough to trip up a worker on the roof (see Figure H234C3d(1)). The change to 600 mm (2 ft) was inadvertent and was not identified until after the 1990 vote had been taken and the Code books printed. It should be noted that the NEC applies a further limit—requiring the roof to have a height/run slope of not less than one-third. This effectively prohibits use of the 900 (3 ft) EXCEPTION over flat roofs and shallow sloped roofs. On the other hand, the NESC limits application of this EXCEPTION to roofs not accessible to pedestrians, but the NEC does not. Where the service drop is installed and maintained by a utility, the NESC applies, not the NEC.

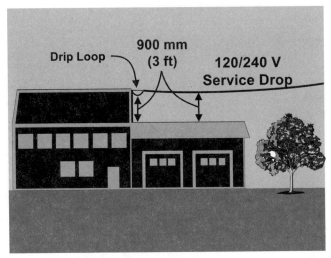

Figure H234C3d(1)
EXCEPTION where roof is not accessible to pedestrians

EXCEPTION 2 to Rule 234C3d(1) of the 2002 and prior Editions (now part of EXCEPTION 1) applies to service drops attached to through-the-roof masts; it was revised in the 1993 Edition to coordinate with concurrent NEC changes recommended by the NESC/NEC Coordination Working Group. Now up to 1.8 m (6 ft) of cable can overhang the roof of the 450 mm (18 in) clearance level (to allow re-roofing) to serve a through-the-roof weatherhead. NESC dropped the former limitations to being over the overhang portion of the roof, but the NEC has yet to do so. The former limitation effectively prohibited serving a through-the-roof mast by coming across a corner of the building and crossing a wall portion. The 1997 Edition revised the EXCEPTION editorially and added the drip loop into Rule 234C3d. Rule 234C3d(2) prohibits service drops within 900 mm (3 ft) of windows (except above window), doors, fire escapes, etc. EXCEPTION 2 to Rule 234C3d(2) does not require the full 900 mm (3 ft) clearance to windows *or portions of windows* that do not open (see IR 541 issued 16 December 2005).

For the house with a balcony over a carport, like that shown in Figure H234C3d(2), service drops cannot be placed in the shaded areas.

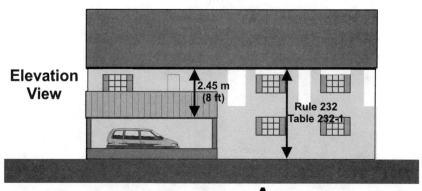

- Service Drops prohibited in shaded area
- Ground Clearance must also be met

A 2002 & Prior : 2.45 m (8ft)

A 2007 & Later : 3.0 m (10ft)

Figure H234C3d(2)
Potential locations for service drops on side of house

Rule 234C allows communication conductors and cables to be attached to buildings or other installations. Supply conductors and cables are *not* allowed to be attached to buildings or other installations, *unless* they are attached for the purpose of an entrance.

Prior to 1997, there was no stated prohibition against attaching a guy to a building or other installation. Albeit unusual, there is no prohibition against attaching a guy to a building foundation or other structural point, providing that adequate strength is available (see Rule 012C, Rule 217B, and Rule 261B). Clearances for grounded and ungrounded guys to buildings and other installations are given in Table 234-1.

Table 234-1 does not apply to clearances of lines from trees. Trees are not considered to be obstructions for clearances purposes. Vegetation clearances are covered by Rule 218 (see IR 537 issued 3 June 2004). Rule 217A4 requires supporting structures to be kept free from climbing hazards, such as vines.

234D. Clearance of Wires, Conductors, Cables, and Unguarded Rigid Live Parts From Bridges

The 1997 Edition revised the title, rule, and Table 234-2 to appropriately address communication clearances to bridges. Previously, the title and rule referred to supply lines, but the table contained requirements for open communication conductors.

The clearances given are designed to prevent contact of conductors with bridges by swinging in the wind or by sagging with ice or high temperature. They are also intended to provide adequate clearances (1) for painters and others who may have to work about ordinarily inaccessible parts of bridges and (2) others who occupy ordinarily accessible parts of bridges.

NOTE: Bridge span support members composed of completely enclosed, hollow segmental box girders that enclose electric supply and communication cables are generally considered to be tunnels as far as the utility facilities are concerned (see Part 3—*Underground*, particularly Rule 314B, Rule 323B, Rule 340B, and Rule 391A3).

Footnote 5 to Table 234-2 covers the situation where conductors passing under bridges are (1) adequately guarded against contact by unauthorized persons and (2) can be de-energized for maintenance of the bridge. In this case, the question of damage to persons is removed and the bridge assumes the characteristic of any other supporting structure. The additional increment of clearance (equal to one-half the final unloaded sag of the conductor at the point of clearance) was added to provide adequate clearance at every point, whether the crossing is made with or without attachment to the bridge. Other footnotes cover the situations where portions of the bridge move or overhang other thoroughfares.

During the major revisions of the 1977 Edition, the scope of this rule remained essentially unchanged. Table 234-2 replaced Table 5 of the Sixth Edition. The table was rearranged to correspond with the format of using voltage classifications as column headings that was established in that edition. The voltage classifications were modified to agree with the ranges found in tables elsewhere in the Code. An editorial gremlin caused a column heading in Table 234-2 to include neutrals; these are exempted by the EXCEPTION to Rule 234D1. Although the errata sheets for the 1977 Edition show the needed correction, the heading was inadvertently retained in the 1981 Edition. The 1984 Edition shows the correct headings. As in other rules of this section, wind displacement of the conductors (except guys, etc.) must be considered, even when the conductor is attached to the bridge (see Figure H234D). Also see the requirements for safety signs in Rule 217A1c when attaching conductors to bridges.

Rule 234D2 applies to current-collection systems using trolley poles. Pantographs are not covered in this rule. In the absence of specific requirements, please see Rule 012C.

Clearances requirements for rigid live parts to bridges were contained in Rule 286F prior to moving to Rule 234J in the 1990 revision. The 1990 Edition also included modification to some clearance requirements to meet the coordinated uniform clearance system. Table 234-2 was also modified in 1997 to include clearances of rigid live parts and to appropriately specify clearances for ungrounded guys.

Figure H234D

Attachment of conductors to a bridge

Note that Footnote 5 to Table 234-2 is not limited to trolley contact conductors; it applies to any conductors or cables near ordinarily inaccessible portions of bridges. Where the conductors can be de-energized for maintenance of the bridge, the clearances in Table 235-6 for conductors to surfaces of support arms plus one-half of the final unloaded sag of the conductor at the particular point may be used as the clearance (see IR 518 issued 13 September 1999). Generally such conductors are supported on short spans and have little movement in the wind. However, if the spans or longer or reduced tensions are used that yield greater sags and greater movement in the wind, Rule 012C may require greater clearances.

234E. Clearance of Wires, Conductors, or Cables Installed Over or Near Swimming Areas With No Wind Displacement

Rule 234E of the 1977 and later editions was necessitated by the increase in the number of outdoor public and private pools and the development of recreational areas near lakes and rivers. Clearances are given for three water recreation areas: (1) swimming pools; (2) beaches and waterways restricted to swimming; and (3) waterways subject to water skiing.

Spas, whirlpools, hot-tubs, jacuzzis, and similar installations that are not suitable for swimming are not considered as swimming pools and are not covered by Rule 234E. The clearance requirements of Rule 234E and Table 234-3 are based upon the expected use of pool vacuum *skimmer* poles and rescue poles and/or diving from platforms. The typical hot-tub type of installation is so small that it is normally not expected to be cleaned with a skimmer pole nor require the use of a rescue pole.

CAUTION: Where such installations are located adjacent to a swimming pool for which skimmer poles are used, it is not unusual for the skimmer pole to be moved over to clean out the hot-tub after cleaning the pool. As a result, Rule 012C may require pool clearances.

It is not practical to prevent electric supply lines from being over or near swimming areas. As population densities have increased, so have the demands for effective use of land; residential lots have grown smaller and supply-line rights-of-way have been increasingly used as general recreation areas. This rule covers clearances for overhead facilities over or near swimming areas; underground facilities are covered by other rules.

The formation of the new clearances over swimming areas in the 1977 Edition consumed a considerable amount of effort. Accident data was analyzed in detail to determine the errant actions involved and to develop practical countermeasures. Almost all of the accidents involved the use of long vacuum skimmer poles, with lightweight aluminum handles, used in cleaning debris from the water surface. The accidents generally involved either (1) transporting the pole from storage on a nearby fence or (2) using the skimmer close to the edge of the pool and thereby raising and backing the extended rear portion of the handle up into a conductor while only paying attention to the location of the skimmer end. After examination of common sizes of pools and their cleaning and rescue poles, the Clearances Subcommittee concluded that the clearances required by Table 234-3 and Rule 234E are practical to achieve and provide adequate safety.

The clearances for swimming pools allow for the normal use of skimmers and similar maintenance tools with a maximum length of 4.9 m (16 ft). The clearances to diving platforms, slides, towers, etc., recognize that the principal activity on or near these items is human movement, usually either a single person jumping or diving or one person sitting or standing on the shoulders of another.

The rule was revised in the 1984 Edition to recognize appropriate applicability of the rule to conductors and cables that generally pose only a mechanical problem. These cables are now all included in Table 234-3, which is normally applicable within 7.6 m (25 ft) of the edge of a swimming pool, but EXCEPTION 2 limits the applicability to areas within 3 m (10 ft) of the edge of a pool, diving platform, or diving tower. (Water slides and other fixed, pool-related structures were added to this list and to Row B or Table 234-3 in 2007.) (see Figure H234E) This change matches Rule 680-8 of the NEC.

The clearances were revised in the 1990 Edition to meet the requirements of the coordinated uniform clearance system and to recognize reports of interference of overhead service drops with use of rescue poles and skimmer poles. In the 1993 Edition, the requirement to provide the clearances under high wind conditions was deleted, since divers are not expected to use the boards during 50-mph winds.

Prior to the 1997 Edition, clearances for rigid live parts were not specified in Rule 234E. As a practical matter, the clearances for conductors were used for energized equipment bushings, service conductors were connected to them. The 1997 Edition specified clearances for rigid live parts in both Rule 234E and Table 234-3. In addition, Table 234-3 was modified in 1997 to specify clearances to ungrounded guys.

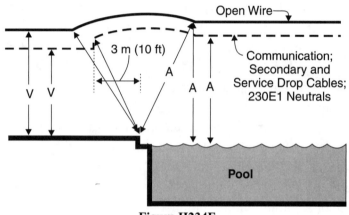

Figure H234E
Clearance above swimming areas

234F. Clearances of Wires, Conductors, Cables, and Rigid Live Parts From Grain Bins

(This rule was added in the 1990 Edition; previous Rules 234F–H were renumbered to 234G–I.)

The 1990 Edition culminated over a decade of effort in examining the expected operations around grain bins. The special Working Group of the NESC Clearances Subcommittee was aided by efforts of the NESC Committee of the Rural Electric Power Conference, the North Central Area REC Consulting Engineers Association; numerous farm-related organizations, Federated Rural Electric Insurance Corporation, U.S. Department of Agriculture, the American Society of Agricultural Engineers Task Force on Clearance of Power Lines, the Texas Farm Bureau Safety Department; and numerous individual utilities, individual engineers, and associations of utilities. The first proposal published for public comment was published in the 1985 Preprint for 1987 NESC revisions. Over 25 formal comments were received of such varying nature that the proposed rules were held out of the 1987 Edition for further study. This massive work effort culminated in the addition of clearances near grain bins to the 1990 Edition.

It should be noted that the Clearances Subcommittee intentionally limited the application of these rules to grain bins; however, the system used here may be useful under Rule 012 when determining appropriate clearances to similar facilities that are not specifically covered by this rule.

Rule 234F recognizes that there are two different problems associated with grain bins: (1) moving the loader into position and (2) using a probe to measure internal temperature or to sample the grain.

The first problem only occurs around bins that are loaded by portable augers or conveyers, but the temperature probe and sample probe problems may be present on any bin. As a result, all bins are required to have vertical clearances above probe and fill ports sufficient to allow the expected temperature-measurement or grain sampling activity; otherwise, those bins that are loaded with permanent loading systems are generally treated as a normal building (see NESC Figure 234-4(a)).

It is the bins to be loaded by portable loading system to which the special provisions of this rule are addressed. NESC Rule 234F2 and Figure 234-4 (b) require the wire exclusion zone to be maintained on all loading sides at a height above grade equal to the probe clearance (bin height plus the 5.5 m (18 ft) probe clearance above the highest port) for a distance *H* equal to the same amount. Only at that point can the limit of the wire exclusion zone start to drop; the allowed drop is at a slope of not more than 1 ft drop for every 1.5 ft run from the point where *H* ends (see NESC Figure 234-4(b)). This gentle drop slope provides room for the lower portions of longer augers used on taller bins to be maneuvered to serve shorter bins. Note that all sides are considered to be loading sides, unless natural terrain features or an agreement prevents loading on one or more sides of grain bins. The line depicted in Figure H234F does not meet the requirements of the rule.

Figure H234F
Grain bin clearance less than required by Rule 234F

The temperature-measurement/sampling probe clearance is based upon the expected use of a probe inserted vertically from the top probe port down into the grain with the aid of a sectional rod system. It is recognized that the use of taller rods is so unwieldy as to be a physical balance safety hazard in itself, due to the precarious footing on the top of a grain bin. Such use is not normally encountered nor reasonably anticipated. There are actually two probe-related problems. The insertion and removal of the probe is covered by the 18-ft dome over the probe ports. The second problem, dropping the probe and having it bridge across from the bin to an energized conductor is covered by the requirement for a horizontal clearance of not less than 4.6 m (15 ft) between the bin and open supply conductors. The 2007 Edition exempted an effectively grounded neutral beside a grain bin loaded by permanent augers, elevators, or conveyors from the 15-ft horizontal clearance—this matched the previous EXCEPTION for neutrals near bins loaded by portable augers or conveyors.

It is also recognized that some bins or rows of bins may be expected to be loaded from any side, while others may be expected to be loaded from only one side or only from an area that is restricted by terrain features or agreement.

These clearances recognize that portable grain loaders are not intended to be transported in the upraised mode; they are too susceptible to overturning and are a hazard to themselves as well as to personnel and equipment along the transportation route. However, there is limited movement expected on the loading side of the bin(s) to maneuver the drop tube into place above the center of the bin. These clearances provide for that required maneuvering room, as well as the room required to move the loader to an adjacent bin in a row of bins. The requirement to provide these clearances under the high-wind conditions was deleted in the 1993 Edition, since grain loading is not expected to occur in 50-mph winds.

Rule 234F allows agreement on a nonloading side. At least one nonloading side is required, if the service drop to supply power for the fans and motors of the grain bin is desired to be aerial. If there is no nonloading side, the service must be underground.

The 1997 Edition added a new Figure 234-4a and revised the rule to require a 5.5 m (18 ft) clearance above *each* probe port of a grain bin served by a permanent auger or conveyor system. Previously, the requirement for 5.5 m (18 ft) above the *highest* probe port had been confusing to some who thought that the intention was to require this level to be maintained horizontally over the whole bin (as required for a bin loaded by a portable auger or conveyor). The new figure depicts the intended clearances. Previous Figure 234-3 was renumbered to Figure 234-4b in 1997. The 2002 Edition split part of the rule into subrules 1 and 2 for clarity.

The clearances of NESC Figure 234-4(b) are required around grain bins that are filled with portable augers. The 5.5 m (18 ft) clearance required above the top probe port is the same as that used for all probe ports on grain bins filled by permanent augers. However, the ground clearance must be maintained at that level on all loading sides of a bin filled by a portable auger for a horizontal distance equal to the height of the bin plus 5.5m (18 ft) from the side of the bin to provide vertical room to maneuver the portable auger up to the bin and remove it from the bin. Since the auger extends upward from the ground at an angle, the clearance envelope has a slope of only a 1 unit drop in a 1.5 unit run. This provides room for the rear of a long auger to be used on a short bin. If the ground slopes, the portable auger will follow the slope of the ground and, thus, the vertical clearance should follow the slope of the ground (see Figure H-234F.)

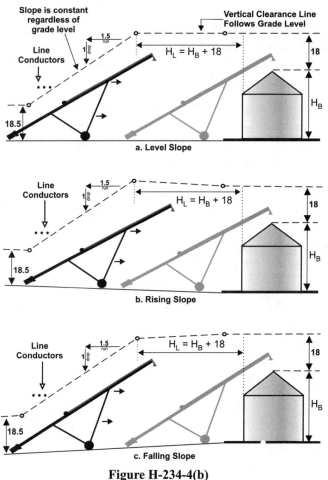

Figure H-234-4(b)
Vertical clearance above flat or sloping ground
near grain bin filled with portable augers

234G. Additional Clearances

(This rule was added in the 1977 Edition as Rule 234F and moved to Rule 234G in the 1990 Edition when the new clearances to grain bins were placed at 234F.)

The additional clearance requirements are consistent with those applicable to Rule 232 and Rule 233; see Rule 232B. Figure H234-1 and Figure H234-2 show the application of these increased clearances.

The requirements of Rule 234G3 (Rule 234F1c of the 1977–1987 Editions) to consider electrostatic effects may be met by a number of methods, including, but not limited to, grounding the object; of course, any related electrolysis effects on the integrity of grounding systems or grounded objects should be considered when choosing methods of limiting electrostatic effects.

The sag adders of Rule 234F2 (1987 and prior editions) apply only to the vertical clearances, not to the horizontal clearances. These requirements were deleted when maximum sag conditions were specified in the 1990 Edition instead of the previous 15 °C (60 °F) conditions. In general, the effects of extreme vertical loadings will not occur simultaneously with extreme wind loadings (see Figure 233-2).

The EXCEPTIONS to Rule 234F2c(1) and Rule 234F2d(1) of the 1981–1984 Editions (Rule 234F2c(1) and Rule 234F2d(2) of the 1987 Edition) were added in the 1981 Edition to recognize that people do not normally use swimming areas during icing conditions.

CAUTION: The sag adders for icing conditions in former Rule 234F2 were useful in meeting Rule 012C for conductors crossing above winter spas where outdoor pools were heated.

The electrostatic limitation of 5 mA rms is a steady-state value. The application of the electrostatic requirements of Rule 234G3 to direct-current systems was deleted in the 1993 Edition.

234H. Alternate Clearances for Voltages Exceeding 98 kV Alternating Current to Ground or 139 kV Direct Current to Ground

(This rule was added in the 1977 Edition as Rule 234G and was moved to 234H in the 1990 Edition.)

The alternate clearances are consistent with those applicable to Rule 232 and Rule 233 (see Rule 232D). Some of the values shown in Table 234-4 changed slightly in 2007 as a result of the coordination of decimal places and rounding of calculation results.

234I. Clearance of Wires, Conductors, and Cables to Rail Cars

(This rule was added in the 1977 Edition.)

Rule 234I establishes clearances to rail cars from conductors that parallel railroad tracks. Assumptions were made concerning the maximum height and width of cars. Clearances are based on the largest car normally in current service. The 6.7 m (22 ft) dimension coordinates with a similar dimension in Rule 231C based upon bridge clearances. However, notice that the actual height of short cars is to be subtracted from 6.1 m (20 ft), not 6.7 m (22 ft), when computing reduced clearances above mining railways and other limited-height railway environments. The long-time value of 6.1 m (20 ft) was changed to 6.7 m (22 ft) in the 1990 Edition as a part of the general coordination effort. However, after further review, a Tentative Interim Amendment was passed, and the 1993 Edition was accordingly revised to go back to the more conservative 6.1 m (20 ft) value that has shown good service.

234J. Clearance of Equipment Mounted on Supporting Structures

(Rule 286F of the 1977–1987 Editions was revised and moved in 1990.)

In the general revisions of clearances (and, especially, clearances to buildings) of the 1977 Edition, Rule 286F was created out of previous portions of Rule 286 and new requirements complementing changes in other rules. Clearances of rigid live parts were then specified here. From 1977 through 1987, Rule 286F had required rigid live parts to have the same clearances as conductors of the same voltage. When Section 28 was removed in the 1990 Editions, former Rule 286F was revised and moved to 234J. At the same time, Table 234-1 and Table 234-2 were revised to specify clearances of rigid live parts, based upon the coordinated clearance system (see NESC Appendix A).

235. Clearance for Wires, Conductors, or Cables Carried on the Same Supporting Structure

(This section was completely renumbered, revised, reworded, and expanded in the 1977 Edition. Various parts of Section 238 concerning separation of crossarms and separation of conductors on crossarms were moved to Section 235 and now concern themselves with clearances of conductors.)

As in other areas of the Code, the 2007 Edition revised the applicable clearances tables of Rule 235 to clearly specify the expected classification of effectively grounded neutrals. Additional revisions were made (1) to aid users in selecting the correct clearance requirements between specified utilities and (2) to reflect the general coordination of decimal values and rounding in tables and calculations.

Before considering vertical clearance requirements at midspan, care should be taken to determine what conditions apply to each conductor for the edition of interest. In some cases, both wires, conductors, or cables will be at the same temperature. In others, they may be the same ambient air, but the upper one will be influenced by ice or thermal loading.

The horizontal clearances required by this rule may be intended to be measured under the often used NESC conductor conditions: 15 °C (60 °F) conductor temperature at final unloaded sag. The conditions are different in various editions and the effects of these conditions can be dramatic. The effects of both storm loading and long-term creep are to be considered in development of final unloaded sags to be used in determining both vertical and horizontal clearance requirements.

The requirements of Rule 235 are generally intended to limit the opportunity for midspan contact between conductors and cables carried on the same supporting structures. Thus, the clearances at the structure are generally to the surface of a conductor or cable, not including ties, but may also need to include the effect of the increased diameter of weights or other attachments out in the span. This rule is also coordinated with the climbing and working space requirements of Rules 236–239 so that the clearances required at the structure in Rule 235 match or complement those of Rules 236–239.

The clearances of Rule 235 are from the closest surface of the wires, conductors, or cables of interest; they do not apply to support brackets. In contrast to Rule 235, Rule 238 *does* consider the mounting brackets in its clearance requirements. It should be noted that the provisions of Rule 235 and Rule 238 are independent; an installation must comply with both rules.

> **Example:** Rule 235C requires 750 mm (30 in) of clearance between a neutral meeting Rule 230E1 and a communication cable, and Rule 238B requires 750 mm (30 in) between the neutral bracket and the communication conductor, the former clearance may have to be increased in order to meet the latter requirement.

Care should be taken when comparing the clearance required by Rule 235 in different editions of the Code. For example, the 1984 Edition of Table 235-5 differed in several respects from the requirements of the 1987 and later editions.

> **Example:** Consider a supply line with 115 kV overbuilding a 12.5/7.2 kV grounded-wye distribution circuit.
>
> (1) *Table 235-5 (1984)*: the basic requirement was 40 inches; since the upper circuit exceeded 50 kV, a voltage adder of
> (115 x 1.05/1.732 − 50 = 19.72 kV)(0.4 in/kV) = 7.89 inches was required. Total clearance is 47.89 in.

(2) *Table 235-5 (1987):* the basic requirement is 16 inches. The voltage adder of (the phasor difference voltage (see Rule 235A3) less 8.7 kV) times 0.4 in/kV
= ([115 x 1.05/1.732= 69.72] + 7.2 – 8.7 = 68.22 kV)(0.4 in/kV)
= 27.29 in. Total clearance is 43.29 in.

235A. Application of Rule

(The individual parts of Rule 235A of the Sixth and prior editions were renumbered in the 1977 Edition as follows: A1—235A; A2—235B; A3—235E; A4—235G; and A5—235F.)

Rule 235A1—*Multiconductor Wires or Cables* is repeated from the Sixth Edition. Unfortunately, the 1977 and later editions contain a typographical error. The rule should refer to "whether single or *grouped*," not "whether single or *grounded*." Rule 235A2 is the same as in previous editions, but Rule 235A3 is not. The 1977 Edition clearly indicated the difference between "clearance" and "spacing" in Rule 230B; as a result, Rule 235A1(c) of the Sixth Edition was dropped.

Rule 235A3 was added in the 1977 Edition to indicate that the maximum voltage potential between the conductors involved was to be used to determine the clearance requirement; the wording was clarified in the 1981 and 1984 Editions. This rule is used only where a subrule or table in Rule 235 does not specify phase-to-ground voltage or voltage between conductors. Note the greater of phase-to-ground or phasor difference is required when this rule applies (see Figure H235A3). Rule 235A3 was revised for clarity in the 1990 Edition and the NOTE concerning using a phasor relationship of 180° if the actual relationship is unknown was added for information.

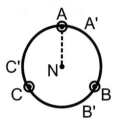

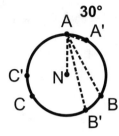

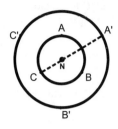

| a. | Two circuits of same voltage exactly in phase:
AN > AA';
AB' = AB' > AN | b. | Two circuits of same voltage 30° out of phase:
AN > AA';
AB' > AB > AN | c. | Two circuits of different voltages and unknown phasor relationship:
A'N > AN';
phasor difference voltage between any conductor of one circuit and any conductor of another circuit = A'N + CN |

Figure H235A3
Relative voltages between a conductor of one supply circuit and a
conductor of another supply circuit on the same supporting structure

235B. Horizontal Clearance Between Line Conductors

(This rule is essentially Rule 235A2 of the Sixth and prior editions.)

In the 1977 Edition, EXCEPTION 2 of Rule 235A2(a) of the Sixth Edition was deleted. A new EXCEPTION was added to permit lesser clearances for higher-voltage circuits where the maximum switching-surge factor was known.

In the 1977 Edition, the term *separations* was changed to *clearances* for clarity. The values specified in Table 235-1 for the clearances apply where the spans are short and the sags small. Where the sags are greater, increased clearances are required at the structures to provide sufficient clearances in the span when the conductors swing in opposition to each other (see Table 235-2 and Table 235-3).

Rule 235B of the 1990 Edition clarified that the use of conductor-to-conductor voltages was intended except for railway feeders. Where the conductors operate at voltages in excess of 8700, the clearance is increased by an increment that is determined by the flashover distance in air; this distance is not directly proportional to the voltage but the increment has been made proportional, in order to simplify computations and to provide a working value. Over the years, this value has proven to be a practical increment, especially since an alternate clearance requirement was allowed to be applied under certain conditions for the larger voltages in the 1977 and later editions.

The conductor clearances determined according to sags, which are shown in Table 235-2 and Table 235-3, are intended to provide sufficient space for workers on structures and to prevent swinging contracts between the conductors (except for some of the very smallest conductors, which swing about more in the wind because of their relatively large sags and light weight; with today's wire sizes, this is rarely a problem). These tables show specific values for the formula in the rule. It continues to be practical to adhere to a comparatively simple rule for clearances and to make clearances depend on voltage, wire size, and sag. Experience has shown that, where followed, these clearances do provide appropriate horizontal clearances based upon the amount of sag of the conductors. Over 50 times from 1978 to 1995 these clearances tables were challenged by utility personnel who said that (1) they installed conductors by these tables (235-1, 2 or 3) and (2) they had wires banging together in storms. Each time, the actual span was shown to have been installed with excess sag or, occasionally, the wrong table had been used (i.e., they stopped when they saw the title of Table 235-1 and didn't read the rule and get to Table 235-2 and Table 235-3).

This problem points out two well-known facts. First, conductors tend to look like they have less sag than they actually have, so the natural tendency is to leave the wire too slack, unless sag or tension is measured. Second, if the wire is installed the same on a cold or hot day as it should be on a 15 °C (60 °F) "blue bird" day, the result can be catastrophic. Pulling too tight on a hot day causes structural problems on cold days. Pulling too loose on a cold day produces too much sag on a hot day. Both problems are particularly acute with compact construction designs. Such designs require shorter spans and/or higher tensions to keep the conductors from banging together in the wind. *Wires, conductors, and cables should be checked at installation to assure appropriate sag and tension.*

In the 2007 Edition, Rule 235B1b was revised to clarify that the horizontal clearances of this rule are to be measured from the surfaces of the conductors themselves (not the surfaces of armor rods, tie wires, or other fasteners), since the issue is prevention of midspan contact between swinging conductors. The specification of the horizontal clearance calculations for Rule 235B 1b were also revised to clearly indicate that the voltage to be used is the voltage in kilovolts between the conductors involved—the voltage may be phase-to-phase, phase-to-ground (when a neutral is involved), or some different phasor-difference voltage (when two different circuits are involved). In addition, the layout was changed to make it more obvious to the reader that the voltage adder above 50 kV of Rule 235B1b(4) applies to both the customary unit calculations and the SI unit calculations. In addition, the values in Tables 235-2 and 235-3 were revised in 2007 to reflect the use of the coordinated decimal and rounding system. As of 2007, the clearance calculations are rounded up; the inch tables (and the results of inch calculations) all now use whole inches.

When suspension insulators are used and are not restrained from motion, such conditions as changes in temperature and ice loading can cause the free end of the insulator to move in the direction of the line. A movement of only a few inches of the free end of the insulator can, in some instances, increase the sag of the conductor by as many feet. The minimum clearances of conductors attached to suspension insulators are those clearances at the extreme position to which the insulator is displaced.

In the Fifth Edition, one string of insulators was required to be displaced at 45° because a 26.8 m/s (60 mph) wind blowing at right angles to the line could, under some conditions of loading, swing the insulator

45° from the vertical position. The values in Table 6 were required to be met when suspension insulators are used and are displaced 45°. In the Sixth Edition, the swingout requirement was reduced to 30°. In the 1977 and later editions, whatever displacement occurs under 290 Pa (6 lb/ft²) wind with the conductor at 15 °C (60 °F) and final sag is required. This maximum design swing angle could be 30° for large conductors and 60° or more for small conductors under comparable conditions.

The 2007 Edition added the statement at this location, as it had earlier in others, that trees are not considered to shelter a line from horizontal wind loading.

The language of Rule 235B2 does not indicate whether *loaded* or *unloaded* sag is intended. Since this rule is intended to link to Rule 235B1, which uses final *unloaded*, it would appear that Rule 235B2 is also intended to be used with unloaded sag.

The general discussions of Rule 232B and Rule 232D apply to this rule for additional clearances above 50 kV. No value is specified for clearances between line conductors of the same circuit rated above 50 kV.

As in other rules, alternate clearances are allowed under certain conditions.

235C. Vertical Clearance Between Line Conductors

(Rule 235A4 of the Sixth and prior editions was moved to Rule 235G in the 1977 Edition. Rule 238A and Rule 238B of the Sixth Edition were combined and moved to Rule 235C in the 1977 Edition.)

Rule 238B of the Sixth Edition specified a required vertical separation between conductors indirectly by reference to the required separation between horizontal crossarms in Rule 238A. The required vertical clearance between conductors on the same supporting structure is now specified in Table 235-5 and does not specify how they might be supported.

In 2007, Rule 235C1 was respecified to clearly direct the user to the appropriate rules and tables for vertical clearances between different kinds of utility conductors and cables. In particular, this was done to direct users to Rule 235H for clearances between communication cables that are all located within the communication space. All other clearance specifications are located in Table 235-5 of Rule 235C, including supply circuit to supply circuit, supply circuit to communication circuit located in either the supply space or communication space, and communication circuits located in the supply space. In addition to adjusting required table values consistent with the 2007 coordination of decimal points and rounding, EXAMPLES were revised and augmented. As in other areas of the Code, the 2007 Edition clearly specified the expected classification of effectively grounded neutrals in the applicable clearances tables of Rule 235.

No value is specified for clearances between line conductors of the same circuit rated above 50 kV; thus Rule 012C applies. Flashover characteristics of lines above 50 kV vary significantly with configuration and with various base impulse level (BIL) control methods. Therefore, it is not appropriate for the NESC to specify clearances between conductors of the same circuit as the voltage exceeds 50 kV. The 50-kV clearance would be the base clearance plus the appropriate voltage adders.

There is often a question about what voltage to use for high-voltage transmission lines. Such lines are almost always wye-connected at the source, with the center point grounded with a high-impedance connection to limit ground fault currents. The maximum ground fault voltage is phase-to-ground voltage, not phase-to-phase. Note that Table 235 columns and rows use phase-to-ground voltages, but the calculation within the table uses phasor difference voltages (see Rule 235A3).

CAUTION: *The clearances and voltage adders of Rule 235C and Table 235-5 do apply to clearances between circuits of any voltage and any other circuit, regardless of ownership.*

The general discussions of Rules 232B and 232D apply to this rule.

The basic vertical clearance required between high-voltage conductors that are operated by different utilities is greater than the vertical clearance required between those operated by the same utility. The lack of familiarity of the employees of one utility with the property of another often necessitates a greater clearance.

It may be necessary to increase these vertical clearances under some conditions, such as when conductors on different support arms have materially different sag increases under load or high temperatures. The values given in Table 235-5 (Table 11 of the Sixth and prior editions) are minimum values, except as covered in the NOTES to the table.

Where supply conductors of the same circuits are arranged vertically on separate crossarms, the vertical clearances are determined by the highest voltage concerned or, after 1987, by the *greater of* the phase-to-ground or phasor-difference voltage (see the discussion of Rule 235A3).

Although Table 235-5 requires, in some cases, a greater vertical clearance between two conductors in different consecutive voltage classifications than between two conductors of the higher voltage classification, it should not be interpreted as applying to the condition shown in Figure H235-1, where the conductors of different voltages are on opposite sides of the pole. In this arrangement, the vertical clearance is that for the higher voltage.

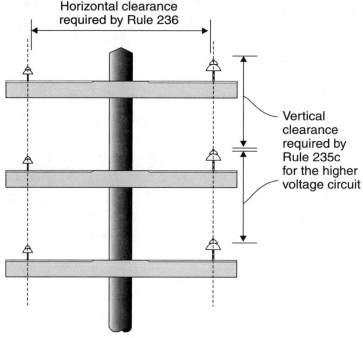

Figure H235-1
Vertical arrangement of circuits

On joint-use structures, a *communication worker safety zone* of 1 m (40 in) between communication and supply conductors of up to (1) 8700 V to ground for effectively grounded circuits or (2) 8700 V between conductors for other circuits is generally considered an appropriate value. The *communication worker safety zone* terminology has been in long use and was codified in the 2002 Edition. The communication worker safety zone is only needed if the communication utility chooses to use communication work rules and equipment. Experience has shown that, with span lengths of 45 m (150 ft) or less, such as are commonly found in urban joint-use construction, a 1 m (40 in) clearance at the structure will generally minimize the possibility of accidental contacts between the usual types of supply conductors and communication cables in the spans, even when the supply conductors are loaded with ice. This clearance is also generally sufficient to limit contact in situations where ice may fall or be jarred off communication cables in the lower position while the supply conductors are still under load. Such clearance also provides a clear working space between the two types of facilities so that (1) line workers working on supply wires at about waist level will have clear leg room below such wires and (2) communications workers will be provided with clear headroom while working on their facilities. Increased clearances are required with increased voltage.

Experience indicates that adequate clearance at the supports is a fundamental requirement for safety where joint-use construction is employed. While the rules provide for a minimum clearance of 1 m (40 in), greater clearances are required where spans exceed 45 m (150 ft) in length and for higher voltages. For application of Rule 235C2a, the calculation of voltage is intended to require the two circuits to be considered as being 180° out of phase, as in all similar calculations in the Code.

Where direct-current feeder circuits of voltages in excess of 750 V to ground are installed above communication conductors, particular attention should be given to the sags. Because of their size and weight, it is somewhat difficult to deadend them under some conditions and they are often given large sags. Consequently, the vertical clearance between these trolley feeders and communication conductors at the supports should be increased over what is usually provided for supply conductors of equal voltage.

EXCEPTION 2 of Rule 235C1 was added in the Sixth Edition solely to encourage the use of common crossing poles for communication service drops crossing under supply lines. EXCEPTION 2 applies only where a communication drop from one *line* crosses under an effectively grounded supply neutral of another line and is attached to the structure of the other line. It was intended to recognize that many existing supply lines built solely for supply facilities would not have sufficient height to allow both the normal supply/communication clearances and the required ground clearances at the same time. It was concluded that, since multigrounded neutrals do not ordinarily represent a safety hazard, and since relatively few operations on such service drops would be required by communications workers, the greater safety of a joint-crossing pole justified the reduced clearance allowed in this special instance. EXCEPTION 2 does not apply to joint-use or colinear construction. EXCEPTION 3 was added in the 1981 Edition.

EXCEPTION 3 of Rule 235C1 was added in the 1981 Edition to reflect appropriate standard practice.

The 1981 Edition modified Rule 235C3 to show that it applied when one or both of the circuits exceeds 98 kV to ground.

Table 235-5 was extensively revised in the 1987 Edition. Phase-to-ground voltage values are normally used in the column and row headings to enter the table. However, where a calculation is required within the table, Rule 235A3 applies and the greater of phasor difference voltage or phase-to-ground voltage is used. This recognizes that the worst case for conductors of similar voltage and phase relationships may be when one line is turned off and grounded for maintenance.

The vertical clearances of Table 235-5 are from the horizontal plane of the lowest surface of the upper conductor at its attachment point. This is a "square box" concept; vertical clearances are intended to be exactly that; they are not diagonal clearances (see Rule 235D).

A new EXCEPTION under Rule 235C2b(1)(a) was added in the 1987 Edition that allows neutrals meeting Rule 230E1 to be attached with a clearance from communication of 750 mm (30 in) at the structure *if* it maintains a clearance from communication of 300 mm (12 in) or more at all points in the span. This change was coordinated with Rule 238. The requirement that the neutral be bonded with the communication messenger was added in the 1990 Edition. New Rule 230F of the 1990 Edition allows certain fiber-optic supply cables to be treated in the same manner. If the fiber-optic supply cable meets Rule 230F1b, bonding of the messenger to the communication messengers is not required. The 2007 Edition augmented EXCEPTION 1 of Rule 235C2b(1)(a) to add the items allowed by Rule 230F to be treated the same as a neutral meeting Rule 230E1 for clearances purposes into EXCEPTION 1 for easier use. Now fiber-optic cables meeting either

Rule 230F1a or 230F1b, insulated communication cables located in the supply space and supported on an effectively grounded messenger are shown directly in EXCEPTON 1.

Rule 235C2b(1) is the so-called *75% rule*: it requires midspan clearances to be not less than 75% of that required at the structure at any time. Thus, for wires, conductors, or cables of greatly differing sag characteristics or electrical loadings, significant increases in vertical clearances are required at the structure to keep them apart at midspan. For example, if Figure H235C2b represents the closest approach condition when the supply secondary has a maximum sag of 1270 mm (50 in) and the communication below has a minimum sag of 254 mm (10 in), the clearance at the pole cannot be less than 760 mm (30 in) (i.e., 75% of value in Table 235-5 for clearance at the structure) plus 1.02 m (40 in) (i.e., difference in sags) = 1.78 m (70 in) at the pole. Because this assumes no errors in stringing sags, designers would typically add an appropriate amount to account for such errors.

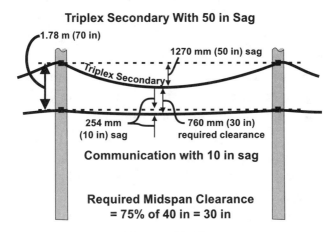

Triplex Secondary With 50 in Sag

Figure H235C2b
Clearance at pole based upon closes-approach midspan clearance

A requirement was added in the 1990 Edition to consider ice loading in Rule 235C2b(1). Both summer and winter sag mismatches must be checked. For a check of the potential midspan mismatch in sags during the winter (in icing areas), the upper conductor is assumed to be loaded with the radial ice required by Rule 250B for the loading district, while the lower conductor has dropped its ice and has no electrical loading. Both are assumed to be in the same ambient air (temperature). It should be noted that, if the upper line has enough electrical loading to heat the conductor up to 0 °C (32 °F) while covered with ice during a period of –18 °C (0 °F), –9 °C (15 °F), or similar ambient air temperature below freezing, the greatest mismatch will occur with the upper wire electrically loaded with temperature at 0 °C (32 °F) about to melt the ice off, with the ambient air temperature (and, thus, the lower conductor) at –18 °C (0 °F), –9 °C (15 °F) or other applicable temperature.

This requirement was added because of problems in much of the upper east coast, portions of the Rocky Mountains, and other areas which have sunny days following ice storms. In many areas, the combination of reflected solar energy and radiant energy from the earth will often melt ice off the lower cables and conductors before the upper ones. If the clearances do not plan for this, the conductors can touch (leading to burndown after repeated contacts over time) and cable lashing wires can be damaged enough to drop cables. Depending upon the relative conductor sizes and span lengths, it only takes from 2.5 mm (0.1 in) to 7.5 mm (0.3 in) difference in radial ice thickness to place conductors at the same level that are installed at the vertical values of Table 235-5 without considering icing differentials.

An EXCEPTION to Rule 235C3b(1), including this requirement to consider ice on the upper conductor and not on the lower conductor, was added in the 1997 Edition. The EXCEPTION applies only to conductors of the same size, type, sag, tension, and ownership. This EXCEPTION only works where significant differentials in ice do not occur or where the original vertical clearance was 1.2 m (4.0 ft) or more, not at the 410 mm (16 in) value of Table 235-5. The 2002 Edition revised the rule for clarity creating a new Rule 235C2b(1)(c) and modifying the EXCEPTION so that it does not apply in areas that experience differentials in ice between conductors.

The 2007 Edition modified the sag temperature conditions contained in Rule 235C2b(1)(c) for the summer sag check. Previously, the upper conductor was required to be at maximum conductor operating temperature under electrical loading and the lower conductor at the same ambient air conditions (air temperature, insolation, and wind speed) without electrical loading. Whereas Rules 232 and 234 require the use of a minimum temperature of 50 °C (120 °F) for the maximum operating temperature even if the maximum operating temperature could not exceed that value, Rule 235 did not do so prior to 2007. As of 2007, the same 50 °C (120 °F) becomes the base floor for maximum operating temperature for the *upper* conductor in Rule 235. This change reflects common usage in the industry; many electric supply utilities have traditionally used 50 °C (120 °F) in this calculation to assure that appropriate clearance would be installed to allow for a small amount of electrical heating due to splitting of single-phase transformer return current or imbalanced three-phase loads between the earth and the neutral between transformer locations or other grounding points. Given that communication messengers bonded to the effectively grounded supply neutral share that load to some extent (steel messengers have a higher impedance and, thus, will carry a lesser portion of the neutral current—but steel heats more than copper or aluminum for the same level of current), many utilities have also traditionally used 50 °C (120 °F) for the minimum summer design condition temperature for a communication cable on a messenger in the upper position. Of course, in some areas of the southwest, such cables may exceed 50 °C (120 °F) without electrical loading on hot summer days.

In the 1993 Edition, Table 235-5 was revised to be consistent with the changes to Rule 224 and 230F. Specifically, the requirements for the location of communication cables in the supply space or the communication space on a joint-use pole is predicated upon the qualifications of those working upon the cables and the limits on the voltage that might be present. What kind of signal or data is carried is not a safety issue.

Note that a fully dielectric fiber-optic cable carried on a nonmetallic messenger is considered as a supply neutral meeting Rule 230E1 (if located in the supply space) or an ordinary communication cable (if located in the communication space). Such cables must be located either in the supply space or the communication space, *not* in the safety zone between the two spaces.

Footnote 9 of the 2002 Edition of Table 235-5 was added as Footnote 10 in the 1993 Edition to recognize the lack of voltage potential between neutral conductors meeting Rule 230E1 and effectively grounded communication messengers located in the supply space. Electrically, a grounded communication messenger is part of the supply neutral and requires no clearance thereto. While a 900 mm (1 ft) clearance is often used to limit the opportunity for a lashing machine to damage the neutral as it spirals around the communication cable and messenger, no clearance to a supply neutral is specified when a communication cable or grounded messenger is located in the supply space.

Footnote 10 in the 2002 Edition of Table 235-5 was added as Footnote 11 in 1997 to allow entirely dielectric fiber-optic supply cables (i.e., located in the supply space) meeting Rule 230F1b to have no specified clearance to supply cables and conductors. This allows entirely dielectric fiber-optic supply cables to get close to supply cables and conductors, with less clearance than would otherwise be required. However, this is not intended to allow them to be so close as to interfere with each other. Entirely dielectric fiber-optic cable meeting Rule 230F1b can be "wrapped around" or can be part of an energized supply conductor. While Footnote 10 states that no clearance is specified between such fiber-optic cables and supply conductors, the intent of the rules is that they should be either:

(1) cabled together or otherwise constructed without separation, or

(2) separately supported far enough apart so as to not physically contact each other in the span during expected wind and sag conditions.

As a practical matter, they should not be installed closer than would be allowed by Rule 235G. If lesser clearances are desired, consideration should be given to directly attaching the fiber-optic supply cable to the supply cable or conductor, as allowed under Rule 230F1d.

Note that if the communication cable wire is to be located in the communication space, Rule 238 and Rule 235C2b(1)(a) EXCEPTION allows a clearance between the supply neutral and the communication cable of 750 mm (30 in) at the pole and a 300 mm (12 in) in the span if the communication messenger is bonded to the supply method. Figure H235C shows the basic clearances from Table 235-5 and the EXCEP-TION in rule 235C23b(1)(a). Two locations are shown for fiber-optic supply cables (FOSC). Fiber-optic cables generally have much less sag than supply secondary conductors or cables. Thus FOSC Position B is deliberately sagged in with greater sag than normal to match the secondary sag and maintain the minimum clearances of Rule 235G to prevent mechanical contact. Often, heavy messengers are used and/or weights are added to achieve the desired sag. From a sag standpoint, FOSC Position B is usually better, but is often less desirable because the FOSC workers (using supply work rules to meet Rule 224A) must climb past the supply secondary.

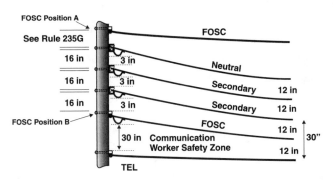

Figure H235C
Two possible positions of fiber-optic cable in the supply space

In 2002, Footnote 2 of Table 235-5 was deleted (1) because it is not necessary with the restrictions now placed on communications installed in the supply space by Rule 224A and Rule 235 and (2) for consistency with recent changes in other rules that deleted clearance differentials based on which was above the other.

Footnote 5 of Table 235-5 was modified in 2002 (Footnote 6 of 1997 and prior Editions) to add entirely dielectric fiber-optic supply cables to the list to save the code user from having to go back to Rule 230F1b. The remaining communication items in the supply space allowed by Rule 230F to be considered as a neutral (meeting Rule 230E1 for clearances purposes) were added into Footnote 5 in 2007. As in the rule calling out Table 235-5, Footnote 5 also specifies that bonding is required between the neutral or effectively grounded messenger of all specified items in the supply space and the communication messengers in the communication space for the reduced clearance of 30 in to be allowed—bonding is not required for entirely dielectric messengers in entirely dielectric cables meeting Rule 230F1b.

When sample calculations were added to Rule 235C2b(1)(b) in 2007, the sample calculations previously in Footnotes 7 and 8 of Table 235-5 were removed.

In 2007, the former values of 0.41 m (16 in) in the last column and last row of Table 235-5 (up to 50 kV over up to 50 kV with different owners) were increased to 1.00 m (40 in) to match other treatments in the table.

Rule 235C4 was added in the 2002 Edition to clearly define and name the "communication worker safety zone" between communication and supply facilities. The communication worker safety zone has always been required if communication is not placed in the supply space under Rule 224A and Part 4, and the name has been in common use, but its name now appears in the code. If communication utilities choose (1) to equip their workers with insulated buckets and insulating gloves, sleeves, tools, etc., and (2) to train their workers to use supply work methods, the communication can be installed in the supply space under Rules 224A and 235, and no separate communication space is required. However, if communication utilities choose to use normal noninsulated communication equipment and communication work methods, a separate communication space is required—and it must be separated from the supply space by a communication worker safety zone to provide head room for communication workers.

235D. Diagonal Clearance Between Line Wires, Conductors, and Cables Located at Different Levels on the Same Supporting Structure

(Rule 238C of the Sixth Edition was moved here in the 1977 revision.)

In essence, no diagonal clearance reduction is allowed. The clearance envelope is squared-off at the junction of horizontal and vertical clearance requirements to properly reflect the needs of workers in the vicinity of these conductors at the structure and the action of the conductors themselves in midspan (see Figure H235D).

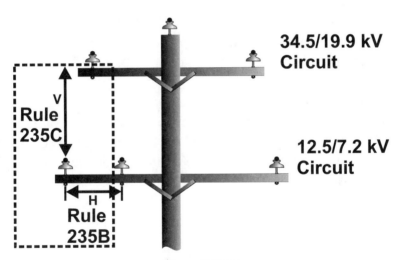

Figure H235D
Vertical and horizontal clearances on same pole line

235E. Clearances in Any Direction From Line Conductors to Supports, and to Vertical or Lateral Conductors, Span or Guy Wires Attached to the Same Support

(This rule was numbered Rule 235A3 prior to the 1977 Edition; Table 235-6 was numbered Table 9 in prior editions.)

Where a conductor is supported above a crossarm by a pin (or other) insulator, the required clearance between the conductor and the crossarm is the straight-line distance between the conductor position and the crossarm surface; i.e., wider skirts on pin insulators do not allow a shorter pin to be used. The voltage to be used is the phase-to-phase voltage.

The minimum conductor-to-structure clearances given in the rules are considered necessary to protect the public from flashover to structures, where persons nearby may be subjected to step potentials or induced voltages in adjacent metallic objects. The general discussion of Rule 232B and Rule 232D and earlier portions of Rule 235 apply to this rule.

The Footnotes to Table 9 and Table 235-6 have been augmented in successive editions to reflect continuing experience with various operating systems. The values of the table have changed little over the years.

Calculations of alternate clearance requirements under Rule 235E3 are similar to those of Rule 232D. Notice that the *Configuration Factor* is different and the built-in safety factor is 1.0.

The title of the rule was clarified in the 1981 Edition.

Footnote 1 of Table 235-6 was revised in the 1984 Edition to allow reduced clearances between an insulated or effectively grounded guy and a communication cable, if appropriate abrasion protection is provided. Footnote 8 was revised in the 1990 Edition to recognize the lack of need for a stated clearance of covered facilities at the pole when clamped to the pole and restrained from movement and the possibility of abrasion. A new Footnote 11 was added in the 1984 Edition to Table 235-6 to recognize appropriate clearances to insulating sections in guys; reduced clearances are allowed to portions of the insulating section as long as full clearances are maintained to its metallic end fittings and the guy wires.

The 2007 Edition added NOTES under Rule 235E1 to direct users to the appropriate rules for clearances to communication antennas in the supply space (Rule 235I) and communication antennas in the communication space (Rule 236D1). A reminder that trees do not shelter a line was added to Rule 235E2.

The values in Table 235-7 showing results of the alternate clearance calculations of Rule 235E were adjusted to reflect the 2007 coordination of decimal places and rounding.

235F. Clearance Between Circuits of Different Voltage Classifications Located in the Supply Space on the Same Support Arm

(Prior to the 1977 Edition, this rule was numbered 235A5.)

In many cases, because of a lack of vertical space on the structures or the necessity for stringing additional conductors, it is impossible to install more support arms in order to provide proper vertical clearance between the conductors of different classifications. In order to provide safe construction under these conditions, the requirements of this rule will permit two circuits or sets of conductors to occupy the same support arm in the five cases listed, provided a sufficient clearance is maintained (see Figure H235-2). The first two cases may be applied to communication circuits used in the operation of supply lines.

Rule 235F is limited to clearances between supply circuits of different voltage classifications on the same support arm; thus the rule is concerned with horizontal clearances and conductor arrangements where such conductors are on the same support arm. IR 519 issued 7 July 1999 clarified that, if the circuit includes a neutral conductor, the neutral conductor is part of the supply circuit (see NESC definition of *circuit*). Note that, if a common neutral is used, the common neutral will be associated with more than one supply circuit. The answer to IR 519 continued to clarify that, with the language of Rule 235F and

Table 235-5, the neutral does not carry a different voltage classification than the associated phase conductors of the same circuit.

Rule 235F and Table 235-5 first appeared as Rule 235A and Table 11 in the Fifth Edition in 1941. Table 11 did not contain a separate category for neutral conductors. The revisions in the 1977 Edition introduced neutral conductors meeting Rule 230E1 into the classification scheme. This change was intended to affect the vertical clearance requirements of Rule 235C and its associated Table 235-5, but it was not intended to affect the operation of Rule 235F.

The 2007 Edition modified Rule 235F to clearly state that the neutral conductor is considered for purposes of this rule to have the same voltage classification as the circuit with which it is associated. Rule 235F does not restrict the relative position of any of the conductors associated with a common circuit, including either phase or neutral conductors. A neutral conductor may be placed in either an inner or outer position, assuming that other requirements are met. In addition, Rule 235F does not apply to the phase and neutral conductors of the same circuit where that circuit is the only circuit on the arm.

The 2007 Edition also removed the requirement from Rule 235F5 for either (1) two communication circuits or (2) a communication circuit and a supply circuit of less than 8.7 kV phase-to-ground to both be owned by the same utility, if they are to be located on the same arm. Rule 224A provides for the safety issues involved; the workers on the communication must be qualified to work in the supply space—ownership is not the issue in this case.

The classification reference for 2002 and prior editions was to Table 235-5, with 750, 8700, and 22 000 V being the division points between classes. For the 2007 Edition, the voltage classifications were added directly into Rule 235F.

The arrangement of conductors shown in Case 4 of Figure H235-2 is not permitted for ordinary constant-voltage distribution circuits. It is intended to provide only for series-lighting and similar circuits that are normally dead during the day and that would, therefore, not present a hazard to workers working on the lower-potential circuits beyond them

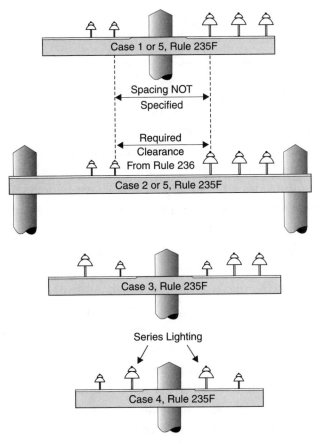

Figure H235-2
Permissible arrangements of supply circuits
of different consecutive voltage classifications
on the same crossarm

during daylight hours. Where it is customary to test series arc circuits during the day, it may not be advisable to employ this type of construction, unless the workers take proper precautions.

The intent of the rule was clarified in the 1981 Edition.

As of the 2002 Edition, communication lines in the supply space are allowed on the same arm with supply under specified restrictions.

235G. Conductor Spacing: Vertical Racks, or Separate Brackets

(Prior to the 1977 Edition, this rule was numbered 235A4.)

In many localities it is customary to install the low-voltage secondary conductors on racks attached directly to the structures. Such construction facilitates the connection of services and of branches and simplifies the wiring on the poles. However, the climbing space cannot be maintained continuously on one side of the pole. It is therefore necessary to supply sufficient lateral working space both above and below the racks to permit the workers to work around them.

Where conductors are supported by racks, the vertical clearances specified in this rule are considered satisfactory values for voltages of less than 750 V. Note that, prior to 2007, the rule used the term *separations*; however, *clearances* is the appropriate term. It is assumed that due care is exercised when the conductors are installed in order to have the same spacing in the spans. It is appropriate to give caution about two situations that can occur with inappropriate stringing of secondary conductors. If the neutral conductor is below the phase conductors, the latter may sag into the neutral under high-temperature operation. If the neutral is above the phase conductors, it may sag low enough under icing conditions to contact a phase conductor, if the ice comes off the warmer phase conductor first. The latter case is extremely rare since the conductors are so close together that (1) they both receive essentially the same heating from the sun and radiant heating from the earth and (2) rising heat from the phase conductors also helps to warm the neutral above.

Rule 235G serves as a guide for clearances of items in Table 235-5 for which no clearance is specified, if they are to be separately installed and without being bundled together. This limits the opportunity for mechanical damage due to contact during storms.

Rule 235G was originally written to specify clearances between secondary conductors mounted on multiple-spool racks. It also applies to secondary conductors mounted on separate brackets on the same side of the pole. Modern construction often uses multiplex cable for secondary, instead of open-wire construction. When a secondary must be added after a line is built, it has been common practice for many electric utilities to attach the neutral of the secondary multiplex cable to the same spool used to attach the existing neutral, with the insulated energized conductors positioned away from the bracket (see Figure H235G4). Because of the greater weight of the cable, there was no potential mechanical conflict between any energized conductor of the multiplex cable and the neutral at the pole or in the span. This construction has proved to be economical, convenient, and safe. The basic intentions of Rule 235G were met: i.e., (1) conductor insulation would not be damaged by physical contact between the different conductors during storm conditions and (2) differences in loadings would not cause uninsulated conductors to touch.

Unfortunately, the rather simple language of Rule 235G implied that the vertical clearances required by the rule were to exist all through the span, not just at midspan. The original language of Rule 235G had not been designed with attachments of multiplex cable neutrals to open neutral brackets in mind. The answer to IR 523 issued 31 July 2001 concluded that this construction was not allowed by the language of the rule.

The 2007 Edition reversed IR 523 by adding Rule 235G4 to explicitly allow the long-time, safe option of attaching multiplex cable neutrals to open neutral brackets, so long as the traditional clearances required by the newly named Table 235-8 are maintained in midspan. (The values of the previous, unnumbered table did not change.) The 2007 Edition also allowed application of the rule to wires, conductors, and cables of different ownership by agreement between the parties involved. The 2007 Edition also clarified the EXCEPTION to Rule 235G to apply only to open wire conductors; cables meeting Rule 230C do not have requirements between the individual conductors of the cable (see Rule 235A1).

See Rule 220B for a discussion of jumpers.

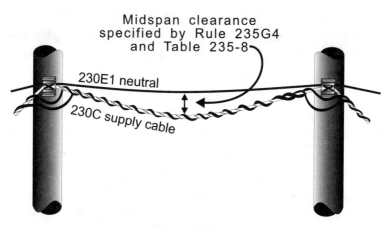

Figure 235G4
Multiplex secondary cable attached to neutral bracket

235H. Clearance and Spacing Between Communication Conductors, Cables, and Equipment

(New in the 2002 Edition.)

Normal spacing between communication cables is 300 mm (12 in) to allow a lashing machine to be pulled along one messenger without damaging the lashing wire on another messenger. Where space is at a premium, communication utilities can, by agreement, install cables closer than a 300 mm (12 in) spacing. This often requires moving a cable out on a temporary attachment arm to work on it.

With the proliferation of new communication attachments after the 1996 Telecommunications Act came serious problems with some later attachers installing cables and equipment so close to existing cables and equipment as to interfere with the ability to safely and efficiently work on the existing facilities. Except by agreement, no communication utility can install conductors, cables, or equipment closer than 100 mm (4 in) to the conductors, cables, or equipment of another utility *anywhere in the span*, including at the pole. Icing differentials may need to be considered, especially where strong winds often follow ice storms.

Appropriate clearance should be maintained to avoid damage to lashing wires due to mechanical contact during high winds.

235I. Clearances in Any Direction From Supply Line Conductors to Communication Antennas in the Supply Space Attached to the Same Supporting Structure

(New in the 2002 Edition.)

Clearances are specified to communication antennas installed in the supply space. Work on these devices requires the use of insulated tools, equipment, and personal protection under the supply work rules of Section 44. Clearances to the antennas are the same as for line conductors under Row 1b of Table 235-6.

Such communication antennas mounted on utility structures are essentially considered as rigid, vertical, open-wire communication conductors. Of concern is degradation of the flashover characteristics of the structure if an antenna is introduced between the structure and an energized high-voltage conductor. If flashover occurs due to overvoltage, the flashover should go the structure, not to the antennas and/or its related circuit. The designer should consider the good practice required by Rule 012 when positioning the communication antenna on the structure so that the air gap clearance between the antennas and the nearest energized conductor should exceed the equivalent dry arcing distance of the shorter (or lesser insulating) string of insulators by an appropriate amount to ensure flashover to the structure. This good practice is intended to limit the exposure of communication equipment to overvoltages, if a conductor-to-ground flashover should occur.

The equipment case supporting the antenna is considered as part of the structure under Row 4a of Table 235-6, as long as it is effectively grounded. The jumpers going to the antenna are considered as vertical or lateral conductor under Rule 239. Note that, if an antenna is mounted in the inverted position on a street light bracket, and the bottom tip of the antenna is the lowest item in the supply space, the communication worker safety zone starts there. The antenna cannot protrude into the communication worker safety zone. At best, it is a mechanical hazard to the eye. If the radiant power of the antenna is such that approach distances greater than those of Table 431-1 or Table 441-1 are necessary, (1) workers must be appropriately trained and (2) appropriate safety signs should be installed in a prominent position on the antenna case.

The 2002 radio frequency specification of 0–750 V in Rule 235I2 was erroneous; this was corrected to 3 kHz–300 GHz in the 2007 Edition. A new NOTE was also added in Rule 235I2 to refer users to the associated work rules in Rule 420Q.

236. Climbing Space

(Rule 236 has remained essentially as it was in the Third Edition. Rule 236B, Rule 236C, Rule 236D, and Rule 236F are short and self-explanatory and are not further illustrated herein. Rule 236D was clarified in the 1977 Edition.)

This rule applies only to the portions of structures that workers ascend. The specified climbing space is not required where it is the unvarying rule that such a structure, or such portion of a structure, will be worked from aerial lift equipment and will not be climbed. However, where such structures or portions of structures are sometimes expected to be climbed (such as when major storm damage has occurred), climbing space is required.

236A. Location and Dimensions

Climbing space may be thought of as an imaginary box whose width, depth, and height dimensions are specified by Rule 236E, Rule 236F, Rule 236G, and Rule 236I. It is only required to be provided on one side or "corner" of a structure. Structures may be considered to be divided into sides (by the line) and, further, into quadrants (by the crossarms). The term "corner" means any quadrant. The climbing space may be shifted to any other side or corner, providing that appropriate transfer room is provided.

236B and 236C. Structure Components in Climbing Space.

With the exception of the special considerations for buckarms in Rule 236F, structure components should not be located in more than one side of the climbing space. Where practical, support arms should be located on one side of the structure to allow climbing space on the other (see Figure H236C). Note that there is no prohibition against climbing over crossarms, but it is preferable not to have to do so.

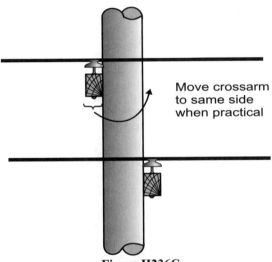

Move crossarm to same side when practical

Figure H236C
Crossarm location relative to climbing space

236D. Location of Equipment Relative to Climbing Space

(Rule 286B of the 1987 and prior editions moved to 236D in the 1990 Edition.)

The selection of a transformer location involves other factors besides the question of load center. Structures that carry complicated wiring or circuit junctions generally are not appropriate transformer locations. Since maintenance work, such as replacing fuses or exchanging transformers, frequently has to be done at night or in stormy weather, the less wiring there is in the vicinity of the transformers, the safer the working conditions will be.

In any case, it is important to provide adequate climbing space all the way up the pole so that (1) it is not necessary for workers to climb around the ends of crossarms, and (2) they will not injure crowded equipment with their tools or spurs when climbing up the structure.

This rule is essentially unchanged since the Fourth Edition.

In the 2002 Edition, clearances for ungrounded luminaires were moved from Rule 232B4a, where they had been inappropriately placed when Section 28 was disbanded in 1990.

In the 2007 Edition, antennas were added to Rule 236D1 to the long list of supply and communication equipment prohibited from the climbing space.

236E. Climbing Space Between Conductors

On structures where it is expected that workers will be required to climb through lower-level circuits and work on circuits at a higher level, adequate clearances must be maintained between conductors of the lower-level circuits.

The intent of this rule is to provide a minimum space of 600 mm (24 in) for the workers to pass between conductors. If conductors are to remain energized and be covered with a temporary protective covering suitable for the voltage, a minimum clear space of 600 mm (24 in) between conductor cover devices must be allowed for safe passage of the workers. The clearances given in Table 236-1 are, therefore, intended to allow a minimum clear space of 600 mm (24 in) when conductors are temporarily covered with electrically and mechanically adequate coverings. No clearances have been given for open-wire circuits through which workers may climb while the conductors are energized and uncovered; such a practice is unsafe and should be avoided.

Table 236-1 was expanded in the 1977 Edition and covers four voltage classifications above 15 kV. The classifications reflect currently popular operating voltages such as 25, 34.5, 46, and 69 kV. As in other areas of the Code, clearances were effectively reduced in the Sixth

Edition for some classifications of wye-connected circuits. The voltage classifications were further revised in the 1977 Edition and some clearances were reduced to reflect the capabilities of modern protective equipment and operating practices.

When both supply and communication conductors are attached to the same pole, the same climbing space is required for communication conductors as for supply conductors immediately above them, up to a maximum of 750 mm (30 in). This requirement is not so much intended to limit any hazard due to the communication conductors alone; it is primarily intended to limit the hazard that might exist if a supply conductor were elsewhere in contact with one of the communication conductors. In this case, a high potential might exist between the two pole-side conductors of the communication circuit; this could cause a serious accident to a line worker required to crowd through communication conductors having a reduced climbing space. Other considerations are that supply line workers (1) will not get their feet in contact with communication conductors and (2) will not injure those conductors in climbing through.

Wherever a primary supply circuit is installed on the same poles with communication conductors, so as to provide sufficient space for the installation of a secondary arm between the two, the intent of the rule is met if the communication conductors have a spacing at the poles corresponding to the secondary voltage. However, where the clearance between the primary and communication arms is not sufficient for the insertion of a lower-voltage arm, the climbing space through the communication conductors should correspond to that of the primary voltage.

Communication line workers, in general, are not accustomed to working near supply conductors. It is therefore desirable to allow liberal free working space for these workers when communication conductors are on the same structure and above supply conductors. This will tend to prevent accidental contact with supply conductors from occurring when the communication worker's attention is on the work at hand.

EXCEPTION 1 of Rule 236E recognizes that there may be certain types of structures or pole configurations where climbing is not permitted regardless of what other structures characterize the general line. This EXCEPTION was clarified in the 1981 Edition.

236G. Climbing Space for Longitudinal Runs Not on Support Arms

It has become common practice in many localities to place the low-voltage conductors vertically on racks or brackets close to the poles, thus practically cutting the climbing space in half. While such construction provides comparatively easy and simple methods for the attachment of services, it requires readjustment of other construction to avoid obstructing the workers when they climb up and down the pole. Unless other arrangements in the locations of the adjacent conductors are made, this constitutes a hazard.

In order to comply with the provisions of the rules without variation, these racks are occasionally placed on extension pieces. In lieu of this, the nearest supply conductors on crossarms may be 1.2 m (4 ft) above or below the rack, or the conductors on the adjacent arms may be so installed as to provide the full climbing space on one side of the rack. Where attachment of conductors close to the pole seems advisable, the racks should generally be on only one side of the pole for uniformity, and the climbing space should generally be carried vertically at the other side. The climbing space between any two wires is required, however, by the rule, to be carried vertically at least 1 m (40 in) above and below them, and any shifting of the climbing space from side to side must, therefore, be done in steps not less than 1 m (40 in) apart.

Prior to 2007, longitudinal runs on racks and cables on messengers were explicitly *not* considered to obstruct the climbing space *if all the wires were either covered by insulating protective equipment or otherwise guarded as an unvarying rule.* This simple specification provided appropriate safety for most installations—indeed, it was required before the work rules (1) were completely revised and augmented in the 1973 Edition and (2) were continually refined since that time, including the requirement for installing insulating coverup when climbing past secondary that is now contained in Rule 441A2. However, the old specification was too simple to work well with installations that have many spools on racks or many single cables installed close together, because it did not provide foot room for climbing past the group of wires or cables—especially when draped by insulating blankets. The 2007 Edition removed the previous specification and replaced it with "if the location, size, and quantity of the cables permit qualified workers to climb past them" which, along with requiring insulating coverup, has been one of the real concerns for the entire history of the rule.

The description of cables in EXCEPTION 1 was changed in the 1977 Edition to reflect the cable definitions in Rule 230C and Rule 230D.

In the 1987 Edition, an EXCEPTION and Figure 236-1 were added under Rule 236G for a different shaped and sized climbing space where certain supply service drops take off from the structure at a small angle to the line. It requires the unvarying practice to protect against contact with such facilities when climbing past them.

236H. Climbing Space Past Vertical Conductors

This rule makes it clear that, when the climbing space is changed from one side to a corner of the pole, as illustrated in Figure H236-1, neither the pole itself, nor conductors attached to the pole and enclosed in a conduit or protected by a molding, are considered as an obstruction. Note, however, that this illustration was taken from the Discussion of the Fifth Edition; it may not be applicable where large conduits are used.

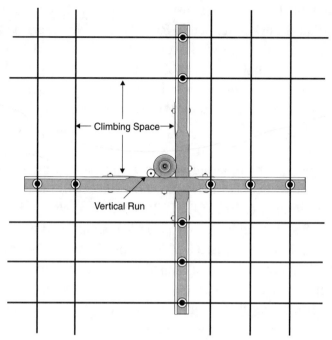

Figure H236-1
Climbing space

237. Working Space

(Rule 237A, Rule 237B, and Rule 237C are essentially unchanged since the Third Edition. Rule 237E and Rule 237F were moved here in the 1990 Edition from their previous location in Rule 286C and Rule 286D; they are essentially unchanged since the Fourth Edition except for the rewording for clarity in the 1977 Edition.)

Sufficient clear working space must be provided between the conductors supported on adjacent support arms to permit line workers to work safely upon the conductors supported by a structure. The vertical and horizontal clearances required in the rules are generally between conductors rather than between pins or crossarms (see Figure H237-1). However, the Discussions of the early Code editions have indicated that, in cases where the crossarms fulfill the vertical-clearance requirements but, owing to the use of different types or sizes of insulators or different manners of attachment, the clearances between the conductors themselves are slightly reduced, the requirements of the rule will be considered as having been met. It is apparent that the reference in Rule 238A to "crossarm spacing" influenced these statements. Rule 238B allows reduced conductor clearances if the crossarm clearances are met.

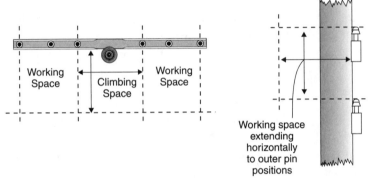

Figure H237-1
Working space

Since those Discussions were written, support apparatus and methods have changed and varied widely enough that it would be appropriate to consider carefully the effect of any such reduction. In addition, clearances in Rule 238 in the 1977 and later editions are between conductors, not crossarms. It should be clear that the intention is for workers not only to have room to work but also room to work safely.

The requirements of this rule are to ensure that the proper dimensions of the working space are maintained at all times. During reconstruction, or when new apparatus (such as a transformer or switch), is being installed, there will be a tendency to place taps or leads in the working space unless the matter is given proper attention. Such connections can generally be placed on the opposite side of the pole from the working side. If this is impossible, it is necessary to install additional arms or other means to support the conductors in order to provide the proper clearances.

The use of buckarms on poles carrying a considerable number of wires makes it difficult to provide normal climbing and working spaces. Some concessions have been made in the rules in order to make the use of buckarm construction practical. Even if a pole were to be specially designed to provide the normal clearances, general levels would be disturbed where the buckarms were numerous, as at a junction pole.

The rules require the provision of adequate climbing space, in accordance with Rule 236, under all circumstances. To accomplish this, an EXCEPTION to the general requirement for horizontal clearances between wires at supports is made by Rule 236F under certain conditions. For voltages not exceeding 8700 V, an EXCEPTION was added in the Fifth Edition to permit a 900 mm (12 in) working space, instead of an 450 mm (18 in) working space. To qualify, the construction must involve no more than two sets of line arms and buckarms, and certain prescribed safety measures must be practiced. Where crossarms have the usual 600 mm (2 ft) spacing and the 450 mm (18 in) working space is provided, the buckarm is placed close to one of the line arms, as shown in Figure H237-2. This should be the line arm carrying the conductors that are connected to conductors on the buckarm. The vertical and lateral conductors will then not obstruct the free 450 mm (18 in) space that constitutes a reduced working space. A line worker can then have access to one set of conductors from below and the other from above. Note that, if buckarm deadend construction is used, clearance to the deadend suspension insulators must be considered, as in Figure H237-3.

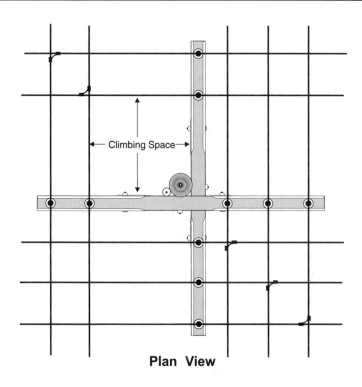

Plan View

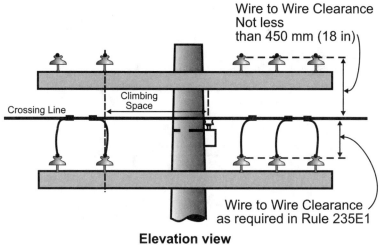

Wire to Wire Clearance
Not less
than 450 mm (18 in)

Crossing Line

Climbing
Space

Wire to Wire Clearance
as required in Rule 235E1

Elevation view

Figure H237-2
Obstruction of working space by buckarm
construction for crossing line

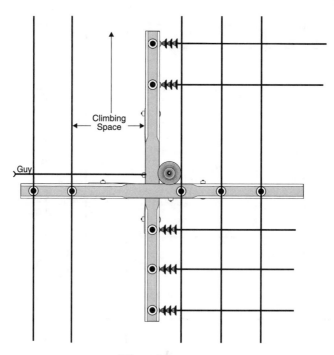

Plan View

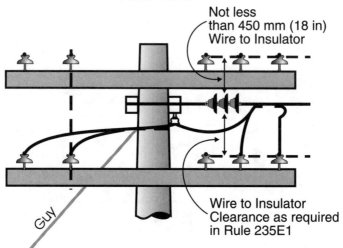

Not less
than 450 mm (18 in)
Wire to Insulator

Wire to Insulator
Clearance as required
in Rule 235E1

Guy

Elevation view

Figure H237-3
**Obstruction of working space by buckarm construction for
deadend tap line**

Even if such current-carrying parts are on the opposite side of the pole or above the climbing space, as with some pole-top fixtures, they should either be (1) suitably enclosed and arranged for adjustment without opening the enclosure or (2) so located that, in adjusting them, it is not necessary to put a part of the body near either current-carrying parts at different potential or a grounded part. In the 1977 Edition, the requirements of Rule 286D (now Rule 236F), were broadened to cover all parts of the body, not just the hands.

The 2002 Edition clarified that Rule 237D on buckarm clearances applies to *crossing* or *tap line conductors*. The previous term lateral conductors was left from earlier codes where it referred to a line at right angles to the main line. Today, lateral conductors are jumpers entirely supported on the same structure.

237E. Guarding of Energized Equipment

(Rule 286C of the 1987 and prior editions was moved to Rule 237E in the 1990 Edition.)

It is not appropriate for current-carrying parts of equipment to be located in the climbing space. This rule is essentially unchanged since the Fourth Edition; it was reworded in the 1977 Edition.

237F. Working Clearances From Energized Equipment

Even if such current-carrying parts are on the opposite side of the pole or above the climbing space, as with some pole-top fixtures, they should either be (1) suitably enclosed and arranged for adjustment without opening the enclosure or (2) so located that, in adjusting them, it is not necessary to put a part of the body near either current-carrying parts at different potential or a grounded part.

This rule has been in its current general form since the Fourth Edition. The requirement was broadened in the 1977 Edition to cover all parts of the body, not just the hands. The rule was revised in the 1987 Edition to include the relationships to supporting platforms. In 2002, luminaires and support brackets were added to the list of items which may require increased clearances to meet Rules 441 or 446. In essence, the approach distances of Part 4 are required to exposed live parts from locations where line workers are expected to climb or work.

238. Vertical Clearance Between Certain Communications and Supply Facilities Located on the Same Structure

(Rule 238A, Rule 238B, Rule 238C, and Rule 238D of the Sixth and prior editions dealt with clearances between conductors. They were moved to Rule 235 in the 1977 Edition for better organization. Rule 238E of the Sixth Edition became Rule 238 in the 1977 Edition. Rule 238 now concerns itself with clearances between wires, conductors, or cables and noncurrent-carrying parts of equipment. All references to "street lights" or "lamps" were changed to "luminaires.")

Rule 238 creates a communication worker safety zone on joint-use supporting structures for communication lines that are not installed in the supply space in accordance with Rule 224A and Rule 235.

Rule 238 of the 1977 and later editions is composed of short and generally self-explanatory paragraphs. In the 1977 Edition, the minimum clearance between ungrounded, series-fed luminaire brackets and communication crossarms or messengers located above were increased to improve safety for the communications worker. Because it was considered to be design information and not a safety requirement, the 1977 Edition eliminated the old Rule 238A requirement for a clearance between supply and communication crossarms of 4 ft. The 40-in clearance between conductors and cables and "equipment" was retained in Rule 235 and Rule 238. As a practical matter, the minimum attachment spacing is often still 1.2 m (48 in), so that 200 mm (8 in) is left for brackets and jumpers. Even then, neatness is required to keep jumpers within the 200 mm (8 in) dimension (see Figure H238A). The titles to the rule and Table 238-1 were reworded in the 1984 Edition to reflect clearly the content and application of this rule.

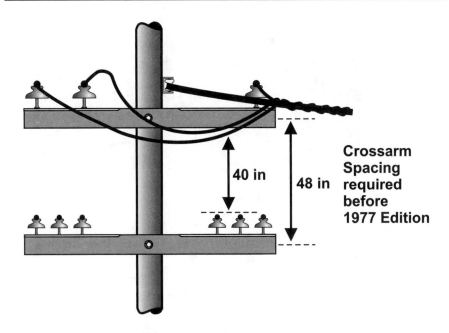

Figure H238A
Older NESC requirements

The vertical clearances of this rule are required vertically from the horizontal plane of the lowest surface of the lowest supply conductor or bracket to the horizontal plane of the highest communication bracket or cable. The values given are vertical, *not* diagonal (see Figure H238B-1).

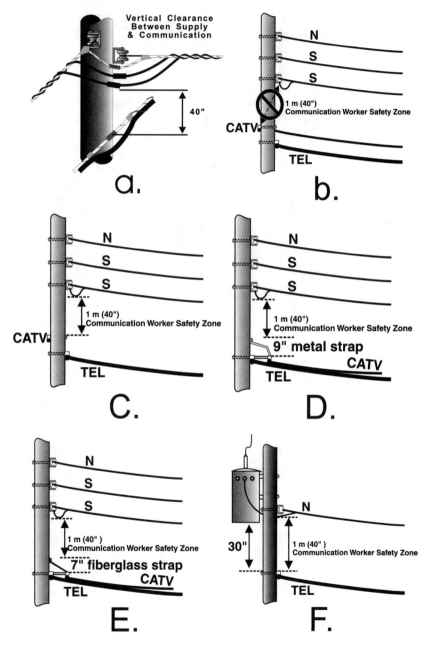

Figure H238B-1
Basic required vertical clearances required
by Rule 235C, Rule 238B, and Table 238-1

For their safety, it is intended that communications workers will not work on communication conductors, cables, or brackets located less than 1 m (40 in) below supply conductors, cables, or brackets. The exception to this general rule is that, where communication lines and equipment are found to be installed too close to supply lines and equipment, communications workers may relocate their lines or equipment if they do so in accordance with the supply work rules. Note that Rule 432 prohibits communication workers from placing any portion of their body above the lowest supply facility.

Clearances between grounded noncurrent-carrying parts of supply equipment and communication conductors may be reduced to 750 mm (30 in) when the former is effectively grounded, although this is conditioned by the requirement that this be a uniform practice over a well-defined area (see Figure H238B-2). Supply neutrals meeting Rule 230E1 were added in the 1987 Edition to this list; note that the EXCEPTION to Rule 235C2b(1)(a) requires the neutral to be bonded to all communication messengers below for this reduction to apply. In the 2002 Edition, the existing Footnote 1 to Table 238-1 was rewritten to clearly specify conditions of application of the reduced clearances from communication in the communication space to grounded supply conductors and equipment. Bonding to the communication messengers is now required. Rule 230F of the 1990 Edition also allows some fiber-optic supply cables to have these reduced clearances. Under these conditions, the potential for communications workers to get shocked from grounded supply equipment is practically nonexistent. However, communications workers cannot be expected to determine by inspection whether supply equipment is grounded. It is expected that areas where grounding of supply equipment is practiced will be *well defined* and *made known* if the lesser clearances permitted by Footnote 1 are to be employed.

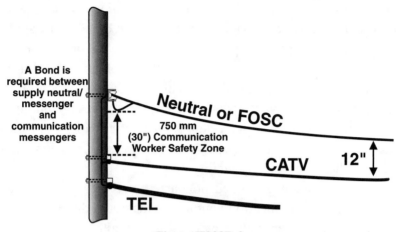

Figure H238B-2
Clearances allowed by the EXCEPTION to
Rule 235C2b(1)(a) and Table 238-1

Rule 238D allows a reduced clearance of 300 mm (12 in) between drip loops of luminaire brackets and communication cables when the drip loop enters the luminaire directly from the surface of the structure (see Figure H238D-1). The 1984 Edition added one EXCEPTION to allow clearance of only 75 mm (3 in) if the drip loop is covered by a suitable nonmetallic covering that extends at least 50 mm (2 in) beyond the loop. In essence, 50 mm (2 in) extra covering is needed on each end to ensure that any of the loop within 300 mm (12 in) of the communication bolt is covered with the auxiliary nonmetallic covering; see Figure H238D-2. This rule is intended to recognize that some communities require certain luminaire heights that would ordinarily violate the communication space requirements. Traffic signal bracket were added in the 1990 Edition. These reduced clearances only apply to luminaires and traffic signal brackets under the stated conditions; no other supply conductors are allowed such clearances by this rule—even if they meet the same conditions.

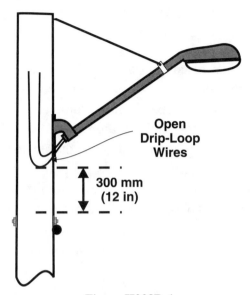

Figure H238D-1
Clearance of communication bolt to
open luminaire drip-loop wires

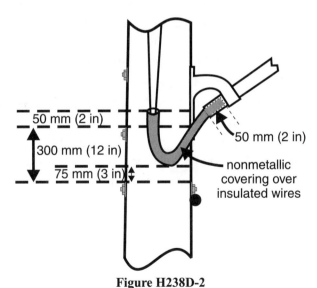

Figure H238D-2
Clearance of communication bolt to open luminaire
drip-loop wires with supplementary nonmetallic covering

238E. Communication Worker Safety Zone

(New in the 2002 Edition.)

See the companion discussion of Rule 235C4 and Rule 224A. The communication worker safety zone is required between the supply space and communication space when communication workers use communication work rules, tools, equipment, and methods. The communication worker safety zone creates headroom for communication workers. When communication workers keep all parts of their bodies below the lowest supply facility, this zone allows safe working space for communication workers observing communication work rules.

If communication workers are *authorized* to work in the supply space; *use supply work rules and methods, insulated buckets, insulating tools and insulating personal protective gear; and otherwise meet Rule 224A*, there is no requirement for a separate communication space and communication worker safety zone.

However, if communication is not authorized in the supply space or communication workers choose not to use supply work rules and methods, insulated buckets, etc., a separate communication space is required. No supply or communication lines or equipment are allowed in the communication worker safety zone except those allowed by Rule 238C, Rule 238D, or Rule 239.

Although a luminaire or traffic signal bracket or span wire is allowed in the safety zone by Rule 238C and Rule 238D (when required to meet operation height requirements), packet radio antennas (as shown in Figure 238E) cannot be mounted on such devices when they are in the communication worker safety zone; they must be mounted outside of the communication worker safety zone.

Figure H238E
Antennas not allowed in communication worker safety zone

239. Clearances of Vertical and Lateral Facilities From Other Facilities and Surfaces on the Same Supporting Structure

In many places throughout this rule, the word "structure" was substituted in the 1977 Edition for the word "pole" to make the rule apply to poles, towers, H-frames, etc. EXCEPTION 1 in the beginning of this rule in the 1987 and prior editions was relaxed to permit the use of conduits other than iron pipe, since there seems little need for the special properties of iron pipe in this situation. EXCEPTION 4 in the 1987 and prior editions was changed to agree with the change in Rule 220B2.

The overall portion of Rule 239 was revised for clarity in the 1990 Edition: the previous wording was in effect before the use of plastic conduits and had not been updated to reflect modern practice. This led to an even more extensive revision in the 1997 Edition to increase the clarity and streamline the rule.

239A. General

(This rule was created in 1990; former Rule 239A was moved to Rule 239B.)

A new Rule 239A was added to place the former EXCEPTIONS to Rule 239 in positive terms and increase the understandability of the intention of the requirements. The 1997 revision added surge-protection wires, insulated communication conductors and cables, and insulated supply cables of 0–750 V to the list of items allowed to be directly attached to the structure. Conduits enclosing conductors (or empty) may also be mounted directly on the structure. However, all of these items may have to meet other requirements for bonding or covering specified in later subrules. The 1997 Edition clarified the use of the term *nonmetallic covering* in this rule; a cable jacket does *not* meet the requirement.

The 2002 Edition added Rule 239A6 to clearly allow either a conduit or U-guard for vertical riser protection. In order to prevent exposure of the enclosed cable, a backing plate is now required for a U-guard, unless it fits snugly to the structure. In 2007, the backing plate requirement of Rule 239A6 was moved to the revised Rule 239D2 to require a backing plate wherever guarding is required by other rules. Generally guarding would be required within 2.45 m (8 ft) of the ground or other areas readily accessible to the public (see Rule 239D1), but it could also be required in other situations, such as for cable risers on poles near buildings or other structures (see Rule 234C2).

The 2007 Edition revised Rule 239A2 to state in specific terms (1) which types of cables are allowed to be installed together in the same duct (raceway) or U-guard (covering) and (2) under what conditions such joint installations may be made.

239B. Location of Vertical or Lateral Conductors Relative to Climbing Spaces, Working Spaces, and Pole Steps

(This rule was moved from 239A in the 1990 Edition; Rule 239B was moved to Rule 239C.)

To facilitate uniformity in the arrangement of conductors and equipment on a pole, it is usual to designate one semicircumference or quadrant of the pole as the climbing space. Where poles are used jointly by supply and communication conductors, it is customary to designate the sidewalk side as the climbing side, leaving the street side clear for the attachment of lamp leads and, where a street railway is also concerned, for the attachment of span wires or brackets. However, where service drops take off from the sidewalk side, at least part of the climbing space may, of necessity, have to be located on the street side. A NOTE referencing Rule 239H was added in the 1997 Edition.

239C. Conductors Not in Conduit

(This rule was moved from 239B in the 1990 Edition; former Rule 239C was moved to Rule 239D.)

Conductors not in conduit naturally require necessary clearances from other live conductors, from grounded surfaces, and from surfaces of structures.

239D. Mechanical Protection Near Ground

(This rule was moved from 239C in the 1990 Edition; former Rule 239D was moved to Rule 239E.)

This rule was completely revised in 1997. It now requires mechanical protection of *all* conductors and cables within 2.45 m (8 ft) of the ground *or other area readily accessible to the public*. Protection is now also required higher on the pole if it is near a window, stairway, etc. Previous EXCEPTIONS 1 (armored cables), 2 (communication cables and conductors), the last half of EXCEPTION 4 (ground conductors on multiground circuits), and 5 (wires used solely to protect structures, not equipment or lines, from lightning) were retained as an EXCEPTION to Rule 239D1. The previous EXCEPTION 2 (lead-sheathed cable in rural districts) was deleted and the first half of EXCEPTION 4 (grounding wires for delta systems used in rural districts) was deleted.

The NESC does not specify either the form of "suitable mechanical protection" to be used or the materials to be used. A cable jacket does *not* meet this rule. Suitable protection for some local conditions may not be enough for others, and vice versa.

The 1997 revisions mirror similar requirements in other parts of the Code. Where guarding is not used, the side of the structure least exposed to mechanical damage (if there is one) should be used (see Figure H239D-1). Guards that are placed over and completely enclose (i.e. conduits, not U-guards) conductors used to ground lightning protection equipment must be either nonmetallic or be bonded to the ground wire at both ends. This 1997 rule differs from Rule 93D in that it requires bonding all metallic conduits to the lightning protection wire at both ends—even if the metal was a nonmagnetic variety. In essence, Rule 93D has been superseded by Rule 239D in this respect. Neither rule requires a double bond on metal U-guards (see Figure H239D-2).

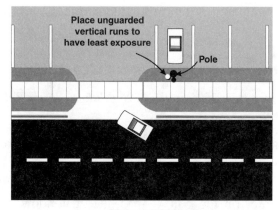

Figure H239D-1
Locating unguarded vertical runs
to have least exposure to mechanical damage

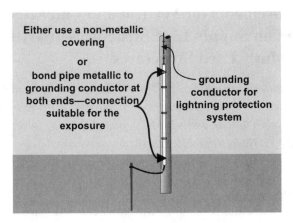

Figure H239D-2
Bonding of metallic pipe to grounding conductor

Grounding wires that become broken by traffic, riding lawnmowers, or other cause may lose or reduce their ability to protect the circuits or apparatus to which they are connected. Thus, mechanical protection is essential in certain instances to guard against such abrasion or breakage. The rule was changed in the 1977 Edition to make it clear that equivalent *mechanical* protection is required when materials other than wood molding are used.

NOTE: Questions are frequently asked of the Interpretations Subcommittee for guidance in this area. Most frequently, the questioners refer to PVC conduit and want to know if Schedule 40 PVC is "enough." The Subcommittee is prohibited from approving specific company standards or materials, or otherwise acting in a consulting manner, and will not do so. The Subcommittee has reminded us, however, that there are several "Schedule 40" PVC pipes listed by the American Society for Testing and Materials (ASTM), and they vary by polymer, impact strength, modulus of elasticity, and deflection under load. Diameter also affects crush-resistance. The variation in attributes is so great that one "Schedule 40" PVC conduit might be better for a specific application than another "Schedule 80" PVC conduit, due to the differing characteristics and different local conditions.

239E. Requirements for Vertical and Lateral Supply Conductors on Supply Line Structures or Within Supply Space on Jointly Used Structures

(This rule was moved from Rule 239D in 1990; former Rule 239E was moved to Rule 239F. A major revision of this rule occurred in 1997. Former Rule 239E2d was combined with Rule 239E2a, and Rule 239E2c was called into the new EXCEPTION to Rule 239E2a.)

The only persons concerned when supply conductors pass through the space occupied by supply conductors, or on structures occupied only by supply conductors, are line workers who are, or should be, entirely familiar with the hazards incidental to the voltage concerned. The requirements of Rule 239F are, therefore, modified by this rule so that conductors that are not enclosed may be either supported at such a distance from the structure, or insulated and directly attached in such a manner, that there is little likelihood of contact with them by workers on the structure.

The voltages specified in Table 239-1 (Table 13 of the Sixth and prior editions) are deliberately on a conductor-to-conductor basis, rather than a voltage-to-ground basis, for effectively grounded circuits; the clearances are quite small and are between conductors in some cases. The words "all voltages are between conductors" were inadvertently omitted from the first printing of the Sixth Edition of the Code.

The reduced insulation requirement of 600 V in Rule 239F2(c) in the Sixth Edition was also intended to apply to the 1000 V insulation requirement of Rule 239D2(b).

The 1977 Edition added a reference to Table 235-6, which also covers some of the clearances mentioned in this rule. Table 239-1 was expanded to three voltage classifications for easier use and limited so as to cover only clearances between vertical or lateral conductors and surfaces of supports, span, guy, and messenger wires. Clearances between vertical or lateral conductors and line conductors now appear only in Table 235-6.

In the 2002 Edition, a Footnote 5 was added to allow reduced clearances to guy insulators. This is similar to Footnote 11 on Table 235-6.

Rule 239D2c of the Sixth Edition was changed in the 1977 Edition to allow the use of 600 V insulation with conductors of 300 V and less, to conform with industry practice. This rule was renumbered to Rule 239E2c in 1990 and incorporated into Rule 239E2a in 1997. In the 2002 Edition, a clearance is now specified from vertical conductors to the pole surface and not the former spacing to the pole center. The clearance now does not vary with pole size. The clearances of Footnote 5 of Table 239-2 were revised to be consistent with Rule 239E2a, since they are now clearances to pole surface, rather than spacing from pole center.

Table 239-2 was developed in the 1977 Edition from the language of old Rule 239D2e. However, it was felt that the old provision of 20 in from the pole center is inadequate to cover voltages between 16 and 50 kV to ground. Rule 239D2d was changed in 1977 by deleting grounded metal conduit as a means of construction within the zones of Table 239-2, in order to provide a safer installation.

The 1984 Edition recognized that low-voltage 230C3 cables are widely used at a lateral conductor to a building or intermediate pole and that they must often pass near a guy on a supporting structure. Since the cable is insulated and the guy is either grounded or insulated, a clearance is needed primarily to limit mechanical abuse rather than electrical problems. As a result, a new NOTE 5 was added to Table 239-1 to allow a clearance of 50 mm (2 in) for such installations.

A new NOTE 6 was added to Table 239-1 in the 1984 Edition to be consistent with Rule 235E, Table 235-6, NOTE 10. New NOTE 6 requires the phase-to-neutral voltage to be used to determine clearances from phase conductors to the support when a neutral meeting Rule 230E1 is used.

Rule 239D2d(3) was revised in the 1984 Edition. It applies only to "supply" grounding conductors (and, thus, is not in conflict with Rule 239G3, which prohibits bare communication grounding conductors in the supply space) and it serves as a reminder about the Rule 239C mechanical protection requirements. In 1997, the grounding conductor attachment to a pole was moved into Rule 239A1.

As a part of the extensive revision of Rule 239 in 1997, this rule was greatly streamlined by combining former Rule 239E2a, Rule 239E2c and Rule 239E2d into a new Rule 239A1 and new Rule 239E2a and its EXCEPTION. Rule 239E2b was revised to allow open supply wires to be run directly to the head of luminaires on joint-use poles, if the luminaires are located completely in the supply space (see Figure 239).

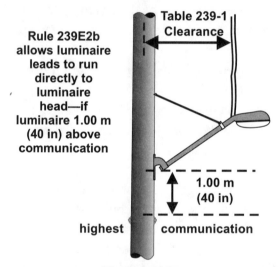

Figure H239E
Wires run directly to luminaire head

239F. Requirements for Vertical and Lateral Communication Conductors on Communication Line Structures or Within the Communication Space on Jointly Used Structures

(This rule has remained relatively unchanged since the Fourth Edition. It was moved from Rule 239E in 1990; former Rule 239F was moved to Rule 239G.)

Rule 239F provides mechanical room for maintenance and provides consistency with the requirements of Rule 235 and Rule 238. The 1997 revision of Rule 239 modified the title and content of the former rule for internal and external rule consistency, but did not create new requirements not previously applicable through rules.

239G. Requirements for Vertical Supply Conductors and Cables Passing Through Communication Space on Jointly Used Line Structures

(This rule was moved from Rule 239F in 1990; former Rule 239G was moved to Rule 239H. In the 1997 extensive revisions of Rule 239, this rule was completely reorganized and shortened. Previous Rule 239G1 was titled Cables Meeting Rule 230C, *but other rules referred to it for the extent of guarding above communication facilities. New Rule 239G1 now specifies guarding for all supply conductors and cables. Former Rule 239G2 became part of EXCEPTION 1 to new Rule 239G1. Former Rule 239G3 was renumbered to 239G2. New Rule 239G3 is developed from (1) the former portion of Rule 239G1 specifying nonmetallic conduit or covering for protection of workers under various conditions of exposure to ungrounded conductors or parts plus (2) portions of former Rule 239G5. New Rule 239G4 was created from the former 239G7. Rule 239G5 is the former 239G6.)*

The 1990 Edition revised the rule for conformance with modern materials and associated good practice. Rule 239G1 requires vertical supply conductors or cables attached to the structure (see Rule 239A for items that can be directly attached) to be physically covered to protect those conductors and cables (and, of course, the line worker) from being gaffed by a line worker climbing the pole. This is a physical covering that is not part of the cable construction; cable jackets do not meet this requirement. The intent is to ensure that neither a cable jacket nor its internal conductor(s) will be breached by climbing line workers. EXCEPTION 1 allows (1) 230E1 neutrals, (2) 230C1 supply cables, and (3) jacketed multiple conductor supply cables to omit the covering, only if they are not located in the climbing space. Rule 239G3 allowed the use of coverings and specified an option: (1) of bonding conductive conduits or coverings to the grounded communications facilities or (2) of covering such conductive facilities with nonconductive ones to limit the opportunity for two voltage potentials to exist on "grounded" facilities within the communication space.

Vertical supply conductors carried through a space occupied by communication conductors require special protection, especially where the voltage is high. Line workers who make repairs or extensions to communication circuits cannot easily avoid coming into contact with such supply conductors; therefore, such supply conductors must be protected where they are liable to be touched by communication line workers. The distance to which the insulating or grounded enclosure extends below the communication conductors is determined by the position of the line worker's spur when working on the wires.

EXCEPTION 2 to Rule 239G1, which allows a supply grounding conductor to go down through the communication space without a covering was further restricted in the 2002 Edition by requiring it to be bonded to the communication messengers at that structure. This is similar to the treatment of metallic riser coverings in Rule 239G2.

Rule 239G (Rule 239F of the 1987 and prior editions) was rewritten in the 1977 Edition to clarify its meaning and to minimize the presence of exposed grounded metal near ungrounded light fixtures.

Another change was made to clarify the type of cable and the method of its attachment where secondary conductors are run through communication space on joint poles. Jacketed multiple conductor cable is specified for vertical runs of secondary when not in conduit, in order to provide some protection for the conductor and insulation, and to keep the conductors from separating. Duplex, triplex, and quadruplex cables meeting Rule 230C3 or Rule 230D do not have an outer jacket over the internal, insulated conductors and, therefore, are not considered as jacketed multiple conductor cables.

Using vertical runs of supply cable on pins and insulators as a means of running vertically through communication space is largely outmoded; that provision was deleted in the 1977 Edition.

Rule 239G4 was added in the 1984 Edition (Rule 239F7 in 1984 and 1987; and Rule 239G7 in 1990 and 1993) to recognize the use of nonmetallic U-guards, but now refers only to the relationship of aerial supply services to communication on the same structure.

It should be noted that Rule 239G3 [Rule 239F5a(2) of the 1987 and prior editions; Rule 239G5a(2) in 1990 and 1993] does not *require* grounding of luminaires and traffic signal attachments, but it limits what can be done within communication space unless certain grounding requirements are met.

The 1997 major revision consolidated like requirements and coordinated this subrule with the remainder of Rule 239 and with Rules 235–238.

239H. Requirements for Vertical Communication Conductors Passing Through Supply Space on Jointly Used Structures

(This rule was moved from 239G in the 1990 Edition.)

Communication conductors passing through a space occupied by supply conductors require an insulating protection to limit contact by supply line workers. The reference to wood molding was deleted in the 1997 Edition, in favor of the more general language *suitable nonmetallic material*. In the 2002 Edition, EXCEPTIONS were added to no longer require covering over communication cables in the supply space on metallic or concrete supporting structures, since the structure itself is conductive.

239I. Operating Rods

(This rule was added in the 1990 Edition.)

Operating rods for supply switches and equipment can have a voltage upon them that is different than that on the grounded communications facilities, even when both are grounded. As a result, they are required to be located outside the climbing space to avoid potential voltage differences as well as mechanical impediment problems.

239J. Additional Rules for Standoff Brackets

This new 1997 rule clarified the capability of using standoff brackets to support vertical runs away from the structure to aid line workers in climbing. Nonmetallic conduit may be used, but it cannot be used to meet basic cable insulation requirements. A cross-reference to Rule 217A2 is given to remind the user of the required clearances *between* standoff brackets.

Only three types of cable (communication, supply cables of any voltage meeting Rule 230C1a, and supply cables less than 750V) may be supported directly without an enclosing conduit. However, Rule 239D1 requires conduit within 2.45 m (8 ft) of the grade to protect the supply cables.

Section 24. Grades of Construction

240. General

(Rule 240 remained untouched from the Third through the Sixth Editions. The 1977 revision included the requirements of previous Rule 241B within Rule 240A in the 1977 Edition. Rule 240B was added in the 1977 Edition. The 1997 Edition completely revised Sections 24–26, merging Grade D with Grade B and further splitting the load factors out from the older overload capacity factors.)

Grades of Construction are required to differentiate between the relative degree of strength and expected performance required of different constructions, especially where a line of one classification of line is constructed near or over another line or a major transportation right-of-way. In a few cases, the relative increase in exposure in urban areas requires superior construction. Differences in voltage classifications of facilities also affect required Grades of Construction. The Grades of Construction are generally more restrictive where conductors of higher voltages cross or conflict with those of lower voltages.

The failure of a supply conductor that crosses above another of lower voltage, whether or not on a common crossing structure, may subject the equipment of the lower-voltage system to abnormal electrical strain. Should this result in failures of low-voltage apparatus or wiring, then operatives and consumers may be exposed to conditions with which they are neither familiar nor prepared to meet.

Furthermore, the falling of any conductor across the signal wires used for controlling train movements may cause serious accidents if it inhibits the use of the signal system. Adequate strength, as indicated in the succeeding rules, is therefore necessary to maintain the clearances specified.

Different requirements are appropriate for different degrees of potential problems. Three different degrees are recognized for supply lines; the corresponding graduations in the minimum standard for construction apply mainly to the strength of the supporting structures. The current Grades of Construction applicable to supply lines are designated as B, C, and N; for the first two of these grades, specific strength requirements are provided. Grade B represents the strongest construction.

Prior to the 1997 Edition, four different Grades of Construction were applicable to communication lines: B, C, D, and N. For communication lines at crossings over railroad tracks, the Grade of Construction was designated as D, which varied so slightly from Grade B that Grade B was substituted in 1997.

In the Fourth Edition of the Code, two other Grades of Construction were designated: A and E. Experience with the rules in that edition indicated that certain of the strength requirements for Grade A could reasonably be modified to agree with those for Grade B. In order to simplify the Fifth Edition, former Grades A and B were combined into a new Grade B. Grade E was also eliminated by appropriate changes in and additions to the rules for Grade D.

The loadings that must be assumed for Grades B and C are contained in Section 25. No specific loading requirements are provided for Grade N. The strength requirements for the various Grades of Construction are specified in Section 26.

No requirements for provision of insulating coverings for conductors in overhead lines of any voltage have been made. While such coverings sometimes aid reliability by limiting burnouts due to conductors contacting one another or grounding out on trees, the reduction of hazard derived from their use is problematical. Their use may even cause an added hazard for the higher voltages. They deteriorate after being in service for several years, and their use in this condition may give rise to a false feeling of security. The provision of appropriate clearances, spacings, and Grades of Construction have generally been shown to be more effective than the use of covered conductors in reducing problems associated with overhead utility construction.

Adding a covering around a conductor increases the diameter and weight of the conductor. When ice-loaded or wind-loaded, a covered conductor transmits a greater loading to the supporting structure than does an uncovered conductor; stronger structures may thus be required. Taller structures may also be required as a result of increased sag caused by the additional mass of the covering; the covering both adds weight and reduces conductor cooling under thermal load.

Where two or more conditions define the Grade of Construction required, the supporting structure itself is required to meet the highest of these Grades of Construction. However, it should be recognized that several different Grades of Construction for conductors may be applicable to different parts of the structure, such as support arms and attachments.

> *NOTE*: Section 24 specifies the Grade of Construction that is *required*. Various utilities initially install a Grade of Construction higher than that required for safety purposes to increase reliability (such as for transmission lines) or increase the life of the line. When such occurs, the NESC would not require other items to be increased in Grade of Construction to match, but this is often required by the utility so that the overall installation can achieve the increased reliability or longevity.

241. Application of Grades of Construction to Different Situations

(In the 1977 Edition, former Rule 241B was moved to Rule 240. Although slightly reworded, the other requirements of Rule 241 remain essentially unchanged.)

241A. Supply Cables

Where the conductors of a circuit are all in a cable, well-insulated from each other, and enclosed in a grounded metal sheath, the danger of shock from contact is greatly reduced; the likelihood of a high potential on such conductors being communicated to another wire coming in contact with the metal sheath is likewise reduced. Such conductors are consequently not required to be of as high a Grade of Construction as open high-voltage wires. For further discussion of this subject (see Rule 261I).

241B. Order of Grades

This rule details the restrictive nature of the grades, with Grade B being the highest and most restrictive Grade of Construction for supply facilities (see Rule 240).

241C. At Crossings

When an overhead line crosses in one span over two other lines, two problems can occur. One problem is the possibility of the accidental contact of the higher wire with one of the others. The second problem is a possibility that, by falling upon both, the upper line may bring the two lower lines into electrical connection. The Grade of Construction required for the higher line is, therefore, required to be not less than that required if one of the lower lines crossed the other, since the same possibilities are involved.

A new paragraph was added in the 1984 Edition to define "At Crossings" and specifically to indicate that joint-use or colinear construction is not in itself to be considered in the crossing category.

241D. Conflicts

A distinction was made in earlier editions of the Code between conductor conflict and structure conflict, as can be seen by referring to the definitions. A structure conflict imposed requirements only upon the supporting structure and not upon those conductors not involved in the conflict. Conversely, if a conductor alone was conflicting, only the conductor was required by the Sixth and prior editions to meet the corresponding obligations; the structure that carried it was allowed to be of a lower grade. This rule was revised in the 1977 Edition and, in concert with other changes, the distinction between conductor conflict and structure conflict was eliminated by deleting conductor conflict. See the discussion of Rule 221 for more explanation of conflicting lines.

242. Grades of Construction for Conductors

It must be recognized that the several parts of a structure (including support arms, pins, and insulators) may be required to comply with several different Grades of Construction. In addition, the minimum sizes and sags of different wires or sets of wires on the same pole line may have to meet the requirements of different grades. However, the Grade of Construction for the structure is required to be not less than that of the highest Grade of Construction for any of the supported facilities (see Rule 243A).

For reasons already stated (see Rule 240), a distinction was historically made in the requirements for urban districts and rural districts. However, as time has progressed, the distinction has become more difficult to ascertain, and the results no longer appropriate, particularly because some areas previously considered as rural often become urban in a very short time as development takes place. In many areas, the urban development may not be contiguous to existing urban areas, such as when whole farms are subdivided (sometimes with golf courses, stores, and offices mixed in with the residential areas), new vacation lodging and destinations are developed, etc. As a result of changes in the growth habits of urbanized areas throughout the United States in the last few decades, few rural lines can reliably not expect to have one or more convenience stores erected at intersections or residential developments of some sort somewhere in their length during their expected lifetimes.

Growth often occurs so fast or in such places that the utilities have no advance warning when designing lines in what are presently rural areas. Further, using Grade N in some circumstances can produce inappropriate results, such as a rural constant-potential upper supply conductor crossing over a constant potential supply conductor in excess of 8.7 kV (which could be transmission voltage). As a result, the 2007 Edition removed the distinction between urban areas and rural areas in both Table 242-1 (electric supply) and Table 242-2 (communication). This essentially limits using Grade N Construction to supply service drops, some aerial high-voltage cables, and some communication cables.

I   In each case in Tables 242-1 and 242-2, the degree of potential hazard is determined by the voltage of the circuits concerned and, when circuits of different voltage are placed on the same supporting structure, by the arrangement of the circuits with respect to each other.

When lines are upon private rights-of-way, the relative potential for a hazard to the public may be less than for lines upon well-used, public rights-of-way. It should be considered, however, that some locations on private rights-of-way could present a relatively greater problem potential, depending upon the character of the land and its use.

The term *limited-access highway* is used to distinguish between areas requiring Grade B versus Grade C construction. The term has caused difficulty in interpretation, primarily because of the lack of specificity of which highways do (and do not) fit the meaning of partially controlled access. Based on the code language and American Association of State Highway and Transportation Officials (AASHTO) documents, an official Interpretation was issued concluding that the term *limited-access highway* is intended to refer to both "fully controlled access" (where grade crossings are prohibited and entrances are carefully controlled) and "partially controlled access" (where some grade crossings are allowed and entrances are restricted, often using collector roads for local access), where entrances and crossings are controlled for the safety and improved flow of traffic. After further review of these issues, the 1997 Edition added a definition of limited-access highway that limited the use of the term to "fully controlled access." This is a practical term to apply. In addition, so much extra capacity is required for Grade C construction that there is little concern about structural problems over any kind of highway. There has never been an intent to require Grade B over ordinary streets and highways, and that was made clear in 1997 in footnote 11 to Table 242-1 and footnote 5 to Table 242-2.

The 2007 Edition added navigable waterways to railroad and limited-access highway crossings in requiring the use of Grade B Construction. Recent problems in using waterways for rescue access due to fallen lines after hurricanes and other natural disasters prompted the change.

The Grade of Construction requirements for supply conductors crossing over or overhanging communication cables may increase to Grade B if the requirements of Footnote 7 or Footnote 8 are not met. For Grade C construction under the 1990 and earlier editions, supply crossing over communication requires the use of a higher *overload capacity factor* in Section 26; for this purpose, joint-use colinear construction is *not* considered to be a crossing.

Footnote 7 of Table 242-1 was revised in the 1984 Edition to allow the note to apply to 4800 V delta systems, as it had been in the Sixth Edition.

The "contact" referred to in Footnote 8 to Table 242-1 is electrical contact. Self-supporting communication cables having an insulated messenger can be used with this rule if (1) such insulation would prevent effective electrical contact or (2) if electrical contact is made, the prompt de-energization requirements will be met. *Prompt de-energization* is not defined in the Code. Footnote 8 (2) of Table 242-1 should be considered in conjunction with Rule 223 (Rule 287 of the 1987 and prior editions). If communication conductors meet the requirements of Rule 223, then the requirements of Footnote 8 (2) are also met. Typically, such measures as those listed in Rule 223 (Rule 287 of the 1987 and prior editions) are used in conjunction with bonding of the communication cable messenger(s) to the neutral of the electric supply system; this limits the voltage that can be impressed on the communications facilities to a level at which those measures can protect customers' premises.

242A. Constant-Current Circuit Conductors

"Constant current" circuits are arranged and operated so as to intentionally vary the system voltage in order to maintain the system current at the regulated level. "Constant potential" circuits, on the other hand, are the normal class of supply circuits where the voltage is held more or less at some predetermined level, such as within the tolerances of ANSI C84.1, *Voltage Ratings for Electric Power Systems and Equipment (60 Hz)*.

Constant-current circuits, if located above streets and alleys in urban districts, cause the same general hazards to traffic below, or to other supply circuits near them and at lower levels, as do other supply circuits of the same voltage. Under such conditions, therefore, the same Grade of Construction is required. Where such circuits, however, expose communication conductors by crossing above, conflicting with, or being located above and on the same poles with the communication conductors, the relative hazard to the communication conductors, considering existing methods of communication-conductor protection, may be very different from that of a constant-potential circuit of equal voltage.

The circuit protection provided by existing types of communication protectors is sufficient to prevent circuit interruption from contact of the communication circuits with the common types of constant-current circuit in which (1) the current does not exceed 7.5 A and (2) the arrester will, in general, withstand discharge up to this limit. A ground resistance of even 15 Ω will not then raise the voltage at the communication instrument high enough to present a serious hazard.

On the other hand, if higher-rated protectors and arresters were to be used, the danger to communication subscribers from constant-potential circuits would be increased. It is, therefore, appropriate to limit the size of the circuit protection for communication circuits exposed to constant-current circuits. If, however, the communication circuit should be interrupted, the inductive character of the series circuit might sustain an arc and present some degree of hazard to equipment and possibly to personnel.

242B. Railway Feeders and Trolley-Contact Circuit Conductors

The relative hazard potentials of supply wires are due to the voltages at which they operate; trolley feeders must be considered hazardous for the same reason. This is particularly true where the trolley feeder is bare and placed below communication conductors on joint-pole construction. This position is practically necessary because of the relatively greater sag of the trolley feeder and the need to avoid vertical runs through communication conductors. The fall of a communication conductor may, in this case, cause damage. The necessary climbing space should be provided in spite of the extra crossarm strength and bracing required by the usually heavy feeders.

242C. Communication Circuit Conductors Used Exclusively in the Operation of Supply Lines

(This rule has remained essentially unchanged, except for the inclusion of all types of communication conductors located in the supply space in the 1997 Edition.)

Rule 242C was changed in the 1997 Edition to reflect changes in the clearances rules in Rules 224 and 235 of the 1993 Edition. The former rule only applied to communication used exclusively in the operation of supply lines. Rule 224 places limitations on what communication lines can be located in the supply space and required qualifications for those who work on such communication lines. Note that, although this rule allows communication meeting Rule 224A2 to have Grade N construction, Grade B will be required, depending upon the location.

242D. Fire-Alarm Circuit Conductors

In this rule, consideration has been given to the adverse consequences that would result from a broken fire-alarm circuit conductor. Early official Discussions of this rule stated the desirability of such conductors being strung with sags considerably greater than those then specified as minima in Rule 262I4. As of 1997, these conductors are considered as communication circuit conductors and are required to have the Grade of Construction appropriate for their location.

242E. Neutral Conductors of Supply Circuits

(This rule was added in the 1977 Edition.)

If the neutral conductor meets the grounding requirements and circuit-voltage-limitation requirements of Rule 230E1, the neutral can be located below the distribution primary voltage phase conductors and either above or below the distribution secondary phase conductors. As a result, the voltage rise from contact by the secondary voltage would be small. However, neutrals that do not meet Rule 230E1 are required to be treated the same as a primary voltage phase conductor for both clearance and strength purposes.

242F. Surge-Protection Wires

(This rule was moved here from prior Rule 261F3 in the 1977 Edition.)

The title was changed in the 1984 Edition to reflect the general change to the use of "surge" instead of "lightning" protection.

243. Grades of Construction for Line Supports

Where there are a number of sets of conductors on the same pole on different crossarms, the longitudinal strength of the crossarms, pins, and fastenings supporting each set of conductors is determined by the Grade of Construction required for that particular set. However, the transverse and longitudinal strength of the supporting structure is determined by the highest grade carried, except for the specific cases listed.

When fire alarm conductors began to be treated as other communication conductors in the 1997 Edition, prior Rules 243A2, 243B2, and 243C2 were deleted.

Rule 243A4 (Rule 243A5 prior to 1997) requires that the Grade of Construction of a conflicting structure be that required for a crossing of its conductors over those of the other circuit. It should be noted that, where track rails serve as a part of the railroad block system (which is usually the case), such rails are considered to be conductors of another line. If an overhead line parallels the tracks close enough alongside to meet the definition of *structure conflict*, then Rule 243A5 will require Grade B construction.

The requirements of this rule, although reworded in the 1977 Edition, have remained essentially unchanged since the Third Edition. Rule 243C5 (243C6 prior to 1997) was added in the 1977 Edition.

Section 25. Loading for Grades B, C, and D

(Section 25 was revised in 1997 by adding new Rule 253)

It is, of course, generally impractical to design overhead structures to withstand the *most severe* weather conditions that may occur anywhere within such a large area as a loading district. Experience has shown that this is not necessary in order to provide a very high degree of safety, since coincident combinations of extreme ice and wind conditions occur very infrequently, and then only in relatively restricted areas. Data on climatic loading have been collected for a number of years by various wire-using organizations, and these data were carefully reviewed in connection with recent revisions to Part 2. Both the climatic data and the extensive experience of the wire-using companies were used as a basis for the selection of the loading assumptions contained in Section 25, as well as for the delineation of the loading districts.

The strength requirements in the rules for Grades B, C, and D provide a degree of safety in keeping with the conditions under which each of these Grades of Construction is required.

In the Second Edition of the Code, different degrees of loading were specified for different types of situations, even in the same loading district. In later editions, a single set of loading assumptions was used in each loading district, but different allowable stresses were specified in different rules for the same materials. This recognized different degrees of potential problems. This latter method simplified the rules but, because of the choice of relatively severe transverse-loading assumptions, necessitated the use of allowable stress values in some cases that were considerably out of line with those used for the same materials in other fields of engineering. Studies made between the issuance of the Fourth and Fifth Editions showed that wind pressures of the level formerly assumed for transverse loading seldom occur concurrently with the assumed ice conditions, and then only in restricted areas.

As in the Fourth Edition of the Code, a single set of loading assumptions was specified in the Fifth and later editions for each loading district for all types of situations covered. The transverse-loading assumptions (1) fall well within the range of weather experience and (2) permit the use of allowable stress values in keeping with usual engineering practice. The vertical-loading assumptions in the Fourth Edition have been retained. While the method of specifying conductor loading in the Fifth and later editions differs from that in the Fourth Edition, substantially the same conductor-loading assumptions were retained by the addition of the constants given in Rule 251. Some of the constants were combined in the 1977 Edition.

At the time of preparation of the Fifth Edition of the Code, considerably more data and experience had been accumulated by the wire-using companies in the United States than were available when earlier editions were prepared. These data were considered carefully at that time. While there are a number of factors involved in the strength of an overhead line, it is not possible to include all of these items in the Code if workable safety rules are to be prepared; the Code is not intended to be a design manual. The principal factors have been included, however, and carefully considered values have been assigned to them for the various Grades of Construction and loading districts covered. Under these circumstances, assumptions made in the Fifth and later editions of the Code may not, in some cases, represent actual pressures and loadings encountered over a period of years in actual practice; they are, however, a much closer approximation than values resulting from the use of rules given in previous editions. When used in conjunction with the allowable stresses specified in the rules, these loading assumptions will provide construction that experience has shown to be on the safe side in situations where Grade B, C, or D is required. For situations other than those for which Grade B, C, or D is specified, the adequacy of line construction can be determined only by examinations of experience and the local conditions involved.

250. General Loading Requirements and Maps

In large portions of the United States, combinations of ice and wind present the greatest structural challenge to be seen by overhead utility facilities (see Figure H250B).

Figure H250B
Ice on pole and supported facilities

The data available from the United States Weather Bureau and from wire-using companies relating to the frequency, severity, and effect of ice and wind storms in various parts of the country provided a basis for dividing the United States into three loading districts. They are shown on the map in Figure 250-1 as heavy, medium, and light.

The stress in a conductor depends upon (1) the pressure of the wind, (2) the thickness of the ice coating carried by the conductor, and (3) the changes in temperature that affect the conductor length and so change its stress, if the supports are fixed. These three factors occur in varying combinations in different districts, and they vary from day to day in the same district. Weather records show that wind velocities of over 36 m/s (80 mph) sometimes occur in districts where ice accumulations over 13 mm (0.5 in) thick and low temperatures are frequent. On the other hand, other districts exist where winds exceeding 18 m/s (40 mph) are unknown and where ice or very low temperatures do not occur.

The maximum velocities recorded by many of the observing stations of the United States Weather Bureau are taken over a period of five minutes; they do not register the maximum values attained in gusts of short duration which, for several seconds, may have velocities far in excess of the average recorded for any 5-minute period. Where the observing stations are provided with instruments giving records of instantaneous values, the instantaneous maxima are found to be considerably in excess of the 5-minute averages for the same stations. Moreover, the Weather Bureau stations are often located in cities and towns that are in low altitudes and sheltered. Such stations do not give a fair indication of conditions that are likely to prevail in the more exposed regions. Buildings, trees, and other obstructions reduce the velocities recorded by the instruments and also reduce the pressure upon overhead wires in these locations.

On the other hand, overhead distribution- and service-voltage supply lines and communication lines are usually nearer the earth's surface than Weather Bureau stations. For moderate wind velocities, this usually means that the winds are less violent, and definite relationships have been published showing how wind velocity increases with distance above the earth's surface. There are significant data that indicate that this relationship does not hold for the wind velocities experienced during storm conditions; the variation with altitude during such conditions is irregular. Observations of wind velocity at various altitudes have shown velocity to increase with height above ground, at times, and decrease, at other times. Under the conditions that place the greatest load upon the line, it cannot be assumed that the wind velocity at the conductors will be any less than at the location of the weather observers' instruments.

The long-used ice and wind loadings of Rule 250B apply only three gradations of loading to the entire nation; it is obvious that different degrees of strength are required in the three loading districts. The loadings specified for the several localities are not intended to represent the actual pressures and loadings to be encountered in those particular regions. The relative values of the loadings cannot be expected to conform to the assumed values given in these rules. They are, however, quantities chosen for use in making computations of working stresses after careful consideration of local weather conditions. The assumed loadings are not chosen to represent the most severe conditions that are likely to be encountered in the various locations, but are values that have been selected after full consideration of both present accepted practice and the influences that tend to modify or diminish the stresses that might be expected to result from the actual loadings.

The western boundary of the Heavy Loading District was shifted eastward in the Fourth Edition so that the entire state of Montana and a considerable portion of Wyoming are now in the medium-loading district.

The substantial further changes in the Fifth Edition were based on studies of the additional weather data and experience accumulated since the Fourth Edition was published. The boundary lines between the loading districts were chosen so that, as far as practical, they follow natural physical dividing lines or the boundaries of major political subdivisions already established and easily recognized.

While general boundaries are indicated in the states of California and Nevada, it has been intended that the detailed boundaries in these states will be as defined by the orders of the regulatory authorities in these states. It is known that storms of heavy-loading intensity occur in certain local areas in Washington and Oregon, and it is the intent that the boundaries of such localized areas be defined in the states themselves.

It must be recognized that weather conditions do not change abruptly at the points that have been fixed as boundaries of the districts. The changes in conditions are generally graded and do not take place at a definite line; they will tend to shift about somewhat from year to year. However, for the purposes of the Code, boundaries are needed to detail the demarcation between loading requirements.

Rule 250B contains loadings for a combination of ice and wind. Rule 250C contains extreme wind loadings on bare facilities. If both Rule 250B and 250C apply, the loading that produces the greatest stress in the structure and supported facilities must be used to determine required strength.

A new Rule 250D was added in 2007 to specify loadings for extreme ice and concurrent wind. The loadings of Rule 250D are based upon freezing rain and concurrent wind. In some localities, the loadings of Rule 250D may be less than those of Rule 250B, particularly where the loadings of Rule 250B are based upon a history of rime ice (hoar frost).

250A. General

There are four general rules. Rule 250A1 requires that the greater loading of Rules 250B (traditional ice/wind combination), 250C (extreme wind), or 250D (extreme ice and concurrent wind) (250D was added in 2007.) be met if two or more are applicable. Rule 250A1 was clarified in the 1981 Edition; the rule in the 1977 Edition had implied unintentionally that the requirements of Rule 250 applied to all structures.

Rule 250A2 was added in the 1977 Edition to recognize that, especially in light-loading districts, some or all parts of structures may experience loadings during construction or maintenance that are greater than those experienced in normal operation. The rule was revised in 2007 to clarify the intent that temporary loads on structure components be considered, in addition to the main structure; bracing or other support (such as a crane) and/or load controls may be necessary to control temporary loads during stringing, worker loads, equipment installation, etc. As a part of the 2007 revision, the former increased vertical load factor for Grade C was reduced to normal relationships with the Grade B vertical load factor.

Rule 250A3 recognizes that there may be places within designated loading areas that have greater or lesser loadings than specified in the rules. No reduction in specified loadings is allowed without approval of the administrative authority. When developing a detailed loading analysis of a specific area, use of statistical methodologies equivalent to those used to develop the maps in Rule 250C and 250D is required as of 2007.

Rule 250A4 was added in the 2002 Edition to recognized that the strength required by Section 25 and Section 26 are sufficient for earthquakes. Generally these structures are flexible enough to withstand earth tremors. Of course, if the earth opens up under a structure, the structure will tend to fail.

250B. Combined Ice and Wind Loading

Rule 250B contains the requirements for combined ice and wind loadings. Table 250-1 was revised in the 1981 Edition to include the temperatures at which tensions are to be measured. The prescribed combination of ice and wind loadings obviously is equivalent to a greater wind load with lesser ice load or vice versa. These combinations were originally selected as representing the worst effective loadings generally found in the respective areas. Since the original development of Rule 250B, additional data on freezing rain ice accumulations with concurrent wind has been developed and included in ASCE Std 7; this data was used to develop an additional ice and wind loading case in Rule 250D in the 2007 Edition. Since Rule 250D loadings are greater than Rule 250B loadings in some areas and lesser in other areas (because Rule 250B also considers rime ice/hoar frost), the greater of the loadings of Rules 250B, 250C, or 250D must be considered, if all apply. Unlike Rule 250C extreme wind loadings, neither the traditional ice and wind loadings of Rule 250B nor the freezing rain and concurrent wind loadings of Rule 250D include a height adjustment factor for wind speed, such as that used in Rule 250C.

The following outlines detail the boundaries between the three *loading districts* of the Fifth and later editions for the benefit of those who may desire to trace the boundary more definitely than allowed by the map.

Boundary Between the Heavy- and Medium-Loading Districts for the Fifth and Later Editions

Beginning at the Atlantic seaboard, follow the 38th parallel of north latitude to Albemarle County, Virginia; follow the eastern boundaries of Albemarle, Nelson, Amherst, Bedford, Franklin, and Henry Counties of Virginia to the southern boundary of Virginia; follow the southern and western boundaries of Virginia to West Virginia; follow the western boundary of West Virginia to the Ohio River; either (1) for the Fifth and Sixth Editions only, follow the Ohio River *down* to the Mississippi River and down the Mississippi to the Arkansas state line, or (2) for the 1977 and later editions only, follow *up* the Ohio River along the Ohio/West Virginia border and then along the northern boundaries of Belmont, Guernsey, Muskingum, Licking, Franklin, Madison, Clark, Montgomery, and Preble Counties to the Indiana state line, and then south along the state line to the Ohio River, and then south along the Ohio and Mississippi Rivers to the Arkansas state line; follow the northern Arkansas state line westward to the Oklahoma state line; follow south along the Arkansas–Oklahoma state line to the Red River; westward on the Red River to the intersection of the eastern boundary of Red River County, Texas; in Texas follow the eastern and southern boundaries of Red River County, the southern boundary of Delta County, the eastern and southern boundaries of Hunt County, the southern boundary of Rockwall County, the eastern boundary of Dallas County, the southern boundaries of Dallas, Tarrant, Parker, and Palo Pinto Counties, the eastern and southern boundaries of Eastland County, the southern boundaries of Callahan, Taylor, and Nolan Counties; north on the western boundaries of Nolan and Fisher Counties; west along the southern boundary of Kent County; north on the western boundary of Kent County to the intersection with the White River; northwest along the White River to the northern boundary of Lamb County; west on the northern boundary of Lamb and Bailey Counties to the Texas–New Mexico state line; north on eastern New Mexico state line to the southern Colorado state line; west on the Colorado state line to the southeast corner of Costilla County, Colorado; follow northward along the eastern boundaries of Costilla, Alamosa, Saguache, Chaffee, Lake, Eagle, and Routt Counties in Colorado to the northern Colorado state line; follow eastward along the Colorado state line to the 106th meridian of west longitude; follow north on the 106th meridian of west longitude to the intersection with the 43d parallel of north latitude; follow east on the 43d parallel of north latitude to the eastern Wyoming state line; follow north on the eastern Wyoming and Montana state lines to the Canadian boundary.

Boundary Between the Medium- and Light-Loading Districts

From the Atlantic seaboard, follow the 33d parallel of north latitude across the states of South Carolina, Georgia, Alabama, and Mississippi to the intersection with the Boeuf River in Louisiana; then southwestward along the Boeuf River to the northern boundary of Caldwell County; along the northern and western boundaries of Caldwell County to the northeastern corner of Winn County; westward along the northern boundaries of Winn, Natchitoches, and Sabine Counties in Louisiana to the intersection with the Sabine River; south along the Sabine River to the northeastern corner of Sabine County, Texas; then in Texas along the northern and western boundaries of Sabine County, and the northern boundaries of Jasper and Tyler Counties to the intersection with the 31st parallel of north latitude; west along the 31st parallel of north latitude to the intersection with the Pecos River; then northwest along the Pecos River to the southern boundary of New Mexico and west on this state line to the ridge of the Guadalupe Mountains; follow the ridge of these mountains to the intersection with the southern boundary of Chaves County, New Mexico; follow the southern boundary of Chaves County, New Mexico; follow the southern and western boundaries of Chaves and Lincoln Counties to the intersection with the Sierra Oscuro Mountains; follow the ridge of these mountains north to the 34th parallel of north latitude; follow west along the 34th parallel of north latitude across New Mexico and Arizona to the southeastern corner of Yavapai County, Arizona; follow west and north along boundaries of Yavapai and Coconino Counties to the intersection with the Colorado River; follow westward along the Colorado River to the Nevada state line; follow north along the eastern Nevada state line to the 38th parallel of north latitude, then westward across the state of Nevada as described in the Rules of the Public Service Commission of Nevada; continue westward along this line to the center of California, then northwestward to the northwestern corner of California.

250C. Extreme Wind Loading

Rule 250C recognizes that tall structures may be subjected to wind-only loadings that exceed the combined ice and wind loadings included in Rule 250B. ASCE Manual 52, *Guide for Design of Steel Transmission Towers*, was one of the references used in this change. This change was made to specifically recognize that the heavy- and medium-loading conditions of Rule 250B do not properly reflect the actual loads applied by wind to structures higher than 18.3 m (60 ft) that carry conductors of 2 cm (0.8 in) diameter or greater. No gust factor was included in Figure 250-2.

The wind loadings of Figure 250-2 are based upon the 50-year mean recurrence interval. It is the intent of the Strengths and Loadings Subcommittee that a loading of that produced by a 50-year wind be used in calculating the loadings to be applied to structures covered by the NESC.

The wording of Rule 250C includes a typographical error which dates to the 1977 Edition; the approved wording of the third sentence begins with "If any portion of a structure *or* supported facilities is located in excess of 18.3 m (60 ft)..."

In Section 26, which specifies *overload capacity factors* to be used with the loads of Section 25, Rule 260C (1977–1987 Editions) allowed the use of an *overload capacity factor* of 1.0 for structures and foundations (1.25 for supported facilities) when the extreme wind loads of Rule 250C are considered; Rule 260C in the 1990 Edition required "at installation" factors to be used for wood structures under the extreme wind conditions. When other wind loads are considered, specific *overload capacity factors* are required by the rules of Section 26.

Rule 250C was extensively revised for clarity in the 1987 Edition. The wind map of Figure 250-2 was updated and Table 250-2 was added. The new table contains translations of the wind speeds of Figure 250--2 into wind pressures on cylindrical surfaces. These wind pressures can be applied to the projected area of conductors and cylindrical structural members without additional force coefficients (shape factors). Where structural members or supported facilities are flat-faced, the force coefficients of Rule 252B are appropriate.

Rule 250C and Figure 250-2 were revised in the 2002 Edition to use the new three-second (3s) gust wind data in revised Figure 250-2 and to include additional factors in the wind loading formula such as gust response factor.

The wind data prior to the 2002 Edition was the so-called *fastest-mile wind speed*. In essence, this was a steady-state storm wind. It was appropriate to apply directly to conductors (which tend to move laterally) before applying significant loads to the structure. It was appropriate to apply a gust factor to wind on a structure, but not to the conductors and cables, since gusts are usually only 20 m (66 ft) or so wide. Since the new data includes a *3s gust wind*, the new data is appropriate for direct application to structures and must be de-rated for application to conductors and cables.

Under storm loadings, waves are continually running up and down a conductor or cable until they hit an attachment point and are reflected back. Maximum transfer of energy to the structure only occurs when the wave hits an attachment point. The new *gust response factors* also consider the diversity of the timing of maximum conductor forces transferred to the structure. As of 2002, the formula for calculating wind pressure now matches that used in ANSI/ASCE 7-98. Included is a *velocity pressure coefficient* k_z that varies with height (see Rule 250C1 and Table 250-2 [Previous Table 250-2 was deleted in 2002.]). Also included is a *gust response factor* G_{RF} that varies with height and span length (see Rule 250C2 and Table 250-3). An *importance factor I* of 1.0 is used for utility structures and supported facilities. The *force coefficient (shape factor)* of Rule 252B is unchanged.

In 2007, the application of k_z was revised so that computations could be made for structure components or large equipment installations at specific heights. For this determination, the k_z for the location of interest is used with the G_{RF} for the total structure height. Some values in Table 250-3 changed in 2007 as a result of the general coordination of the use of decimal places and rounding of calculations.

250D. Extreme Ice with Concurrent Wind Loading

Rule 250D was added in the 2007 Edition. Like 250C, it applies where a portion of the structure or supported facilities extends 18 m (60 ft) above ground or water. Rule 250D also applies where required by Rule 261A1c or Rule 261A2e. This new loading system includes a new freezing rain map overlaid with a concurrent 3-second gust wind speed map in NESC Figure 250-3. This creates another loading case that differs from the original ice and wind case of Rule 250B. Since Rule 250D loadings are greater than Rule 250B loadings in some areas and lesser in other areas (because Rule 250B also considers rime ice/hoar frost), the greater of the loadings of Rules 250B, 250C, or 250D must be considered, if all apply. Unlike Rule 250C extreme wind loadings, neither the traditional ice and wind loadings of Rule 250B nor the freezing rain and concurrent wind loadings of Rule 250D include a height adjustment factor for wind speed, such as that used in Rule 250C.

251. Conductor Loading

(The loadings contained in this rule in the Sixth and prior editions were moved to Rule 250 in the 1977 Edition. This rule was extensively revised in the 1977 Edition to detail the requirements for calculating loads on conductors.)

The 1977 revised wording of Rule 251 specifically details the projection of wind onto ice coverings over conductors and cables. The ice loading is assumed to be that of a hollow cylinder whose inner diameter is equal to the outer diameter of the conductor or cable that it covers; with larger conductors, this is in effect what actually occurs. Generally for the smaller conductors, a lesser volume of ice will actually coat the conductor or cable but, since the roughness factor increases with smaller wire sizes, the rule requirements continue to be practical as the size decreases. The vertical and horizontal components of loading on conductors were also delineated in the 1977 Edition, as well as the total loading to be assumed. These separate specifications are used in calculating required strengths of both a conductor (or cable) and its supporting structure. In the 1981 Edition, appropriate ice loadings for bundled conductors were specifically recognized.

Rule 251A3 was revised in 2002 to require an appropriate mathematical model to determine wind and weight loads on ice-covered conductors and cables. Where available, an appropriate ice accretion model should be used in a qualified engineering study in accordance with Rule 251A4 to determine the expected ice accumulation. The original rule for calculation of ice on conductors and cables is required in the absence of a mathematical model determined by a qualified engineering study in accordance with Rule 251A4.

Rule 251A4 was added in the 1987 Edition to recognize that conductor stranding can affect the "roughness" of a conductor or cable and can affect the ice loading. The loads specified in Section 25 are required to be used unless tests or qualified engineering studies indicate that a reduction is justifiable.

Rule 251B1 recognizes that the weight of spacers, weights, or other equipment on conductors may be significant and needs to be considered.

252. Loads Upon Line Supports

(The loadings contained in this rule in the Sixth and prior editions were moved to Rule 250 in the 1977 Edition. This rule was extensively revised in the 1977 Edition to detail the requirements for calculating loads on line supports.)

252A. Assumed Vertical Loading

The 1977 Edition included a clarification of the requirement that, for vertical loading purposes, the radial thickness of ice is to be computed only upon wires, cables, and messengers—not upon supports. Vertical loads include the weight of the structural components as well as the weight of the supported facilities.

Rule 252A requires computation of radial ice loading on *wires, cables, and messengers,* but the language "but need not be computed on supports" does not specifically require such computation on supports. Consideration of ice loading (and the concurrent wind loading) is not specified in the rules for ice on either the structure or supported facilities. As a result Rule 012C applies, and *accepted good practice for the given local conditions* is required.

The original language of Rule 252A did not consider some installations that are common today, such as flat panel antennas served by 2–4 coax cables running down a tower to the ground. If such installations are located in icing areas and the vertical cable runs bridge over with freezing rain or rime ice (hoar frost) and become a solid mass, they can add significant weight and wind loads to their supporting structures and structural components. If such antenna systems are installed on power transmission line towers, the orientation of the antennas and their vertical cable runs can affect how much of the loads are directed in the transverse direction versus the longitudinal direction. As indicated in IR 538 issued 16 December 2005, such loads may adversely affect the stress on fasteners, support components, and even the supporting structure itself. If so, Rule 012C requires consideration of those loads.

252B. Assumed Transverse Loading

The requirements for application of transverse loads to the structure from conductors and messengers separately from those resulting from wind loads on the structure itself were detailed in the 1977 Edition. The rule continues to recognize that close spacing of conductors may, in some cases, produce a "shielding effect" for wind loads in the heavy- and medium-loading districts. The rule states both the loads that must be considered at angles and the methods by which they shall be determined and combined.

Rule 252B1 requires the transverse loads from conductors and messengers (including the cable loads on messengers) to be the horizontal load determined by Rule 251. Such loads include wind on ice-loaded conductors and cables, if the installation is in an icing area. The effect of ice bridging across between vertical cable runs serving antennas is not specifically required to be considered by Rule 252A for vertical loads. However, there is no such exemption from considering ice loading in Rule 252B1 for transverse loads. It should be noted that wind loads on such installations may be significant, even without ice making a solid panel out of the cables, and can be particularly significant if ice makes a solid wind panel. If such loads are significant, they are required by Rule 012C to be considered. See the icing discussion under Rule 252A and IR 538 issued 16 December 2005 (see also Figure H252B2).

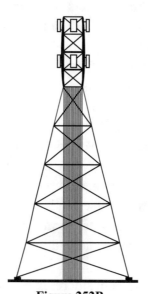

Figure 252B
Wind on vertical cable runs to antenna can be a significant load—
especially if cables are bridged by ice to form a solid panel

Rule 252B1 was rewritten for clarity in the 1987 Edition to state the general rule and provide an EXCEPTION to the rule for open-wire communication with two or more conductors on the same crossarms.

The requirements of Rule 252B2 were clarified in the 1981 Edition by the addition of the specific statement that the wind loads were to be applied to structures and their components. Rule 252B2 was clarified in the 2007 Edition to indicate that the specified force coefficient applies to structures or components having solid or enclosed flat sided cross-sections that are square or rectangular *and have rounded corners*. The structural members used in utility construction meet that criteria; the corners are rounded on the shapes and create smaller turbulence (vortex) loadings than would be the case with sharp edges.

Rule 252B3 details the requirements at angles in the line. As written, it requires the vector sum of the transverse wind load and the resultant load imposed by the wires due to their change in direction. The latter is required to include the effects of wind on the wire tension. At first, this may seem to be a double counting of the wind, since the effect of the wind on the conductors is applied twice—once in conductor tension, and again as a transverse load. To some extent, it is double counting; however, the *angle* used to calculate the resultant is normally the *line* angle, not the increased angle that actually occurs during wind loading. This apparent double counting of wind actually is a proxy that serves to make up for using the shallower angle commonly used when calculating the resultant. If the actual conductor angle under wind-loading deflection, instead of the nominal line angle, is used to calculate the resultant conductor loading on the structure, then it is not expected that the wind should be double-counted. In that case, the actual wind load on the conductors would be translated, along with tension resulting from the weight of the conductors, into the actual amounts and directions of forces applied by the conductors to the structure under wind loading. Then it is no longer appropriate to also apply the transverse wind loading of Rule 252B1. It is necessary, however, to continue to apply the transverse wind loading of Rule 252B2 on the structure itself.

Rule 252B4 was added in the 1977 Edition to recognize appropriately the resultant of wind action on conductors. At the same time, former Rule 252D—*Average Span Lengths* was eliminated.

Rule 252C5 allows temporary measures to be used to hold the forces applied during stringing operations; it does not require the structure to be able to withstand those forces at other times in its life.

The constant of Rule 251B is intended to be used when computing conductor tension under Rule 252B—*Assumed Transverse Loading.*

252C. Assumed Longitudinal Loading

Experience has shown that the placing of guys on wood poles supporting higher-voltage lines may, under many conditions, so reduce the insulation provided by the poles that insulator failures and flashovers are more likely to occur on such guyed poles than elsewhere on the line. On the other hand, line failures of a character such that accidental contacts were prevented at wire crossings by the presence of head guys were much less frequent than formerly anticipated. The rules covering longitudinal strength requirements were accordingly revised in the Fifth Edition to give appropriate weight to these facts.

If a line is built of the same Grade of Construction throughout, it was formerly considered unnecessary to provide special longitudinal strength at intermediate points.

Recent significant lengths of line-structure "domino-type" failures were recognized, however, in the 1977 Edition; Rule 252C6 now recommends that structures with longitudinal strength capability be provided at reasonable intervals along the line. The rule intentionally does not specify the method by which such capability should be achieved, i.e., it does not infer that deadends are necessary; other methods may serve the same purpose.

The recent domino failures included two in Wisconsin: one involved 11.3 km (7 mi) of 345 kV aluminum H-frame structures that toppled between deadends on two different occasions; the other involved similar structures where a 100 km (62 mi) section dropped between deadends. A similar failure in Indiana involved 69 structures between deadends.

At points where there is a change in the Grade of Construction—for example, where a crossing span has been built in a line that is not strong elsewhere—a failure in the weaker portion of the line may affect the stronger portion. Therefore, the longitudinal load at such points is based upon the assumption that certain of the wires may become broken elsewhere. Where wires smaller than AWG No. 2 are carried, it is assumed that two-thirds of them may be broken. However, to ensure protection in cases where there are also a number of larger wires on the pole or where there are no wires smaller than No. 2, a load equal to the loaded tension in two of the largest wires is assumed.

In the case of supply conductors, the actual tension in the conductors corresponding to the existing sag is used in the computation of load. In the case of communication conductors at railroad crossings and limited-access highways, definite tensions in the conductors (in percent of ultimate strength) are assumed for each loading area; such conductors are not ordinarily given larger sags for the purpose of relieving the pull upon supports as is sometimes done with heavy supply conductors.

The term *pull* used in this rule is intended to mean "design tension." Each individual subconductor in a bundled conductor is to be considered as a separate conductor for application of this rule.

The requirements of Rule 252C1 apply to adjacent line *sections* of different Grades of Construction. They do not apply to single points in a line, such as a common-crossing pole.

Rule 252C4 requires the effects of uneven vertical loading or unequal spans to be considered in the structure design. Many companies have produced standards to limit the ration of adjacent span lengths and elevation differences based upon different wire sizes and tensions. Special care is appropriate in mountainous terrain where whole spans may be shaded while others are in the sun; this can produce severe differentials in longitudinal loading if most of all of the ices comes off one span before any comes off the adjacent one.

Except for the requirements of Rule 252C4, which require consideration of differentials in ice loading on wires, conductors, and cables, Rule 252C is silent as to longitudinal loads that may result from ice loading on structure components or supported facilities. Rule 012C may require such consideration on some installations, such as some antenna installations with multiple vertical cable runs located close together on one side of the structure. See the discussions of ice loading under Rules 252A and 252B1 and IR 538 issued 16 December 2005.

252D. Simultaneous Application of Loads

(This rule was moved from 252E in the 1977 Edition when the former Rule 252D—Average Span Lengths was revised and moved to Rule 252B4.)

This rule serves as a proactive reminder that it is the responsibility of line designers to carefully consider the effects of the various conditions that *are expected* to simultaneously load their structures (see Figure H252D-1). Although some designers deliberately overdesign by assuming that the wind at an angle applies perpendicular to each span to get conductor tension and transverse wind loadings, that methodology is not required by the NESC. Others apply the wind perpendicular to the long span and at an angle to the short span; this leads to underdesign. Typically, the NESC design case will be determined by an oblique wind loading along the bisector of the angle (see Figure H252D-2).

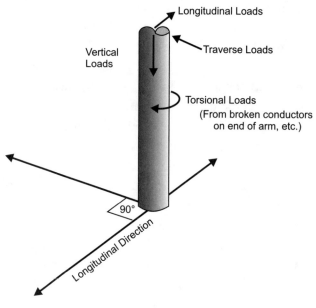

Figure H252D-1
Simultaneous loadings

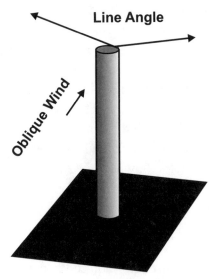

Figure H252D-2
Design wind direction at angles

252E. Simultaneous Application of Loads

(Moved to Rule 252D in the 1977 Edition when the former Rule 252D—Average Span Lengths *was moved to Rule 252B4.)*

253. Load Factors for Structures, Crossarms, Guys, Foundations, and Anchors

(This rule was added in the 1997 Edition.)

This rule was added in 1997 as a part of a general revision of the specification of strength and loadings requirements. Previously, both load factors and strength factors were located in Section 26. As of 1997, load factors for Grades B and C are brought into Section 25 along with the loadings to which they apply.

The load factors of Tables 253-1 and 253-2 apply regardless of material. These load factors are not strictly loading factors; they also include general safety factors, location factors, structure-type factors, and loading direction factors. Distinctions between materials are made in Section 26.

The load factors of Table 253-1 are only to be used with the strength factors of Table 261-1A. This is essentially Method B of the 1993 Edition. The alternate load factors of Table 253-2 are only to be used with the strength factors of Table 261-1B. This is essentially the older calculation method that is still allowed as an alternate method for those who prefer to use it. The 2007 Edition added a sunset clause to phase out use of the older, alternate method by 31 July 2010.

The load factor value in Table 253-1 for Grade C vertical loads on wood structures was revised and increased to 1.90 in the 2002 Edition, and a new footnote 6 was added to keep the value at 1.50 for metal and prestressed concrete. As a result, when strength factors for wood are applied, the total vertical strength required is the same for Grade B and Grade C; this change was made because many structures see their greatest loads when workers are on them. The vertical *overload capacity factors* were the same for Grade B and Grade C before the 1997 change that created Table 253-1.

Support hardware was added to Table 253-1 in the 2002 Edition.

In 2007, the distinction between ***longitudinal loads*** at *crossing* and *elsewhere* was removed from Table 253-1. The practical effect of that change was to retain the *at crossing* values and increase the Grade B load factor for longitudinal loading not at a crossing (i.e., *elsewhere*) from a load factor of 1.00 to the same 1.10 value used at crossings. Also in 2007, Footnote 6 was added to Table 253-1 to clarify that the "at crossings" column under Grade C applies when one line crosses another supply or communication line. Rule 241C and Table 242-1 are both referenced. If a supply or communication line crosses over a limited access highway, a railroad track, or (as of 2007) a navigable waterway, Grade B is required by Rule 241C, Table 242-1, and Table 242-2.

Section 26. Strength Requirements

(Section 26 was extensively revised in 1997. Load factors for Grade B and Grade C were moved to Section 25 to be grouped with their required loadings. Grade D was combined with Grade B in the 1997 Edition.)

Grades of Construction are specified in Section 24 for line conductors and their supports. All lines must meet certain of the requirements of the Code, such as those for clearances. Other requirements depend upon the Grade of Construction; the differences in the requirements for the different grades relate mainly to mechanical strength. They also involve, however, certain other items, such as the electrical strength of insulators.

The Fourth Edition of the Code contained requirements for Grade A, Grade B, Grade C, and Grade N construction. Later experience indicated that Grade A requirements resulted in stronger construction than necessary, in most cases. For this reason, and also in the interest of simplifying the Code, Grade B requirements were amended to include certain of the former Grade A requirements; Grade A was then omitted in the Fifth Edition.

Grade N is the designation given to construction that does not have to meet the requirements of any of the other grades. Section 24 limits the application of Grade N to specific locations. Use of Grade N was further limited in 2007 by the removal of the separate specification for lines in rural areas from Table 242-1. No specified loading requirements, load factors, or strength factors are given for Grade N. There are, however, a few strength requirements for Grade N construction, such as limiting sizes of supply conductors.

The mechanical strength of poles and similar structures is assumed to involve only three considerations: (1) they should be able to support the weight of the conductors when carrying ice of a specified thickness; (2) they should have sufficient strength to withstand the pressure of the wind at right angles to the line; and (3) they should have sufficient strength to withstand the pull in the direction of the line due to any tension in the conductors that is not balanced, such as at a deadend. It is, of course, recognized that actual line failures usually involve complicated combinations of these and other types of loads, such as torsional loads set up by wire breaks, loads due to conductor oscillations and swaying of supporting structures, and many others. However, experience has shown that the strength requirements included in the rules, based on the simple assumption of the three types of load mentioned, will provide adequate overall safety (see Figure H26-1).

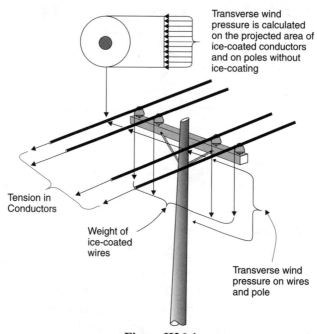

Transverse wind pressure is calculated on the projected area of ice-coated conductors and on poles without ice-coating

Tension in Conductors

Weight of ice-coated wires

Transverse wind pressure on wires and pole

Figure H26-1
Forces producing load on supporting structures

By dividing the allowable stresses given in the rules into the ultimate stresses for the various materials, so-called "factors of safety" may be determined. These factors of safety do not have the same meaning as in many other fields of engineering, where the loads and the resisting strengths of structures against such loads are more accurately known.

Wood-pole lines are essentially flexible structures. Their ability to withstand the varied and irregularly applied loading of wind and ice is proved, by experience, to be in excess of that calculated by the usual methods under Code loading assumptions and strength requirements. In other words, the allowable stresses and loading assumptions contained in the rules are only a convenient means of providing construction that experience has shown to be adequate in the various situations where Grade B, Grade C, or Grade D construction is required.

Prior to 1997, where Grade B, C, or D construction was required, Section 26 required that the material used, when in the position in which it is used, have a strength level equal to or greater than the loads required by Section 25 multiplied by the *overload capacity factor* required by Section 26. This methodology was essentially the standard method used throughout the history of the Code, although earlier editions stated it somewhat differently.

An alternate method for calculating the strength required of wood structures was added in the 1990 Edition as Method B, with the traditional method being named as Method A. The new Method B became the *preferred* method in the 1997 Edition, with the traditional Method A being allowed as an *alternate* method. In 2007, a sunset clause was added to prohibit use of the traditional method after 31 July 2010.

The 1990 introduction of Method B for wood only was the first explicit recognition of the four factors that are included within the Method A *overload capacity factors*:

(1) *Loading Factors*—such as the relative uncertainty of wind versus wire tension loads

(2) *Location Factors*—such as potential conflict with railroads versus cow pastures

(3) *Structure Type Factors*—such as the relative importance of deadends and angles versus tangent structures

(4) *Material Factors*—such as the relative degradability and nonhomogeneity of wood versus steel

The present system (Method B for wood of the 1990 Edition) essentially uses a Strength Factor to limit the allowable stress to be placed upon the wood and uses a reduced load factor (the old *overload capacity factor*) to account for the uncertainty of the load that will be present. Both factors include some recognition of *location factors*. Method B produces results with wood that are generally comparable to Method A. The results are not equal. Either method may be used until 31 July 2010 to calculate the strength required by a rule that allows the use of these alternate methods, but only one method may be used to satisfy the rule for a given structure or structural component. For example, you could not use the alternate method (old Method A) for the wind loading and the present method (old Method B) for the wire tension loading on the same pole. However, one method could be used for the pole and another method could be used for the crossarm.

In 1997, load factors for both methods were moved to Section 25; strength factors for both methods remained in Section 26. The previous Method B became the standard method in 1997; it no longer carries the designation of Method B. The previous Method A was retained as an alternate method for those who choose to use it—until the sunset date of 31 July 2010.

As of 1997, the methodology is essentially to first calculate the stresses in structural components that would occur if the loads of Section 25 were applied with the applicable load factors. These calculated stresses must then not exceed the permitted stress level calculated by derating the material by the appropriate strength factor from Section 26.

See the discussion of Section 24 to determine whether a line is considered to be "at crossing."

Under Rule 012A and Rule 012B, code users are required to install and maintain their facilities to meet NESC requirements. As a practical matter, installations are usually made pursuant to utility-promulgated design standards. Those standards usually start with the NESC as the basic safety requirement and add the effect of additional requirements, including reliability, efficiency of construction and maintenance, etc.

Typically some allowance is included for the effect on structure load-ings and on clearances for normal errors in sags and tension at installa-tion. Properly applied, such standards promote safe, reliable, and economic utility service, as generally mandated by state statutes.

When utility standards are followed, installations are properly main-tained, and structures are not overloaded at later times, structures stand tall and neat and have a long useful life. However, if angles are not properly guyed or carried on special, strong structures; if heavy equip-ment is mounted on the wrong side (relative to other loads); or if too many conductors and cables are added; it is difficult to keep structures in compliance with the NESC (see Figures H260-1 and H260-2).

Figure H260-1
Pole overloaded by unguyed angle service drops, and equipment location

Figure H260-2
Pole overloaded by unguyed communication opposite heavy equipment

As new materials have been placed in use for utility structures and structural components, the NESC has adapted to accommodate those materials. Prestressed concrete and reinforced concrete structures were first addressed in the 1977 Edition. Fiber-reinforced polymer structures and components were first addressed in the 2007 Edition.

260. Preliminary Assumptions

Certain influences that diminish the effect of the actual loadings have received careful consideration by (1) reducing (below what would otherwise be considered proper) the assumed loadings and (2) increasing the allowable stresses which are based upon the strength requirements of the several parts of the line; namely, conductors, fastenings, and pole or tower structures. The computation of stresses is usually made on the assumption that there is no deflection of supporting structures. However, such deflections occur, and the rule permits taking them into account under certain conditions. In this case, the assumed loads may be reduced *when the effects of structure deflection are known but* the designated fiber stress or ultimate stress cannot be changed. The conductors themselves exert a powerful influence in distributing the load along the line and in aiding the stronger structures to help support weaker ones. It should be noted that the 1981 Edition clarified the intention that the flexure of structures and structural supports should be calculated under the loads required by Rule 250 *before* the *overload capacity factors* of Section 26 are applied.

This rule was expanded in the 1977 Edition with the addition of Rules 260B and 260C. The latter rule recognizes a change in the *overload capacity factors* appropriate for use with new Rule 250C. The phrase "other supported facilities" used in Rule 260C refers to such facilities as connectors, insulators, transformer brackets, capacitor racks, etc. The 1.25 *overload capacity factor* does not apply to guys or to support arms or their braces, since they are part of the structure. An "at installation" requirement was added in the 1990 Edition to use an *overload capacity factor* of 1.33.

With the revisions of 1997, former Rule 260A and Rule 260B became Rule 260A1 and 260A2 and remained essentially unchanged. Former Rule 260C became Rule 260B, and the language was revised to recognize the move of the *load factors* to Section 25 and retention of the *strength factors* in Section 26.

Service drops generally have a relatively low tension. As such, they sometimes present a special problem because all too commonly, line workers do not think the tension is enough to worry about. All forces on a pole must be considered. In Figure H260-3, the secondary service drop was installed so tightly to get ground clearance that the pole eventually deformed, thus reducing ground clearance. To meet NESC clearances and strength requirements, care must be taken to ensure assure that forces are appropriately balanced at installation

Figure H260-3
Pole deformed by unbalanced service drop tension

In 2007, a NOTE was added to Rule 260B1 to direct users of fiber-reinforced polymer structures or structural components to methodologies for computing the 5% lower exclusion limit strength for use with the appropriate strength factor.

A NOTE was added to Rule 260B2 in the 2002 Edition to inform code users of various industry design standards, guides, and handbooks that may be useful.

261. Grades B and C Construction

Historical Comments: *Section 24 is used to determine the required Grade of Construction for structures and supported components. If Grade B or Grade C is required, the loads and load factors of Section 25 must be used to determine the expected stresses under the design conditions. That stress cannot exceed the stress levels permitted by these rules.*

The 1977 Edition began to modernize the requirements of the strengths and loadings requirements to match modern practices and to reflect modern materials and types of structures. Prestressed and reinforced concrete was first addressed in 1977. The requirements for wood, metal, and concrete structures were continually refined in the 1980s and 1990s.

In the 1993 and prior editions, the load factors and strength factors now contained in Rule 253 and Rule 261, respectively, were contained in a single number in Section 26. The language and format of Sections 25 and 26 are still being modified to be more compatible with modern load and resistance factor design (LRFD) calculation methodologies. Specificity of requirements for new materials and new loading requirements are also being increased. Requirements for fiber-reinforced polymer structures and components, as well as requirements for loadings from freezing rain and concurrent wind, were added in the 2007 Edition.

Because of (1) the complexity of the requirements, (2) the time required to review recent data, develop and evaluate proposed new requirements, and coordinate new requirements and methodologies with existing requirements and methodologies, and (3) the need to allow time for users to change internal systems to match new methodologies, these changes are being made in a relatively sequential, coordinated manner over time. At the same time, various existing requirements relating to strengths are being consolidated and coordinated. For example, both new materials (fiber-reinforced polymers) and existing requirements (crossarms and braces) were added to Table 261-1A in the 2007 Edition—the former to specify new requirements for new materials and the latter to coordinate with existing rules and limit the opportunity for confusion as to appropriate strength factors to be use. A new Rule 261N was added to specify working loads on steps and attachments used for climbing and fall protection, in order to allow better specification of load factors between Grade B and Grade C.

261A. Supporting Structures

(This rule was extensively revised in the 1977 Edition; several subareas of the previous rule have been regrouped or separated. The basic structure of the rule was retained in 1997, but the language was greatly simplified by the move of load factors to Section 25 and the creation of the new complementary sets of tables: Table 253-1 and Table 261-1A; and Table 253-2 and Table 261-1B.)

261A1. Metal, Prestressed, and Reinforced Concrete Structures

(Former Rule 261A1—Average Strength of Three Poles was moved to 261A2g in the 1977 Edition.)

These structures were grouped in the 1977 Edition when prestressed concrete structures were added. The previous safety factors were retained for reinforced concrete structures and expressed in terms of *overload capacity factors.* These factors are greater than those required of metal structures due to the reduced homogeneity of the concrete material.

Beginning with the Fifth Edition of the Code, the required strengths of steel-supporting structures were specified in terms of an *overload capacity factor* of the completed structure. Beginning with the 1977 Edition, *overload capacity factors* have been applied to concrete structures as well. This makes it unnecessary to consider the stresses in individual members and greatly simplifies both the Code treatment and the administration of the rules.

The *overload capacity factors* for Grade B metal structures were changed in the 1977 Edition to reflect continued experience and to effect appropriate rounding of numbers. Prestressed concrete structures are manufactured with greater quality control and are more homogeneous than reinforced concrete structures. Because the limitations of prestressed concrete structures are essentially the same as those of metal structures, the *overload capacity factors* of metal structures were found to be appropriate for prestressed concrete structures as well.

Former Rule 261A3(e), Rule 261A3(f), Rule 261A3(g), and Rule 261A3(h) were deleted in the 1977 Edition because the Code is not intended as a design manual.

For a discussion of required strength of angle structures, see the following discussion on requirements of wood structures.

The language was revised in 1997 to reflect movement of load factors to Section 25.

In the 2002 Edition, Rule 261A1c (the requirement for considering a gust factor on metal structures) was removed, since it is now included in the extreme wind data of Rule 250C. A new Rule 261A1d was added to recognize reinforcements and splices on metal, prestressed, and reinforced concrete structures, similar to Rule 261A2d for wood structures. Like Rule 261A1c for metal and concrete structures, new Rule 261A2f requires all wood structures to be able to withstand the extreme wind while standing alone without conductors. While this is no problem for poles and most framed wood structures, it can definitely be a problem for some tower-type designs and others that receive guying effects from attached conductors.

261A2. Wood Structures

(Former Rule 261A2—Reinforced Concrete Poles was included within Rule 261A1 in the 1977 Edition. The current rule was numbered as 261A4 in the Sixth and prior editions.)

Where lines carried on wood poles are necessarily heavy, it is usually advisable to install poles giving some margin of strength over that required just to meet the rule. Preservative treatment, butt reinforcement, or other methods may be used to maintain the pole to a high percentage of its initial strength during its life.

It was recognized in the 1977 Edition that the maximum bending moment on a pole may not be at or near the ground line. However, because of the taper of naturally grown wood poles, the allowable bending moment usually decreases faster than the actual strength of the pole as the calculation moves toward the top of the pole. In recognition of the latter, and in recognition of successful good service from wood poles installed under the Sixth Edition requirements, EXCEPTIONS to Rule 261A2b and Rule 261A2c were added in the 1981 Edition and carried forward in Rule 261A2a(2) and Rule 261A2a(3) and Rule 261A2b(2) of the 1990 Edition to allow calculation of required strength at the ground line. This also recognizes that ground-line decay usually occurs faster than decay further up the pole, due to increased moisture. The "at replacement" load factors for wood were determined with this in mind. This EXCEPTION continued in Rule 261A2 of the 1997 Edition.

The extent of the deterioration of a wood pole is often difficult to determine. Where the butt has been subjected to insufficient preservative treatment, rot may develop in the interior of the pole and not be visible from the outside. Although such interior rot does not, for a given loss of material, weaken the pole to the same extent as butt rot on the outside of the pole, the pole may be weakened considerably or its life may be shortened.

Rule 261A2a, EXCEPTION 1 requires meeting this rule at the ground line for unguyed structures and at the guy point(s) for guyed structures (see Figure H261A2a).

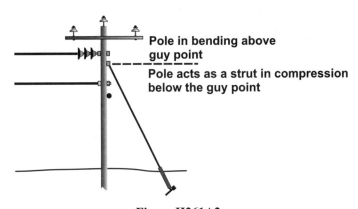

Figure H261A2a
Consideration of portions of a guyed pole

Because of the significant changes in ground-line strength of a wood pole, wood poles are required to be oversized when installed. This allows room for exterior rot at the ground line and ensures sufficient strength up the pole where the point of maximum bending moment and shear stress occurs. When the strength at the ground line can no longer meet the "at replacement" requirements, either the pole must be replaced (in which case the new pole must meet the "at installation" requirements) or the ground-line area must be strengthened by an appropriate method. In the latter case, it is *not* necessary to bring the ground-line strength all the way back up to "at installation" standards.

Rule 261 was revised in the 1990 Edition to add Method B, an alternate method for calculating the required strength of wood structures. The original method is retained as Method A. In Method A, the applicable *overload capacity factor* (OCF) of Section 26 is to be multiplied by the loads of Section 25; the resultant stress cannot be greater than the strength of the material. Method B is similar except that

(1) reduced *overload capacity factors* are used, and

(2) the resultant cannot exceed the permitted stress level (a percentage of the fiber strength).

The two methods yield similar results and *either* method may be used. The methods cannot be mixed for the same component. For example, Method A could be used for the pole and Method B could be used for the crossarm, but Method A could not be used for wind loading if Method B was to be used for other loads. Note that Method B became the preferred method in 1997, with Method A being allowed as an alternate until the 2007 Edition placed a sunset date of 31 July 2010 on the use of the alternate method (the older traditional Method A).

For Method A, the existing rules for the strength of wood structures were moved down one level as illustrated in the following table.

1987 Edition	1990 Edition
261A2	261A2a
261A2a	261A2a(1)
etc.	

Method B became Rule 261A2b and retained the ancillary require-ments of Method A.

Rule 261A2a(4) and Table 261-3A (Rule 261A2d and Table 261-3 of the 1987 and prior editions) must be considered when fulfilling the requirements of Rule 261A2a(5) (Rule 261A2e of the 1987 and prior editions). The column must be designed to take (1) the vertical compo-nent of the wind loading times the OCF of 4, plus (2) the vertical com-ponent of the wire tension loading times the OCF of 2.

The above system was essentially retained in the 1997 Edition. Load factors are now moved to Section 25. The former Method B is the pri-mary method. The former Method A was retained as an alternate for those who wish to continue to use it (until the sunset date of 31 July 2010 added in the 2007 Edition).

As of 1997, the methodology is essentially to first calculate the stresses in structural components that would occur if the loads of Section 25 were applied with the applicable load factors. These calcu-lated stresses must then not exceed the permitted stress level calculated by derating the material by the appropriate strength factor from Section 26.

Rule 261A2b—*Permitted Stress Level* (1993 Rules 261A2a(1)(a) and 261A2b(1)(a); Rule 261A2a(1)(a) of prior editions) recognizes that the values for the strength of wood poles of different varieties are estab-lished in ANSI O5.1, *American National Standard Specifications and Dimensions for Wood Poles*. They are based on figures somewhat lower than the average value of breaking strength for a given kind of pole. This ensures that the actual strength of the majority of poles will fall above the assumed strength. In the 2002 Edition, Table 261-1A and Table 261-1B clearly specify that, when new items are added to existing structures, the original and present structure strength must be great enough to meet the strengths required at installation and at replace-ment, respectively. The values required at installation ensure that the top of the pole is strong enough for expected stresses. The values at replacement ensure that the ground line, which tends to decay faster due to moisture in the soil, will remain in an appropriate condition.

The values in Table 261-1A are "at-installation" values for wood and reinforced concrete and "complete-life" values for metal and pre-stressed concrete. Footnotes 2 and 3 specify "at-replacement" values for wood and reinforced concrete.

If a new load is added to an existing structure, (1) the structure in new condition must have been strong enough to meet the "at-installation" values in the table and (2) the structure in its present condition must still be strong enough to exceed the "at replacement" values.

CAUTION: If the ground-line area is augmented under Rule 261A2d with a splint or other reinforcement to increase both the at-installation and at-replacement strength to add increased load to an existing structure, the top of the structure may become the limiting factor and also require augmentation. See Rule 012C.

Note that the "at-replacement" factors apply to the required strength of the pole, not the installed strength. It is not unusual for a utility to choose to install larger poles than immediately necessary to allow for longer life. Any loads added at a later date decrease the life of the structure.

It is recognized that the strength of an occasional pole will fall below that specified, but such a pole in a line, when flanked by poles of superior strength in spans of ordinary length, is not likely to fail. This latter phenomenon is one of the reasons for Rule 261A2(e)—*Average Strength of Three Poles.* That rule was removed in the 2007 Edition (see the discussion of Rule 261A2(e) later in this section).

Recognition is also given to the use of possible future developments in new materials.

Rule 261A2d—*Spliced and Reinforced Poles* are sometimes stubbed, instead of changed, in order to save costs; if properly done, the results are effective. In earlier editions of the NESC, such ground-line strength enhancement measures were not allowed at certain locations. However, later editions have recognized that it is the strength of the structure that is the key element, not the method of achieving that strength.

Wrapping of the stud joint with wire or steel bands, in addition to bolting, may prevent the bolts from pulling through the pole. The stub should be placed beside the pole, not in line with it, so that it will develop the required transverse load capability (see Figure H261A2d-1). If the butt of the pole is badly decayed, it may be advisable to cut it off and remove it. On the other hand, where a pole is broken or deteriorated above ground and is spliced in order to promptly restore service, the new pole section should be placed directly on top of the base so that the centerline of each coincides with the other; splices should then be placed around the joint to overlap the two sections. Today, a number of channel, plate, or tube systems are available for use in enhancing the strength of a wood pole at the ground line.

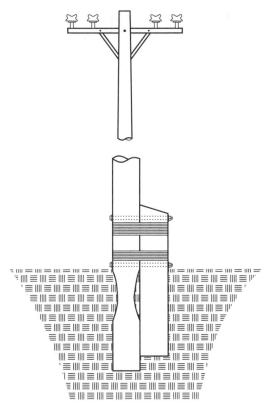

Figure H261A2d-1
One method of stubbing to reinforce a deteriorated pole

Figure H 261A2d-2 shows a stub pole used to rehabilitate a pole damaged by a vehicle fire. Note that, although the pole stub is placed correctly to rehabilitate the transverse strength of the pole, this installation does not meet the requirements of Rule 261A2d because the stub pole does not connect to the top portion of the pole.

Figure H261A2d-2
Stub pole too short

Rule 261A2d does not cover pole extensions; these constructions are considered to be part of the structure and are covered by the normal strength rules.

Figure 261A2d-3 shows a side-by-side splice at the top of a corner pole. The pole is in line with the line going toward the viewer's left but sideways to the line going to the viewer's right. Longitudinal strength is thus strong in one direction, but weak in the other. Good practice would be to (1) use metal splints and set the top pole section directly over the bottom one or (2) if the loads were light (which they are in this particular case), set the top pole outside the corner on the bisector of the angle.

Figure H261A2d-3
Inappropriate side-by-side top pole splice

The cracked pole shown in Figure 261A2d-4 has been repaired with a long metal splint. This pole is located inappropriately, does not meet the requirements of Rule 231B, and should be moved.

Figure H261A2d-4
Damaged pole with metal splint
in location not meeting Rule 231B

The provisions of Rule 261A2e—*Average Strength of Three Poles* Rule 261A2a(7) of the 1990 and 1993 Edition (Rule 261A2g of the 1987 and prior editions) were designed to permit considerable latitude in the construction of wood-pole lines. It was well-known that each pole in a supply line could assist in supporting the poles adjacent to it, particularly with short spans; the conductors themselves act as guys after a pole has deflected to a certain extent. Rule 261A2e allowed a weak pole to remain in place, as long as the specified constraints on average strength, individual pole strength, sag and tension, and span length are met. It was especially important from the standpoint of safety that pole structures of sufficient strength be used at crossings over railroads, communication lines, and limited access highways. Thus Rule 261A2e was not allowed to be used at such crossings. In essence, this rule was not applicable to Grade B construction. Rule 261A2e—*Average Strength of Three Poles* was deleted in the 2007 Edition.

As time has progressed, larger wire sizes with both greater sags (and therefore less support from adjacent poles) and higher tensions (and therefore greater forces) have been used. As a result, Rule 261A2e— *Average Strength of Three Poles* is no longer reliable in many circumstances. Wood-pole strength is based upon the average strength of wood poles. Unlike metal, prestressed concrete, and fiber-reinforced polymer structures, wood has a wide variation in strength, with a typical *coefficient of variation* in excess of 20% of the mean. The former average-strength rule is not applicable for other materials and, after review for the 2007 Edition, was concluded to no longer be appropriate for wood poles. If an existing wood pole is known to be weak, that pole should be spliced or reinforced in accordance with Rule 261A2d or replace or rehabilitated in accordance with Footnote 3 Table 261-1A (the standard method) or Footnote 3 of Table 253-2 (the alternated method that can be used until 31 July 2010).

There are numerous lines that initially required only Grade N construction but which, because of the addition of circuits, were later required to conform to the requirements of a higher grade. For example, if an open-wire supply circuit were the only construction on a line, only Grade N construction would be required for locations. If another line later crosses under this line, however, the Grade of Construction may have to be B or C. If such a line is not originally designed for such requirements, it probably will not conform to that grade after the new line circuit is added. It would be very expensive to rebuild the line to meet the requirements due to the additions made. Since it is very possible for many pole lines to need to change from one grade to another during their lives, lines should generally be originally constructed to comply with the grade that may be required of them in the future.

Requirements such as those relating to minimum pole sizes and freedom from observable defects were deleted in the 1977 revision; these are design considerations, not safety considerations, and duplicated ANSI O5.1 requirements.

The Fourth Edition of the Code was generally interpreted as requiring that poles at angles in a line withstand the arithmetic sum of the following loads without exceeding (at the ground line if unguyed or at the point of guy attachment if guyed) the allowable percentage of ultimate fiber stress specified for transverse loading: (1) wind on pole surface, (2) wind on conductors, and (3) resultant of conductor tensions.

Experience with this Fourth Edition method indicated that it required excessive strength at angle supports, especially for relatively large angles. After considerable study, a modification was included in the Fifth Edition and carried through the Sixth Edition. It was recognized that the three loads listed previously must be taken into account, but that the variability of wire tension is much less than the variability of the wind. It was decided, therefore, to apply the allowable percentage of ultimate stress under transverse loading to the sum of (1) and (2) and to apply the allowable percentage of ultimate stress at deadends to (3), before combining the three loads. To accomplish this, the rule required the *calculated load* to be a combination of (a) the transverse wind loads (times an adjustment factor) and (b) the resultant transverse load of the change in conductor direction (times a factor of 1). The structure was then required to carry this calculated load without exceeding the ultimate stress allowed at deadends. This method effectively accomplished the desired result; the amount of the reduction in overall strength, as compared to the requirements of the Fourth Edition, increases as the size of the angle increases and as conductor tension thus becomes more and more controlling.

The adjustment factors given in the rule (2.0 for Grade B, 1.5 for Grade C) were determined by reducing the two allowable percentages of ultimate stress to a common denominator, using the allowable stress at deadends as a base. For Grade B poles, for instance, the allowable percentage for transverse loading is 25% in Rule 261A2 of the Fifth Edition. This means that the transverse loads discussed previously ([1] wind on pole surface and [2] wind on conductors) must be multiplied by four. The allowable percentage for deadend loadings is 50%, which means that the longitudinal load discussed previously ([3] resultant of conductor tensions) must be multiplied by two. The pole must then withstand these combined loads without exceeding its ultimate fiber stress.

The study of this matter brought out the fact that there was some confusion as to whether combining loads (2) and (3) resulted in error due to twice taking into account the wind pressure on the conductors, since the same wind pressure that causes load (2) also contributes to the conductor tension in load (3). The reason that adding these loads was assumed not to introduce an appreciable error results from the fact that the size of the angle used in the conductor-loading calculations was ordinarily measured by sighting along the line of the supports adjacent to the angle. This is not, of course, the same as the angle in the conductors on the corner support, since the effect of the wind is to displace the conductors out of the vertical plane passing through its two points of support. The angle in the conductors is obviously greater than the angle measured as above by sighting, and by an amount that depends upon the transverse wind pressure on the conductors. It was assumed, therefore, that if the actual angle in the conductors under wind displacement is used in determining load (3), load (2) should be neglected. Otherwise, the wind on the conductors would, of necessity, have to be considered twice, as outlined previously (see Rule 252B for further discussion).

It was permissible, under the wording of Rule 252B6, to take into account the reduction in conductor tension due to the angularity of application of the wind, which is usually assumed to be in the direction of the bisector of the angle. This is seldom done, however, because (1) information as to the amount of the reduction is not available without special computations, and (2) it would have little effect on the overall result in most cases, particularly if the angularity of the wind is considered in determining load (2).

The existing methodology for computing required strength at angles in the line was essentially continued in the 1977 revision, except that equivalent *overload capacity factors* instead of percentages of ultimate strength were used. The tables were expanded in the 1981 Edition specifically to make clear in all cases the intention for reduced *overload capacity factors* to apply to wire *tension* load as opposed to wind load.

In the 1984 Edition, metal portions of a wood structure (not including guys) are allowed to meet the lesser *overload capacity factors* of metal, rather than the higher ones required of wood. Obviously, this means that tests of a composite structure may require additional preparation if the reduced OCFs are employed for metal end fittings; a complete new composite would be suitable for testing the metal portions, but special stronger metal portions would need to be used to test the wood portions that have a higher OCF.

In the 1993 Edition, redundant language in Rule 261A2a(4) of the 1990 Edition was deleted.

261A3. Fiber-Reinforced Polymer Structures

(Former Rule 261A3—Transverse-Strength Requirements for Structures Where Side-Guy Wiring Is Required, But Can Only Be Installed at a Distance was moved to 261A4 in the 2007 Edition.)

This rule was added in the 2007 Edition to specify requirements for structures and structural components made of fiber-reinforced polymers. This includes polymers with glass-fiber reinforcement as well as fiber reinforcement of other materials. The stresses caused by the assumed loads of Rule 252 multiplied times the load factors of Table 253-1 cannot exceed the *permitted load*. The *permitted load* is limited to the 5% lower exclusion limit (LEL) of the strength of the reinforced polymer material (i.e., 95% of the structures or components will be expected to have greater strength) multiplied times the applicable strength factor of Table 261-1A.

261A3. Transverse-Strength Requirements for Structures Where Side-Guy Wiring Is Required, But Can Only Be Installed at a Distance

(Former Rule 261A3—Metal Supporting Structures was included within Rule 261A1 in the 1977 Edition. The current rule was 261A5 in the Sixth and prior editions.)

At many crossings, especially in lines on city streets, (1) it is not feasible to attach side guys to the crossing poles, and (2) the only other method of meeting the strength requirements is to use special structures, such as steel poles or towers. To obviate the additional expense of such construction, the alternative is offered of treating several spans collectively and providing the transverse strength at those poles where side guys can be erected. This treatment is restricted to sections 250 m (800 ft) in length, and the intervening line must be of uniform grade in all other respects.

The justification for this alternative rests in the observed fact that the conductors themselves act as guys to the poles; in some instances, they serve to equalize the load; in other cases, the conductors transfer the load to the resisting structures. The guying is not only longitudinal but, as soon as deflection of a pole begins, includes a transverse component. Instances are on record where the conductors have held up poles that, without their help, would have fallen.

261A4. Longitudinal-Strength Requirements for Sections of Higher Grade in Lines of a Lower-Grade Construction

(Former Rule 261A4 was moved to 261A2 in the 1977 Edition. The current rule was 261A6 in the Sixth and prior editions.)

261A4a. Methods of Providing Longitudinal Strength

Just as in Rule 261A3, where unusual conditions may sometimes require special alternative construction to provide the required transverse strength, special alternative construction may sometimes be necessary to provide the required longitudinal strength. The need to use alternative construction to meet longitudinal strength requirements does not often occur; head guys can generally be installed. Perhaps the principal occasion for its use is where a line crosses both a road and a railroad in the same span. The limiting distance in this case has been made the same as that specified in Rule 261A3. Either at a crossing, or at an end section of high-grade construction, the unbalanced tensions may, under certain given conditions, be divided between two or more pole structures, due to their respective deflections toward the crossing section or other section of strong construction.

It is ordinarily impractical to distribute such loads over more than two or three poles; the pole nearest the weak section or the angle in the line must ordinarily withstand most of the load. The allowance of such distribution as an alternative was discontinued in the 1997 revision.

Usually the use of a crossing structure strong enough to withstand the loads, or the transferring of the load to a sufficiently strong and rigid end structure, will be more satisfactory than to attempt to distribute the load over two or more structures, each of which alone is too weak for the load imposed. Often the computation of the division of loads between such poles is difficult; errors in assumptions may result in unanticipated and dangerous weakness in the crossing or end-section span of the presumably strong construction.

When the assumed load cannot be carried, it must be reduced by increasing the conductor sags. The object of this rule is to make the section of higher grade independent so that, insofar as practical, it can stand even in case of failure of the line at a nearby point. If the entire line is built to the same specifications, this procedure is not necessary.

261A4b. Flexible Supports

This rule serves as a reminder that flexible structures may need to be head-guyed or otherwise reinforced to limit reductions in the clearances required by Section 23.

261A5. Transverse-Strength Requirements for Structures Where Side Guying Is Required, but Can Only Be Installed at a Distance

(This rule was moved to 261A3 in the 1977 Edition.)

261A6. Longitudinal-Strength Requirements for Sections of Higher Grade in Lines of a Lower Grade of Construction

(This rule was moved to 261A4 in the 1977 Edition.)

261A7. Strength at Angles in a Line

(This rule was included in Rules 261A1b and 261A2d in the 1977 Edition.)

261B. Strength of Foundations, Settings, and Guy Anchors

(Guy anchors were added to this rule in the 1993 Edition.)

There is general agreement that foundations of metal poles and towers are, as a rule, the weakest feature of the structure. The fact that foundations are subject to variations in the character of the soil, and are affected as well by moisture and frost, whereas line material is of quite uniform and known properties, is further reason why particular care should be given to the design of foundations. Good workmanship is of no less importance than proper design. Insufficient tamping of the backfill is a common source of trouble and has been the cause of some failures.

Owing largely to their lower cost, earth foundations have been used extensively. In many parts of the country where lines are in inaccessible regions, it is so difficult to secure concrete materials without long transportation distances that the cost precludes their use. There has been considerable objection to earth foundations, owing to the large number of failures resulting from their use. Failures have occurred on a number of different lines that were constructed with metal footings and earth backfill and later found to require reinforcement of the footings with concrete.

Foundations must, in general, be designed to withstand bearing, uplift, and a lateral force tending to slide or overturn them. The downward force need scarcely, if ever, be considered, as foundations designed for uplift will usually develop adequate bearing power. One exception to this is swampy ground, where it may even be necessary to resort to the use of piles to give adequate bearing.

The concrete used for tower footings and foundations should be of good quality and proportions. It is a mistake to use a lean concrete on the assumption that its function is merely that of ballast. Not only is the foundation called upon to withstand shearing and bending stresses, but it also acts as a protection to the metal members embedded in it.

The *overload capacity factors* for foundations are essentially the same as required for metal structures. The general intention is for the foundation to maintain the strength required for the structure. Since the reason for the larger OCFs for wood and reinforced-concrete structures is only because of problems with nonhomogeneity of these materials, foundations for these structures are required only to have OCFs equal to those for metal structures.

The need to design foundations to meet expected earth reactions was recognized in the 1977 Edition. The wording of the 1977 Edition was revised in the 1981 Edition to require that the foundations be designed to withstand the reactions on the foundation that will result from applications of the loadings of Rule 252 multiplied by the *overload capacity factors* of Table 261-4 to the structure. These reactions include those of the earth as well as those of the attachment of the structure to the foundation.

It should be noted that the Code does not require the foundation to match the strength of the structure. This has been used to advantage on at least one U.S. barrier island that is subject to periodic hurricanes. In that case, major poles are oversized for their natural sand foundation. When a hurricane comes along, *all* power is purposefully cut off for the duration of the storm. The poles are strong enough that they will lean over in hurricane winds, rather than break. They can be more easily righted and side-guyed than replaced, and thus power is restored more easily, quickly, and less expensively than would otherwise be the case.

The previous discussion also applies in general to guy anchors, which were added to the rule in the 1993 Edition. Of particular importance are matching anchor type, size, and depth to the soil characteristics at the site.

The 1997 revision placed the load and strength factors for foundations, settings, and guy anchors into Table 261-1A.

261C. Strength of Guys and Guy Insulators

Because of the great flexibility of wood and reinforced-concrete poles, they may deflect considerably before developing much resistance to the transverse loads applied. Guys, when installed properly, are under initial stress and would fail before stretching enough to put much transverse load on the poles. Thus, the strength of a pole cannot significantly share the transverse load with a guy and, therefore, a guy must take the total load. However, if a guy is attached to a line-supporting structure not capable of much deflection, such as some metal structures and prestressed-concrete structures, the strengths of the structure and the guy are additive. It is appropriate to note that some metal structures may be designed with such inherent flexibility that they do not behave significantly differently than a wood pole and would, thus, be appropriately treated only as a strut, letting the guy(s) take the total horizontal loading.

The rule appropriately recognizes that (1) guy strands typically fail after little inelastic deformation, but (2) anchors may continue to serve effectively after considerable inelastic deformation.

In the case of a tangent pole with a guyed tap, such as that in Figure 261A2a, the portion of the pole below the guy is considered as a strut, but the portion above is considered in bending (see Rule 261A2a and its EXCEPTION 1).

As covered under Rule 261A7, the total load at corners (other than those where deadend construction is employed) consists of the sum of the load effects of wind on pole surface, wind on conductors, and the resultant of conductor tensions. These three loads can most conveniently be added by first reducing them to equivalent horizontal loads acting at the point of guy attachment.

The matter of how many guys are appropriate for a given design is a design decision, not a Code decision. The NESC is a performance code that specifies *what* must be done, not *how* it must be done. It is the responsibility of the entity that places a load on a line structure to ensure that the structural components have the required strength after the load is added. In the case of a joint-use line, the NESC does not specify what entity must actually install required guys; such matters are subjects for agreement between the parties involved.

Rule 261C2b was deleted in the 1987 Edition because it was redundant with 261C2a.

In 1997, the rule was changed to recognize movement of load factors to Section 25.

In 2007, Rule 261C3 was added to address guys on fiber-reinforced polymer structures in the same manner as previously used for wood structures.

261D. Crossarms and Braces

(This rule was extensively revised in the 1990 Edition to bring all the requirements for crossarms, regardless of material used, under the same rule. Braces were added to the rule in the 1993 Edition.)

The minimum required crossarm sizes vary with the crossarm length and number of conductors carried, since the length of lever arm and the possible stress due to both vertical and longitudinal (parallel with the line) loads vary with these same factors. The given sizes are those that will withstand, with an appropriate margin of safety, a working load due to an unbalanced longitudinal force of 3.1 kN (700 lb) on the end pin. This load can occur if an outer conductor breaks at one side of the crossarm, and it is the working load that can be withstood by good wood pins. These crossarms will also adequately withstand the total vertical load of all conductors under the assumed maximum ice loading up to spans of 90 m (300 ft) with No. 4/0 conductors on all pins. For larger loads, larger crossarms or double crossarms are often advisable.

IR 530 issued 3 September 2002 clarified that Row 2 of Table 261-1A *Wood and reinforced concrete structures* should be used to determine strength factors for solid sawn wood crossarms.

The forces exerted by conductors of overhead lines on pins, crossarms, and poles in tangent sections of pole lines may be balanced at some particular combination of temperature condition and loading of wind or ice or both. At other temperatures and loadings, the forces will be unbalanced to some extent. In general, the longitudinal unbalancing will not be severe, except at angles and deadends, unless a conductor fails. Transverse wind load is unlikely to break conductor fastenings, pins, or crossarms, even with heavy conductors in long spans. The vertical load at times becomes serious for small crossarms, but not for pins.

Through its design, the insulator will take its load as a crushing force at the tie groove and is usually amply strong. The insulator pin acts as a beam whose length is equal to the distance from the top of the crossarm to the point of attachment of the wire. The crossarm also acts as a beam whose length varies with the conditions and, in the case of a crossarm carrying a single conductor on one side of the pole, is equal to the distance from the pin position to the point of attachment at the pole.

This rule was expanded significantly in the Fourth Edition and remained essentially the same until the revisions of 1990, except that Rule 261D was modified in the Fifth Edition to recognize that general use of ridge pins and metallic brackets attached directly to the pole offers strength comparable to wood, usually with superior durability.

The EXCEPTION to the double crossarm or bracket requirements under Rule 261D5c (Rule 261D4c of the 1990–2002 Editions; Rule 261D5 of the 1987 and prior editions) was modified in the Sixth Edition. It applies to the situation that frequently occurs when (1) a supply line, having reasonably uniform spans, is erected at one date in full compliance with the rules; and (2) a communication line crosses underneath it at a later date. Little would be gained by requiring subsequent installation of double crossarms on the supply structure. Rules 261D3(b) and (c) were deleted in the 1977 Edition.

Bracing is, of course, generally necessary to withstand unbalanced vertical loads, as with oscillating conductors, persons at work, or line equipment carried on the crossarms.

The practice of attaching single crossarms on adjacent poles to opposite sides of the poles is to be commended, since it helps considerably to tie the wires in with the poles and, if a number of wires fail in a span, the crossarms on the several adjacent poles will not be pulled off.

The "number of pins" is used to designate the length of the crossarm; 6 pin is 2.45 m (8 ft), 8 pin is 3 m (10 ft). An 2.45 m (8 ft) arm requires use of the "6 or 8" category in Table 261-6 for supply conductors.

Rule 261D was revised in the 1990 Edition to include Method B for wood crossarms as Rule 261D1b. Method A for Wood was placed in Rule 261D1a, concrete and metal crossarms in Rule 261D2, other materials in Rule 261D3, and the remainder of the rules were placed in Rule 261D4—*Additional Requirements*; the remainder of the requirements were renumbered. Wood crossarms were renumbered to 261D2 and metal to 261D1 in the 1997 Edition.

Stresses resulting from conductor tensions computed at 320 kg (700 lb), according to Rule 261D4a (Rule 261D3 of the 1987 and prior editions), provide small factors of safety that vary according to the grade of the crossarm, dimensions, and allowable fiber stress. Accordingly, the 320 kg (700 lb) loading or its equivalent results in a stress that approaches, but does not exceed, an *overload capacity factor* of unity. No *overload capacity factor* is stated as such when the expected loading is greater than 320 kg (700 lb); Rule 261D4a implies a required OCF of unity where conductor tension is greater than 320 kg (700 lb). This rule applies both to deadend and line poles; the Code requires the designer to ensure that the conductor loading on the crossarms will not exceed its designated fiber stress (or ultimate strength). Note that conductor tension resulting from all sources, including wind and ice loading, must be included. Rule 261D4a(1) (Rule 261D3a of the 1987 and prior editions) requires that, at a minimum, the crossarm be able to hold the forces capable of being withstood by good wood pins—320 kg (700 lb).

Rule 261D5a(2) (Rule 261D5a(2) of the 1990-2002 Editions; Rule 261D3b of the 1987 and prior editions) is provided as a convenience in ensuring that appropriate safety considerations have been met for conductor tensions up to 900 kg (2000 lb) per conductor; the indicated construction allows double crossarms to be used without having to make the general calculation required by Rule 2612a(1). The reference to Rule 261D4a(1) was corrected to 261D2a(1) and the strength requirements involved were correctly limited to longitudinal strength in the 2002 Edition. Note that this particular construction is not required for conductor tensions above 320 kg (700 lb); the designer may specify alternate construction that meets Rule 261D2a(1) (Rule 261D3a of the 1987 and prior editions). For conductor tensions above 900 kg (2000 lb), no particular construction is specified as being considered to meet the requirements of the rules; it is the designer's responsibility to ensure compliance with the fiber-stress limit, in accordance with Rule 261D2a(1).

IR 520 issued 7 July 2000 clarified the relationships of the rules. Wooden crossarms must meet the greater of (1) actual loadings as stated in Rule 261D2a(1) and (2) a minimum of 700 lb applied at the outer attachment point, as stated in Rule 261D4a(1)(a) (now Rule 261D5a(1)(a)). Both rules apply to both Grade B and Grade C Construction. Where conductor tensions exceed 2000 lb, it is *not* intended that the strength requirements of Rule 261D5a(1) be increased on a proportional basis. For conductor tensions above 2000 lb, no particular construction is specified as being considered to meet the requirements of the rules; it is the designer's responsibility to ensure compliance with the fiber stress limit (see IR 376 issued 6 November 1985).

Note that, although braces are included within this rule, the loads on the braces may be nothing more than vertical load imbalances with uneven ice, uneven wind loading, broken conductor partially supported by the ground or other facilities, or similar imbalances. The function of many brace designs is merely to provide rotational stability for a crossarm. With some designs, however, a "brace" is expected to carry normal loading of some value. In these latter cases, they should be treated as any other structural member.

Rule 261D3 *Fiber-Reinforced Structures* was added in 2007 to address crossarms and braces made of fiber-reinforced polymers. (*The existing 261D3 and D4 were renumbered to D4 and D5, respectively.*) The provisions are essentially the same as for the structure.

Rule 261D5b *Bracing* requires that, *if necessary,* crossarms must be braced to support the expected loads *including line personnel working on them.* In many cases, crossarm strength is great enough to carry all the loads expected on the arm and braces are used only to support unbalanced vertical loads, such as worker loads or broken conductor/

differential ice loading imbalances. Flat metal braces that provide tension on the side of the crossarm opposite the unbalanced load (and have little compression capability) are sometimes used for this purpose. The 2007 Edition addressed the latter case specifically by clarifying that such braces need only be designed for the unbalanced vertical loads, not the total loads carried by the crossarm.

261E. Insulators

(The former Rule 261E was moved to Rule 261F in the 1977 Edition when metal crossarm requirements were added and placed here. Metal crossarm requirements were consolidated with similar requirements in Rule 261D in the 1990 Edition. This rule was added in 1997.)

Users are directed to Section 27 for the strength of insulators.

261F. Strength of Pin-Type or Similar Construction and Conductor Fastenings

(This rule was 261E in the Sixth and prior editions. The former Rule 261F became Rule 261G in the 1977 Edition.)

This entire section was reorganized in the Sixth Edition in the interest of clarification. The intent of the previously specified 320-kg (700-lb) strength requirement was explained, and recognition was given to the practice of using single metallic conductor supports in lieu of double wood pins. An EXCEPTION similar to that provided in Rule 261D5 for double crossarms was added for double pins and conductor fastenings.

This rule was rewritten in the 1977 Edition to conform in style and substance with Rule 261D; former Rule 261E1 was deleted; the remaining rules were revised and renumbered. A horizontal post insulator is not considered a pin-type insulator.

The 1997 revision recognized movement of load factors to Section 25.

261G. Armless Construction

(This rule was added in the 1977 Edition to recognize changes in available construction materials.)

The rule serves as a reference to other rules appropriate for use with armless construction.

261H. Open Supply Conductors

(This rule was numbered 261F in the Sixth and prior editions.)

The 1997 revision recognized movement of load factors to Section 25.

261H1. Material

(This discussion applies to the Sixth and prior editions only. See the following section for discussion of later Rule 261H1 requirements.)

In the Sixth and prior editions, conductors were required to be of material that would not corrode excessively under the prevailing conditions. The use of noncorrodible material for overhead conductors was intended to prevent conductors from falling due to deterioration. The rule recommended that hard-drawn or medium-hard-drawn copper be used for new overhead lines rather than soft copper, especially in wire sizes smaller than AWG No. 2. Experience with soft copper indicated that, as long as copper wire remains soft, it will stretch in every considerable storm; other facilities and the public below would then be endangered by the real possibility of contact with fallen or deeply sagged wires. By confining the use of soft copper to the heavier sizes—say, larger than No. 2, including railway feeders—the hazard was assumed to be greatly reduced. Railway feeders and secondary distribution conductors are frequently strung with less than maximum allowable tensions. Serious elongations of such conductors under wind and ice loads are not, therefore, to be expected, even if they are of soft copper.

While the preceding cites copper as an example, other noncorrodible materials, such as those using aluminum as a base, were intended to be studied in similar fashion before installation.

This rule was deleted in the 1977 Edition because it was not so much a safety rule as a design rule. Previous Rule 261F2, which became 261H1, appropriately stated the requirements in terms of strength equivalent. These minimum sizes did not cover all types of wire in modern use and were deleted in the 2002 Edition, because the strength required by the required loadings and related factors for Grade B and Grade C must be met. Other rules require that the conductor maintain required strength and sag characteristics throughout its life.

261H1. Sizes of Supply Conductors

(This discussion applies to the 1977–2002 editions only. This rule was deleted in 2002).

The advantages in using the smallest allowable sizes of conductor are frequently not as great as may appear from the initial saving in cost. A larger size of conductor may often be justified for greater reliability of service, ability to meet load increases, improved voltage regulation, reduced maintenance cost, and allowance of longer spans. Table 262-5 was completely revised for readability in the 1987 Edition. When this rule was deleted in the 2002 Edition, a minimum size was specified in Rule 263E for service drops exceeding 45 m (150 ft).

261H1. Sags and Tensions

(This rule was numbered 261F4 in the Sixth and prior editions. Rule 261F3 was deleted in the 1977 Edition. From 1977 to 1997 this rule was numbered 261H2 and was renumbered 261H1 in the 2002 Edition when former Rule 261H1 was deleted.)

Conductors in the spans between line structures may be required to conform to several Grades of Construction. The sags should be fairly uniform and selected to provide a tension within proper limits for all sizes of conductors in spans where Grade B, Grade C, or Grade N construction are likely to occur simultaneously. Sags should be determined after careful consideration of operating experience in maintaining service and providing safety, including a study of the observed mechanical characteristics of conductor materials under operating and test conditions. The effects of both storm loading and long-term creep are to be considered in development of final unloaded sag. Requirements of construction practice make it necessary that sags in adjacent spans of different lengths should provide approximately equal stresses in the conductor in the different spans at the time of stringing.

While large conductors may theoretically be strung to a lesser sag than smaller ones without exceeding their elastic limit when loaded, urban construction practice often uses greater sags than those employed for the smaller wires. There are several reasons for this: (1) railway feeders and other heavy conductors, if strung to small sags, may impose undue stress on poles and fastenings—particularly at angles, deadends, and other points of unbalanced tension; (2) heavy conductors do not swing in the wind as readily as light ones, and the need for small sag is therefore not as great. Furthermore, (3) where heavy feeders are run on the same poles with other conductors, they usually occupy the lower crossarm where an excessive sag will increase, rather than reduce, the clearance from other wires.

For normal construction, the tension of conductors will change with loading, and three limitations on loading are given by Rule 261H1b as follows:

(1) 60% of rated breaking strength (RBS) under the assumed loading of Rule 251

(2) 35% of RBS at 15 °C (60 °F) initial unloaded tension

(3) 25% of RBS at 15 °C (60 °F) final unloaded tension

For catenary suspension systems where loading is uniformly maintained by an autotensioning mechanism (such as a railway contact conductor system), only the first of these limits would apply, since the conductor is always under final load.

The second and third tension limits were added to limit fatigue failures due to aeolian vibration. These simple limits were based upon observations made decades ago, with the types of wires, tensions, and loading conditions in use at that time. The 2007 Edition revised the NOTE under Rule 261H1b to inform users that these limits may not protect some conductors under some conditions. The configuration and mix of metals used in the wires and conductors, the tensions, and the loading conditions should be considered in determining whether vibration mitigation devices are needed.

261H2. Splices, Taps, and Deadend Fittings, and Associated Hardware

(This rule was numbered 261F5 in the Sixth and prior editions and 261H3 in the 1977–1997 Editions.)

This rule recognizes the need for special care that is required in placement and design of conductor terminations and connections. The requirement for deadend fittings to withstand conductor loadings was clarified in the 1977 Edition. Under the prior wording of the rule, some individuals mistakenly assumed that the requirement was that of 261F, i.e., 3.1 kN (700 lb). In Rule 261H3a of the 1987 Edition, the requirement for using an *overload capacity factor* of 1.65 for splices to match Rule 261H3c for deadend fittings was added. The 2002 Edition added *associated hardware* to the rule, since supply conductor hardware must have the same strength as required for the conductors. The 2007 Edition added Rule 250D loadings to the rule.

261H3. Trolley-Contact Conductors

(This rule was numbered 261F6 in the Sixth and prior editions and 261H4 in the 1977–1997 Editions.)

This rule is unchanged since the Fourth Edition.

261I. Supply Cable Messengers

(This rule was numbered 261G in the Sixth and prior editions.)

Most of these requirements were deleted in the general 1977 revision because they were either unnecessary or covered elsewhere.

The 1997 revision recognized movement of load factors to Section 25. The 2007 Edition added Rule 250D loadings to the rule.

261J. Open-Wire Communication Conductors

(This rule was numbered 261H in the Sixth and prior editions.)

As of 1977, this rule links tensions for communication conductors to those required for supply conductors under similar circumstances and, prior to 2002, gives minimum sizes that were included in Grade D prior to 1997. In the 2002 Edition, the minimum sizes for Grade B and Grade C communication wires and associated Table 261-4 were removed (see the discussion of Rule 261H1).

261K. Communication Cables

(This rule was numbered 261I in the Sixth and prior editions.)

Prior Rule 261—*Short-Span Crossing Construction* was deleted in the 1977 Edition.

The 1997 revision recognized movement of load factors to Section 25. The 2007 Edition added Rule 250D loadings to the rule.

261L. Paired Communication Conductors

(This rule was numbered 261J in the Sixth and prior editions. Prior Rule 261L—Cradles at Supply-Line Crossings was deleted in the 1977 Edition.)

This rule is essentially unchanged since the Fourth Edition.

261M. Protective Covering or Treatment for Metal Work

(This rule was deleted in the 1977 Edition.)

This rule was a design rule, not a safety rule. Strength requirements must be maintained as required. It is the responsibility of the designer and operator to ensure continued compliance.

261M. Support and Attachment Hardware

This rule specified requirements for support hardware in the 1997 Edition. Attachment hardware was added in the 2002 Edition. As a result, the title of Table 261-1A was revised.

261N. Climbing and Working Steps and Their Attachments to the Structure

This rule was added in the 2007 to directly address the strength required for steps, ladders, platforms and the structural members that support them. A variety of existing standards and work practices were reviewed during the development of this rule. Such steps and attachments are required to support not less than twice the *maximum intended load*. The maximum intended load must be not less than 300 lb. IEEE Std 1307™ *Fall Protection for Utility Work* gives guidance in this area.

262. Grade D Construction

(Prior Rule 262H and Rule 262K were deleted in the 1977 Edition. These rules were deleted when Grade D was merged into Grade B in the 1997 Edition.)

The Fourth Edition of the Code contained requirements for Grade D and Grade E construction for communication lines that cross railroads. In the interest of simplification, Grade E was omitted from the Fifth Edition of the Code; the few requirements formerly applying to Grade E construction were made as EXCEPTIONS to Grade D requirements. In concert with changes to the 1977 Edition that eliminated related provisions in Rule 261K, the rules for special short-span construction for Grade D were deleted.

262A. Poles

(Former Rule 262A1 and Rule 262A2 were renumbered 262A2 and 262A3 in the 1977 Edition; Rule 262A3 was renumbered Rule 262C5; and Rule 262A4, Rule 262A5, Rule 262A6, and Rule 262A8 were deleted. This rule was deleted when D was merged into Grade B in the 1997 Edition.)

The reference to designated fiber stresses contained in ANSI O5.1-1979 were added in the 1977 Edition as a new Rule 262A1. Those parts of the former rule that are retained in this rule are essentially unchanged since the Fourth Edition.

262B. Pole Settings

(This rule is essentially unchanged since the Fourth Edition. This rule was deleted when Grade D was merged into Grade B in the 1997 Edition.)

262C. Guys

(This rule was generally revised in the 1977 Edition to better group the requirements and more easily specify related requirements. Rule 262A3 was renumbered Rule 262C5. The requirements themselves are essentially unchanged since the Fourth Edition. This rule was deleted when Grade D was merged into Grade B in the 1997 Edition.)

These rules are the requirements for guying overhead line structures. Although Rule 262C5 construction is not the most desirable way to guy a crossing pole, it is considered a reasonable alternative where it is not practical to guy the crossing pole. Some of the rigidity given to the guyed structure will be transmitted to the crossing poles through the head guy; the amount transmitted will depend on the tautness of this head guy and the rigidity of the guyed structure.

262D. Crossarms

(This rule is essentially unchanged since the Fourth Edition. This rule was deleted when Grade D was merged into Grade B in the 1997 Edition.)

The Fifth Edition added the table of crossarm sizes. The Sixth Edition recognized the standard deadend crossarm as providing adequate strength; it is less likely to split than a standard crossarm because its pins are in line with the pull of the conductors instead of being at right angles. The 1977 Edition revised the rule to recognize that double crossarms are not necessary at crossings if the support arm used is of equivalent strength.

262E. Brackets and Racks

(This rule is essentially unchanged since the Fourth Edition. This rule was deleted when Grade D was merged into Grade B in the 1997 Edition.)

262F. Pins

(This rule is essentially unchanged since the Fourth Edition. This rule was deleted when Grade D was merged into Grade B in the 1997 Edition.)

This rule is one of the many for which the 1977 Edition revision made it clear that the requirement is to meet the strength required by *expected* loads.

262G. Insulators

(This rule is essentially unchanged since the Fourth Edition. This rule was deleted when Grade D was merged into Grade B in the 1997 Edition.)

The strength requirement in the 1977 and later editions refers to expected loads, rather than the ultimate strength of the conductor.

262H. Conductors

(Prior Rule 262H—Attachment of Conductor to Insulator was deleted in the 1977 Edition and the remaining rules were renumbered accordingly. Although extensively rewritten for clarity and expanded somewhat in the 1977 Edition, the requirements of Rule 262H—Conductors are essentially unchanged since the Fourth Edition. This rule was deleted when Grade D was merged into Grade B in the 1997 Edition.)

The sags recommended for communication conductors are limited to copper conductors and will result in tensions that approximate the fatigue endurance of the material (1) at 0 °F in the heavy- and medium-loading areas and (2) at 20 °F in the light-loading area, without wind or ice loading in all three cases. This merely reflects what has been deemed good practice in wire stringing, although it results in consider-able variation as to the maximum stress that will occur under storm loading among the three loading areas. For conductors other than copper, this discrepancy has been eliminated by specifying maximum tensions as 60% of ultimate under storm loading and 20% at 16 °C (60 °F) without external load.

262I. Messengers

Paragraph one was revised in the 1977 Edition to require a minimum breaking strength of 2720 kg (6000 lb) for all messengers. The requirements of paragraph two are often controlling for messengers carrying the heavier cable installation.

263. Grade N Construction

(As in other rules, the various requirements of this rule relating to corrodibility of materials were deleted in the 1977 Edition.)

It is intended that Grade N construction have a design safety factor of one for the loads that are expected in that area. However, the general requirements of the NESC that do not relate particularly to strength should be met in all cases.

263A. Poles

This rule was expanded in the Fifth Edition to include requirements for keeping poles and stubs as far as practical from the traveled portion of state and federal highways, for minimizing the number of wire and cable crossings over such highways, and for the maintenance of lines and equipment that are within falling distance of the traveled portion of these highways.

This rule should *not* be misinterpreted to mean that poles and stubs should be located as far away from the road as *possible*. The practical considerations involved in pole-line placement are many; chief among them is the ability to easily maintain the line. References to pole location were removed in 1997; pole location requirements are contained in Rule 231B (see the discussion of Rule 231B).

263B. Guys

(This rule is unchanged from the Fourth Edition.)

263C. Crossarm Strength

(This rule is essentially unchanged from the Fourth Edition.)

As in similar requirements throughout the Code, this rule was revised in the 1977 Edition to delete the specific poundage requirement for line workers. Such requirements will depend upon the weight of the heaviest line worker expected.

263D. Supply-Line Conductors

263D1. Size

*(Prior to the 1977 Edition, Rule 263 had two subsections: 1—*Material *and* 2—Size.*)*

This rule was expanded in the 1977 Edition to include requirements for common aluminum conductors. The "corrodibility" requirements were deleted at that time. See Rule 261H for a discussion of soft conductors.

The 2007 Edition removed the distinction between urban and rural in Table 263-1; the greater of the previous values is now used in the tables of minimum wire sizes for Grade N supply line conductors.

263E. Service Drops

Service leads of considerably smaller size than line wires of the same voltage are permitted; they are usually strung to much greater sags in order to (1) relieve the poles of unbalanced side loads and (2) reduce the pull on buildings to which they are attached. However, because of their small size and the nature of the attachment at the building, such leads are frequently torn down in storms; many utilities, therefore, find it advisable to use larger sizes generally, except possibly in outlying districts and where the load to be supplied is quite small. Where the service crosses a trolley-contract conductor, the necessity for larger sizes is apparent.

It is considered that supply service leads of more than 750 V should be treated as line wires and be built accordingly for all purposes.

The former sag requirements were deleted in the 1977 Edition; a tension requirement was added.

Cabled service drops are in many cases preferable to individual wires, but care should be taken at the attachments that the wires are properly separated and fastened. The insulation requirement was eliminated in the 1977 Edition (see Rules 230C and D).

263F. Trolley-Contact Conductors

(Former Rule 263F—Lightning Protection Wires was moved to 242F in the 1977 Edition. The present Rule 263F was renumbered from Rule 263G and is unchanged from the Fourth Edition.)

263G. Communication Conductors

(This rule is unchanged from the Fourth Edition; it was renumbered from 263I in the 1977 Edition.)

263H. Street and Area Lighting Equipment

(This rule was added in 1990.)

263I. Insulators

(This rule was added in 1997.)

264. Guying and Bracing

(Rule 282 of the 1987 and prior editions moved to 264 in the 1990 Edition.)

264A. Where Used

If a structure does not have sufficient strength to support its load, the necessary strength should be provided by other means. This applies not only to definite strength requirements in these rules, but also to structures for which no transverse-strength requirements are made. Storm guys for wood structures are accepted practice in most parts of the country. When it is necessary to give additional support to a structure by the use of a guy, the lead of the guy (see Figure H264-1) is an important factor in determining its required strength. Sometimes a head guy may be carried back to the next structure in line.

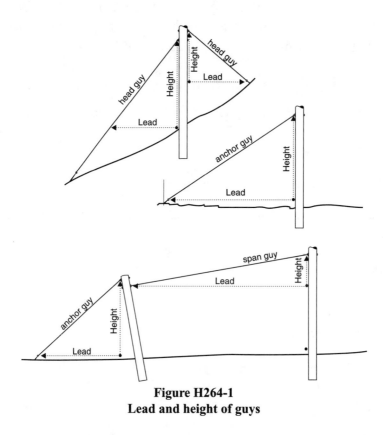

Figure H264-1
Lead and height of guys

It is the usual practice to install guys at angles, corners, deadends, etc., where the forces due to the conductors may cause the allowed stresses of the structure material to be exceeded due to longitudinal and columnar loading. Guys also should be installed on structures carrying very heavy transformers or other similar equipment that would produce such a serious top-heavy condition that the increased load caused by wind pressures on such equipment might cause failure, even though the structure would otherwise be sufficiently strong to support the load due to wind action on the conductors alone. It may also be advisable for storm guys to be installed on extremely heavily loaded lines as an additional precautionary measure, such guys to operate in all four directions.

Where the forces acting upon a structure normally are not balanced, as at angles in the line, the steady pull is likely to gradually displace the pole from the vertical position. This may not lessen its ability to carry its load, but it is objectionable from the standpoint of appearance. More importantly, however, it is objectionable from the standpoint of safety if it leans into the span and effectively shortens the length of span enough to increase the sag of the conductors and enough to significantly reduce the original clearances. In such cases, it is desirable to apply guys in such position as to have the stress in them balance the otherwise unbalanced tension in the wires. This is true especially at sharp corners and at deadends. In the latter case, head guys are, of course, required. Where there is a change in the Grade of Construction, the required longitudinal strength of line supports can often be supplied only by the use of head guys.

Many cases of angles in the line, as well as other instances of unbalanced load, may involve considerable calculation to determine the strength of guy required. If it is not desired to make individual calculations for guying particular structures, then generic calculations should be made and suitable selection criteria should be established to ensure that any guy will always meet or exceed the requirements. The strength of the anchor also needs to be considered and matched to the guy.

This rule has remained essentially as required in the Third Edition.

264B. Strength

A wood pole develops resistance to bending only as it is bent, and the deflection of the top of the pole is considerable before the fiber stress reaches the limiting value fixed by the rules. A very much smaller stretch of the guy will develop its maximum strength, especially as it is normally installed with initial tension. It is evident that the ultimate strength of both cannot be utilized simultaneously. Consequently, when a guy is used on a wood pole, it is required to be strong enough to carry the entire horizontal load. The same applies to flexible steel towers and to concrete poles. The previous requirements relating to flexible structures were deleted in the 1977 Edition. These requirements are in Rule 261C, Rule 262C, and Rule 263B. The 1984 Edition added references to ANSI/ASTM Standards with respect to minimum breaking strength requirements. The 1997 Edition recognized movement of load factors to Section 25.

The 2007 Edition directed users to Rule 217C (the new location for the former Rule 264E requirements) for protection and marking of guys in a new NOTE.

264C. Point of Attachment

This rule essentially is unchanged since the Fourth Edition. It is intended to minimize the bending moment on the pole that is created when guys are located away from the conductors that they are intended to sustain. The insulation value afforded by intervening sections of wood is recognized in the Fifth and later editions.

264D. Guy Fastenings

A high-strength guy can, under stress, do considerable damage if wrapped around a soft wood pole, unless a guy shim is used for protection. After the pole is once cut to any appreciable degree by a guy wire, there is some likelihood of the pole snapping off at the cut under a heavy load. The 2002 Edition changed the reference from rated breaking strength to the true determinant, design load.

Thimbles, guy saddles, pole plates, or their equivalent should be used to attach guys to anchor rods or guy bolts, in order to distribute the load over a greater area.

A 9.1 kN (2000 lb) reference value was added to the thimble require-ment in the 1977 Edition. The 2002 Edition clarified that this referred to rated breaking strength.

In the Sixth and prior editions, this rule required guys to be stranded. The practical use of new, nonstranded materials for straight runs was recognized in the 1977 Edition. These include glass-fiber-reinforced strain insulators used to increase pole-top BIL. Stranded guys now are required only when subject to small-radius bends.

264E. Guy Markers (Guy Guards)

(The requirements of Rule 264E of the 2002 and prior editions were moved to Rule 217C in the 2007 Edition.)

264E. Electrolysis

(This rule was moved from Rule 264F in 2007 when the previous 264E moved to 217C. The title was changed in the 1977 Edition.)

Frequently, anchors for guys are subject to severe electrolysis condi-tions. Anchor rods may be practically destroyed where direct-current railways are in the immediate vicinity. This may be prevented by using suitable insulating blocking between a guy wire and a metal pole, or by using strain insulators in such guys.

This rule was generalized in the 1977 Edition; previously it applied only to anchors attached to metal structures.

264F. Electrolysis

(This rule was moved from Rule 264F in 2007 when the previous 264E moved to 217C.)

264F. Anchor Rods

(This rule was moved from Rule 264G in 2007 when the previous 264E moved to 217C.)

The anchor rod and anchorage are subject to much more rapid deteri-oration than the guy wire; hence, they should be of sufficiently heavy material. In general, anchor rods are of such lengths that their full strength is developed by the anchorage only when installed in solid earth with not more than 300 mm (12 in) of the rod projecting above ground.

When lining up the pull of an anchor guy installed in earth, an error of direction of some magnitude frequently is made and, when installed, the anchor rod will not be in line with the guy. This should not be permitted, as the rod has no holding power in the direction of the strain under such conditions and the guy would soon become slack. This may cause a special problem if the rod is bent sharply along a rock projection; special care in rock areas is often needed.

Where anchor rods are held in the earth by means of wood blocks or pole sections, sometimes called dead men, washers should be installed on the anchor rod of sufficient size to prevent the anchor pulling through the blocks when subjected to the strain for which it is intended. A washer of not less than 25.8 cm^2 (4 in^2) in is recommended.

Anchor rods installed in rock are generally of a special type and are placed at right angles to the direction of the strain, thus securing greater effectiveness.

In the 1984 Edition, the "shall" requirement was relaxed to a "should" in recognition of the fact that it is sometimes not *practical* for joint-anchor construction to meet a "shall" requirement (see Rule 015).

264G. Anchor Rods

(This rule was moved to Rule 264F in 2007 when the previous 264E moved to 217C.)

264H. Grounding

(This rule was expanded and moved to 215C in the 1977 Edition. See also Rule 283B of the 1977 and prior editions.)

Section 27. Line Insulation

270. Application of Rule

This section only applied to Grade A and Grade B construction in the Third and Fourth Editions and Grade B in the Fifth Edition; the Sixth Edition made it applicable to supply lines; the intended application was clarified only to *open-conductor* supply lines in the 1977 Edition. The requirements of Section 27 do not apply to electric supply stations.

271. Material and Marking

The applicability of the requirements was reduced from 7500 V to 2300 V in the Fourth Edition. *All* insulators used on supply lines were required to be made of wet-process porcelain or other equivalent material in the 1977 Edition; the 2300 V break point for the marking requirement was retained.

272. Ratio of Flashover to Puncture Voltage

(In the 1977 and later editions, former Rule 272—Electrical Strength of Insulators in Strain Position was revised and contained within Rule 277—Mechanical Strength of Insulators. See NOTE 1 to that rule. The current Rule 272 is an expansion of former Rule 273.)

273. Insulation Level

(This rule was numbered 274 in the Sixth and prior editions.)

The voltage references were changed to relate to standard voltages in the Fifth Edition. The same voltage references have been retained in the 1977 and later editions. The rule has also been expanded (1) to *allow* lower dry flashover ratings for the stated nominal voltages if based upon a qualified engineering study and *require* such a study for higher nominal voltages, and (2) to *require* the use of insulators with higher dry flashover voltage ratings, or other effective means, where unfavorable conditions exist. The latter requirement was moved from Rule 276C. Note that, as in other cases in the Code where a "qualified engineering study" is referenced, it is generally expected, but not specifically required, that the study will be performed by a Professional Engineer licensed in the state where the facility will be located.

The Code, in essence, recognizes that the required insulating level relates to system BIL. The selection of insulators for a particular system must be coordinated with system BIL, surge-arrester application, and other apparatus and equipment. Where the effects of these factors have been developed by qualified engineering studies, they may be used; otherwise, the minimums listed in the Code are required. Just as the Code requires special calculations for clearances to the higher voltage parts, so does it require special care in insulator selection, although the voltage breakpoint for *requiring* such special calculations differs between the rules. Specifically, this rule allows upgrading the voltage on existing lines on the basis of qualified engineering studies and where other requirements are met.

This rule also permits the use of semiconductive glazing on insulators, sometimes used in high-contamination areas, as long as the appropriate flashover requirements are met.

The values in Table 273-1 and AIEE Std 41 (March 1930) correspond to the bottom end of the 60 Hz test voltage ranges for insulators in common use at the voltages indicated. They are essentially the same as those required since the Third Edition in 1920.

The Strengths and Loadings Subcommittee recognizes that "dry flashover" may not be the best test, but it has been used for many years with reasonable success. The value of considering "wet flashover" is recognized, but consensus agreement on any such method has not been reached.

274. Factory Tests

(This rule was numbered 275 in the Sixth and prior editions.)

In the Sixth and prior editions, a factory test of insulators was required to be used above 15 000 V. Such tests are required in the 1977 and later editions for insulators to be used at or above 2300 V. The specific test requirements also were deleted in the 1977 Edition and the use of ANSI standards was required.

275. Special Insulator Applications

(This rule was numbered 276 in the Sixth and prior editions. Former Rule 276A and Rule 276B have been retained in the 1977 and later editions and former Rule 276C was combined with current Rule 273.)

This rule was added in the Fifth Edition to specify the practice to be followed in selecting insulators for single-phase taps taken from three-phase circuits, either grounded or ungrounded, where such taps are not made through isolating transformers.

276. Protection Against Arcing and Other Damage

(This rule was renumbered from Rule 277 in the 1977 Edition and moved to Rule 447 in the 2002 Edition because it was a work rule, not a construction requirement.)

277. Mechanical Strength of Insulators

(This rule was introduced in the 1977 Edition.)

In 1977, new Rule 277 expanded the requirements of previous Rule 272. It should be noted that, where a multiple insulator assembly is used, it is the assembly that must meet the requirements of the rule. The critical points in a line are the corners, etc., where, with large angles, lightning voltages of twice the tangent line values may be expected due to wave reflections. The usual practice in distribution construction is to double the insulation at these points since, by adding one extra insulator where only one would be sufficient to meet Rule 277, these critical points can be reinforced at the point where a broken insulator can drop the line. This insurance is inexpensive in view of the protection afforded. If a pin insulator shatters, it will often be held together enough by the tie wire to keep the conductor in the air. In contrast, a damaged suspension unit may drop the conductor; it may also shatter the pole top and drop the conductor to the ground, since there is an unbalanced tension at the pole top that is normally held by the guy and the upper section of the pole.

NOTE: Voltage reflections are a function of the surge impedance of the line; in some cases, they may not double or may be reduced by the surge arresters at nearby equipment locations.

The required limitations on loading insulators beyond specified percentages of strength apply throughout the life of the installation. Appropriate recognition should be given to decreases in insulator strength expected during the life of the insulator as a result of the expected cumulative time duration of high loadings.

Strength requirements for composite insulators were added in the 1993 Edition. These requirements refer to the latest ANSI Standards.

In 2007 the specificity of the requirements was expanded and the allowed percentages of insulator strength ratings and related referenced standards were placed in a new Table 277-1.

278. Aerial Cable Systems

(Former Rule 278—Compliance with Rule 277 at Crossings was deleted in the 1977 Edition. The current rule was added in the 1977 Edition.)

This rule specifies both the electrical and mechanical requirements for insulators associated with aerial cable systems. The different requirements applicable to cables meeting Rule 230C1, Rule 230C2, or Rule 230C3 and other cables, such as spacer cable, are specified.

279. Guy and Span Insulators

(Rule 283 and Rule 284 of the 1987 and prior editions were moved to Rule 279A and Rule 279B, respectively, in the 1990 Edition.)

279A. Insulators

(Rule 283A, Rule 283B, and Rule 283C were moved to Rule 279A1, Rule 279A2, and Rule 279A3, respectively, in the 1990 Edition.)

279A1. Properties of Guy Insulators

The material requirements of this rule were applied to Grade C, Grade D, and Grade N for the first time in the 1977 Edition. The requirements of the rule have been essentially unchanged since the Fourth Edition. In the 1977 Edition, it was clarified that a guy insulator could be composed of two or more units. It should be noted that, where an insulator exists in an anchor guy that is grounded at the structure in accordance with Rule 215C, Rule 279A1a and Rule 279A1b (Rule 283A1 and Rule 283A2 of the 1987 and prior editions) do not apply. However, Rule 279A1c (Rule 282A3 of the 1987 and prior editions) is applicable. The voltage between conductors is to be used for both dry and wet flashover ratings.

The language was revised in the 1993 Edition to require insulators based upon expected loading requirements, rather than upon the size of the guy strand that is actually used. This allows oversized guy strands to be used without affecting the size of the insulator.

279A2. Use of Guy Insulators (2002 and prior Editions)

(Rule 279A2 of the 2002 Edition was moved to Rule 215C; see Rules 215C2, 215C3, and 215C5.)

279A2. Galvanic Corrosion and BIL Insulation

(This rule was added as Rule 279A3 Corrosion Protection *in the 1977 Edition and was revised and renumbered to 279A2 in 2007.)*

Insulators placed in a guy solely to prevent electrolysis are not required to meet the requirements for guy insulators, but they may not reduce the strength of the guy. See Rule 217A (Rule 280A of the 1987 and prior editions).

In 2007, insulators used for BIL insulation purposes were specifically added to the rule, along with a reference to new Rule 215C7. Neither insulators used to limit galvanic corrosion nor insulators used for BIL insulation are required to meed the electrical strength requirements for guy insulators, but both are required to meet the mechanical strength requirements (see Rule 279A1c).

279A3. Corrosion Protection

(This rule was added in the 1977 Edition and was revised and renumbered to 279A2 in 2007.)

279B. Span-Wire Insulators

(Rule 284A and Rule 284B were moved to Rule 279B in the 1990 Edition as Rule 279B1 and Rule 279B2, respectively. In 2007, Rule 279B2 Use of Span Wire Insulators *moved to Rules 215C5 and 215C6.)*

When wood poles carry no conductors or attachments except a lamp or trolley-suspension wire, a single insulator at the hanger may be sufficient, since the wood pole usually provides a long, high-resistance path to ground. The public is endangered only by leakage through the pole to ground, and the workers in this case know the hazards of the devices.

Since it is often necessary for workers on the pole to touch the brackets or span wires supporting a series lamp or trolley wire, the insulating value of a wood pole, especially when damp, may not be dependable. It is general practice to provide double insulation between a lamp or a trolley wire and supporting metal poles to ensure continuity of service. It also would seem reasonable for double insulation to be used where line workers will need to work on other circuits carried on wood poles that also carry lighting or trolley brackets, just as it is to ensure the continuity of commercial service of conductors carried on metal poles.

This rule was expanded in the 1977 Edition to include insulation requirements.

The language was revised in the 1993 Edition to require insulators based upon expected loading requirements, rather than upon the size of the guy strand that is actually used. This allows oversized guy strands to be used without affecting the size of the insulator.

Section 28. Miscellaneous Requirements (1987 and prior editions only)

(These rules were recodified in the 1990 Edition to place them near or within related rules. The following is a cross listing of rules from the 1987 Edition with their location in the 1990 Edition.)

1987	1990
280	217
281	218
282	264
283	279A
284	279B
285A	220D
285B1	*D/R213
285B2	*D/R276
286A	220E
286B	236D
286C	237E
286D	237F
286E	232B3 and Table 232-2
286F	234J2
286G1 and 3	232B4a and b
287	223
288	224
289	225
Table 422-4	Table 441-3
Table 427-2	Table 441-1
Table 427-2	Table 441-2
Table 430-1	Table 431-1

*D/RXXX—Deleted due to redundancy with Rule XXX.

Section 29. Rules for Underground Lines (Sixth and prior editions only)

(These rules were recodified in the 1973 Edition into a new Part 3. The following is a cross listing of rules from the Sixth Edition that were moved to Part 3 in the 1973 Edition.)

Sixth	1973
290A, B	320
291D, E	320
290A	320A
291D, E	320B
291	322A
291D, E, H	322B
292A	323A
292B	323B
292E	323C
292F	323D
292H	323E
292C	323H
292D	323I
295A	333
291E, F	341
292G	341
293	341
295B	342
290A	351
294	352
297B, C	360
297A, B	362
292G	371
298	372
296	373
297D	373
299	383

Part 3. Safety Rules for the Installation and Maintenance of Underground Electric Supply and Communication Lines

(Part 3 of the NESC now includes rules relating to underground installations. These rules were developed in the 1973 Edition from Section 29 of the Sixth and prior editions. The previous Part 3 concerned utilization equipment and was rescinded on 14 January 1970. The NEC contains the previous requirements of Part 3 for utilization wiring.)

Section 30. Purpose, Scope, and Application of Rules

(This section was added in the 1973 Edition.)

300. Purpose

(This rule was added in the 1973 Edition.)

The purpose of this section is to detail the requirements for safe installation, operation, and maintenance of underground or buried supply and communications facilities.

301. Scope

(This rule was added in the 1973 Edition.)

Rule 301 differentiates the scope of this part of the NESC from that of the other parts. In particular, it does not cover installations in electric supply stations. Although these rules *may* be useful when considering particulars not detailed in Part 1, they are not required within electric supply stations. For example, the double-door/barrier requirement for access to exposed live parts above 600 V required by Rule 381G is not applicable in areas qualifying as electric supply stations.

302. Application of Rules

(Rule 302 of the 1973 Edition was deleted when general rules were moved in the 1981 Edition to a new Section 1. The present rule was added in the 1993 Edition for consistency.)

303. (Not used in the current edition)

(Rule 303 of the 1973 Edition was deleted when general rules were moved in the 1981 Edition to a new Section 1.)

Section 31. General Requirements Applying to Underground Lines

(This section was added in the 1973 Edition.)

310. Referenced Sections

(Rule 310 of the 1973 Edition was deleted when general rules were moved in the 1981 Edition to Section 1. The current rule was added in the 1981 Edition.)

Rather than repeating the general requirements for each Part of the NESC, these requirements are contained in one group in Sections 1, 2, 3, and 9.

311. Installation and Maintenance

(This rule was added in the 1973 Edition. Rule 311A of the 1973 Edition was deleted when general rules were moved in the 1981 Edition to Section 1.)

Both safety and reliability are served by accurate knowledge of the locations of underground facilities and cooperation between affected parties.

The 2007 Edition added Rule 311C to match existing Rule 230A2d and recognize that, in an emergency, communication cables and specified supply cables may be laid directly on grade, if they are guarded or located so as not to unduly obstruct pedestrian or vehicular traffic and are appropriately marked. This (1) allows quick restoration of service in less time than would be required for permanent replacement of an underground cable and (2) allows timely scheduling of permanent restoration during regular working hours with regular crews. Note that this rule applies to emergency outages, *not to temporary installations*; see the discussion of Rule 230A2d for further information. A *temporary*, above-ground "underground cable" must be installed in conduit to meet Section 32.

312. Accessibility

(This rule was added in the 1973 Edition.)

Adequate working spaces, facilities, and clearances are required for workers to operate and maintain the system efficiently and effectively, especially under emergency conditions. This requirement is similar to Rule 213 for overhead facilities. Because of the greater opportunity for cramped spaces and limited egress in underground facilities, the design of these facilities to include appropriate working spaces and clearances is especially critical.

313. Inspection and Tests of Lines and Equipment

(This rule was added in the 1973 Edition.)

Rule 313 is a duplication of the same requirements for overhead lines. See Rule 214 for a complete discussion.

314. Grounding of Circuits and Equipment

(This rule was added in the 1973 Edition.)

The *requirements* for grounding of underground facilities are found in Part 3. The *methods* of grounding are located in Section 9. Conductive parts, frames, and cases that may present a hazard to workers or pedestrians, if inadvertently energized, are required to be effectively grounded. Equipotentiality of surfaces both limits electrolysis problems and promotes safety. Grounding requirements for nonneutral conductors were specifically included in the 1990 Edition.

For an extended discussion of problems with using the earth as the sole conductor for any part of a circuit (see Rule 215B). Using an equipment case, lighting standard, etc., as a sole conductor is functionally equivalent to using the earth and presents a greater chance of exposure to workers or the public.

Rule 314B does not apply to metallic handhold covers of pull boxes containing street lights. It does apply to conduits and risers coverings over all supply cables and, if exposed to open supply conductors of greater than 300 V, conduits and riser coverings over communication cables.

The NOTE to Rule 314C4 was added in the 1993 Edition to recognize the practical necessity of operating a bipolar HVDC system in a monopolar operation during periods of some emergencies and some maintenance activities. Monopolar operation is not considered appropriate as a general mode of operation. The NOTE was changed to a Rule in 1997.

315. Communications Protection Requirements

(This rule was added in the 1973 Edition.)

This rule is a duplicate of Rule 223; see that rule for a discussion of the requirements. Application of the rule to steady-state induced voltages was added in the 1987 Edition.

316. Induced Voltage

(This rule was added in the 1973 Edition.)

This rule is a duplicate of Rule 212; see that rule for a discussion.

Section 32. Underground Conduit Systems

(These rules were developed in the 1973 Edition from prior Rules 290, 291, and 292.)

NOTE 1 under the section title includes the special definitions for *duct*, *conduit*, and *conduit system*. A conduit is a structure that contains one or more enclosed raceways called *ducts*. A conduit system includes both the conduit and the associated manholes, etc. Thus, when the term *duct* is used, the rule generally concerns fit, smoothness, etc., to limit damage to cables when pulling. Rules affecting conduits and conduit systems generally are concerned with location, strength, etc., to limit damage to the cables from surface usage or adjacent underground activity.

NOTE 2 was added in the 2002 Edition to recognize the increasing practice of installing so-called "cable-in-conduit" with the methods used to install direct-buried cable. Such installations are frequently made between pad-mounted equipment and do not involve manholes or handholes. As of the 2002 Edition, single-duct conduit of all types that does not run to manholes or handholes is intended to meet the same rules as direct-buried cable in Section 35, instead of the requirements for conduits in Section 32. If a single-duct conduit runs to a manhole or handhole, Section 32 applies, not Section 35.

320. Location

(This rule was developed in the 1973 Edition from prior Rules 290A and 290B, and 291D and 291E.)

320A. Routing

(This rule was developed in the 1973 Edition from Rule 290A of the Sixth and prior editions.)

Municipalities will often prescribe the general location of an underground installation. In many cases, the location of existing piping will be a determining factor in the location of underground supply and communications facilities. If given some freedom, a utility can eliminate significant trouble and expense by a careful study of the existing underground structures, together with those being planned for the future, before installation of new facilities. Such careful planning may permit greater manhole dimensions than frequently are provided in congested districts.

Conduits should be installed in straight lines to reduce pulling strain on both conduits and cables. Especially critical is the proper alignment of conduit sections so as to provide smooth ducts. Rule 230A1c requires the bending radius to be large enough to limit the likelihood of damage to the conductor during pulling (a small radius increases pulling tension and can damage the cable). The 2002 Edition removed the former RECOMMENDATION in Rule 320A1c for a 5° angle limit at joints because (1) this is design information, and (2) this is overly restrictive in some cases. It is recognized that single enclosed raceways (i.e., single-duct conduit) have various allowable bending radii, depending upon materials, before the curvature is enough to cause a structural problem. Likewise, the joints have various allowable angles before becoming potentially dangerous to the conduit or cable. Under Rule 012C, angles at joints must not be greater than appropriate for the materials and design involved. In unstable soils, special care may be required. Where curves are required to meet other rules or to accommodate existing site features, the radius of curvature should be as great as practical. If curves of short radius are used, the cable may be damaged during the pulling operation.

If the intended service of the installation is to be accomplished efficiently, the location of facilities must provide safe, easy access for maintenance. The impact of utility system maintenance on the flow of pedestrian and vehicular traffic can unduly affect the safety of both utility workers and the public if the location of underground facilities requiring maintenance is not planned carefully. A conduit system also must be located relative to other underground facilities or specially constructed such that reaction with its environment will neither damage the installation nor require undue maintenance.

320B. Separation From Other Underground Installations

(This rule was developed in the 1973 Edition from Rules 291D and 291E of the Sixth and prior editions. The title was changed from "Clearances" to "Separation" in the 1987 Edition.)

Conduits should be located as far as practical from other underground structures, especially from water mains and gas mains. Water from a broken main undermines the conduit system, causing it to settle or even break. Leaking gas will often find its way through considerable earth to a manhole; it especially tends to follow pipes, ducts, and similar constructions. Thus, the greater the distance between such systems, the less are the chances of damage.

To arrest the action of an electric-power arc, and to prevent it from affecting communication cables, a barrier wall of concrete not less than three inches thick, or equivalent protection, should be placed between ducts carrying supply conductors and adjacent ducts carrying communication conductors. This same means of limiting damage by cable arcs is often advisable for use between conduits containing large supply feeders used for different classes of service or acting as important tie lines between different stations.

Notice that the Code language generally assumes that conduits will be above sewer and water lines; in general this is practical and preferred. However, in certain areas of Florida and other states where, due to a combination of mild weather and flat terrain, water lines are commonly found near the surface, the only practical location for supply and communication conduits may be below the level of water and/or sewer lines. Depending upon the separation, special structural precautions may be necessary to maintain duct integrity.

As of the 2002 revision, both conduits and gas (or other fuel) lines are prohibited from being in the same manhole, handhold, or vault. Handholes and vaults were added to the existing prohibition for manholes. The 2007 Edition recognized that such lines that carry flammable materials that are not ordinarily considered as fuels are also intended to be covered by this prohibition.

321. Excavation and Backfill

(This rule was added in the 1973 Edition.)

The reliability of conduit systems is directly related to the continued integrity of the system. Care should be exercised in preparing the bed in which a conduit system will lie. Notice that the backfill requirement for conduits are not as rigid as those for direct-buried cables.

322. Ducts and Joints

(This rule was developed in the 1973 Edition from Rule 291 of the Sixth and prior editions.)

322A. General

(This rule was developed in the 1973 Edition from Rule 291 of the Sixth and prior editions.)

There are a number of different duct materials, each of which has different strength, finish, and fireproof qualities. The interior of all ducts should be smooth and free from projections so that cables may be readily installed and removed without damage to the sheath. Where the arc from a damaged supply cable in one duct could damage or otherwise harm a cable in an adjacent duct, either (1) the material and design of the duct or conduit should be constructed to withstand such conditions, or (2) the cables should be installed in single-duct conduits spread sufficiently apart to limit such damage.

The outside forces to which the conduit will be subjected, such as surface traffic loadings, should be considered in conduit design. Design and installation of the conduit to meet such loads will help ensure safe and reliable service.

Impact loading from surface forces can be a major cause of deterioration of a conduit system or misalignment of its components. A

30% increase in the live loads expected above the conduit is required for impact loading. This impact loading can be reduced by 10% per foot of cover; i.e., *impact* loading need not be considered below 3 ft (900 mm) of burial depth *but* the effect of the *live* loads must still be considered (see Rule 323A).

322B. Installation

(This rule was developed in the 1973 Edition from Rules 291D, 291E, and 291H of the Sixth and prior editions.)

The care with which a conduit is designed and installed can be the single largest factor in determining the useful life of the system. Installing the conduit so that it is protected from external forces and contaminants increases both safety and reliability.

If conduits are laid carelessly, shoulders can occur between adjoining sections of the ducts; these may damage the cable sheath and even render it impossible to pull the cable into the duct. By the same token, the conduit must be restrained appropriately to ensure continued alignment of its components during pulling stresses and under its environmental loading. Conduits should be so designed that proper alignment of the duct(s) can be maintained during construction.

Where soil is soft and unstable, suitable foundations for conduits should be prepared from materials. In solid ground, a suitable foundation may be provided by tamping the natural soil securely into place. Workers frequently break into conduits when making excavations in streets. Aside from the property damage, such accidents may injure the cables and their sheathing. As a result, it is sometimes advisable to provide covers over conduits to reduce this trouble. Where ducts are embedded in concrete, it is generally at least 75 mm (3 in) thick.

Where external, corrosion-resistant coatings are used on pipes used for conduit, special care is required during the storage, handling, placement, and fitting of the pipe sections to ensure continuity of the corrosion-inhibiting compound over the exterior of the pipe.

Conduit systems are subject to the entrance of underground gases from decomposition and other sources. Where conduits extend through a building wall, an external seal is required around the conduit and an internal seal is required around the cable(s) inside the conduit to limit entrance of gas into the building. Venting of the conduit to limit gas pressure also may be appropriate. Note that the NESC work rules in Part 4 require testing of the air quality in manholes before entrance and may require positive ventilation while work is in progress.

Conduits should be installed so as to be supported continually at the entrance into buildings, bridges, manholes, etc., to limit damage due to settling. Conduits installed in or on other structures must be designed with matching expansion and contraction capability.

323. Manholes, Handholes, and Vaults

(This rule was developed in the 1973 Edition from Rule 292 of the Sixth and prior editions.)

323A. Strength

(This rule was developed in the 1973 Edition from Rule 292A of the Sixth and prior editions.)

It is not contemplated that every manhole should sustain the heaviest loads; each should provide strength in accordance with the conditions that it is expected to meet. The intention of these rules is to cause careful, reasoned analysis and design of conduit systems to meet the conditions that are expected to exist. This rule specifies both specific and general strength requirements and takes into account the effects of both dead loads and live loads, including impact loading. These requirements are also referenced by Rule 322 for determining the required strength of conduits. In roadway areas, the specific weight and dimension between stress points of Figure 323-1 are required to be used (see Figure H323A).

NOTE: During the process of installation, many utilities require that suitable eyes or hooks be embodied in the concrete or brick walls of manholes. If these are located properly, they can facilitate greatly the installation and removal of cables.

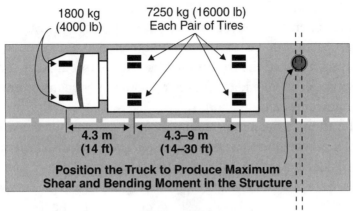

Figure H323A
Underground structure loading

323B. Dimensions

(This rule was developed in the 1973 Edition from Rule 292B of the Sixth and prior editions.)

The dimensions required by this rule can usually be provided in manholes without difficulty. They are the minima required to provide a reasonably safe working space and to provide speedy egress if an accident occurs. The dimensions recognize required actions under different circumstances; lesser values are allowed by EXCEPTIONS where the potential hazard is otherwise limited. The inside dimensions are important especially where transformers are to be installed in manholes. Sufficient space should be provided to safely and easily operate the cutouts necessary with transformers.

323C. Manhole Access

(This rule was developed in the 1973 Edition from Rule 292E of the Sixth and prior editions.)

The increase in average human size since the early Codes were written was recognized in the revision of the 1973 Edition. The width of a round manhole access opening to a supply manhole was increased from

600 mm (24 in) to 660 mm (26 in). The requirements of both round and rectangular openings are recognized. The rule recognizes that one of the purposes of the access opening is quick egress in the case of an accident. As a result, the location of the means of egress is especially important so as not to endanger either personnel or enclosed equipment during emergency actions.

The location of the access openings in highways also can affect the safety of workers as well as the public. Where practical, such openings should be located outside of either pedestrian or vehicle lanes. However, it is recognized that it is all too often necessary to place such facilities within areas exposed to pedestrian or vehicle traffic. In these cases, consideration should be given to locations such that temporary protective railing or other traffic inhibitors can be erected with a minimum of interruption of traffic flow. Such placement will increase the safety of both workers and passersby. Prior planning of the location of all underground facilities in highway areas, where practical, can significantly aid in the maintenance process and reduce potential hazards.

Where practical, personnel access openings should not be located directly over cables or equipment. However, where the personnel access openings would interfere with a curb, etc., if it were located so as not to be *directly* over a cable or equipment, the opening may be located over a cable or equipment—so long as an appropriate safety sign is used to warn utility personnel. The 1997 Edition added the reference to ANSI Z535 safety sign standards.

Rule 323C5 was added in 2002. Manholes deeper than 4 ft (1.25 m) must have ladders or other suitable climbable devices to avoid damage to cables or equipment during entry.

323D. Covers

(This rule was developed in the 1973 Edition from Rule 292F of the Sixth and prior editions.)

The covers of manholes serve two purposes: (1) they limit access to the interior of unauthorized persons, and (2) they form an integral part of the structural arrangement that supports and protects the enclosed facilities.

323E. Vault and Utility Tunnel Access

(This rule was developed in the 1973 Edition from Rule 292H of the Sixth and prior editions.)

The location of vault and tunnel openings affects the safety of both workers and the public. The location of access openings to vaults and tunnels is affected by concerns similar to those expressed for manhole access. In addition, where access doors are accessible to the public, they should be treated similarly to electric supply station gates and remain locked unless under control by a qualified attendant. Similar to rooms in electric supply stations, Rule 323E requires the door to be able to be exited from inside the space, but "panic" hardware is not required by the rule.

This rule recognizes the need to limit access by unqualified people while providing quick egress for any workers who may be inside. The rule is intended to apply to vertical doors, not manhole top covers that happen to be hinged; Rule 323D applies to such covers. As of the 2002 Edition, safety signs are required at entrances to vaults and tunnels containing exposed live parts. See ANSI Z535.2 for sign information.

323F. Ladder Requirements

(This rule was added in the 1973 Edition as 323G. It was moved to 323F in the 1987 Edition.)

This rule recommends specifications for fixed ladders and requires portable ladders to meet the specifications of the NESC work rules. Ladders should be inspected at reasonable intervals to ensure continued safety to climbers.

323G. Drainage

(This is essentially Rule 292C of the Sixth and prior editions. It was moved to 323H in the 1973 Edition and to 323G in the 1987 Edition.)

The apparatus and cables in manholes are not accessible when covered with water. Where water has been present, sediment deposited on the apparatus and walls of the manhole may require extensive cleaning before work can be done in the manholes. As a result, it is important to carefully plan and install appropriate drainage for manholes and conduits. Where drains lead directly into sewers, traps should be used to keep sewer gas out of the manholes.

Sewer gas may be dangerous either because of its toxic effect or because of the lack of sufficient oxygen content. However, its presence generally is evidenced by the sulfide of hydrogen caused by the decomposition of organic matter in the sewers. A poisonous effect is produced by the presence of carbon monoxide and sulfides of ammonia and hydrogen. Carbon monoxide, gasoline vapor, and methane, combined with air, may produce an explosive mixture. All of these characteristics combine to make sewer gas extremely objectionable in manholes.

323H. Ventilation

(This rule was developed from Rule 292D of the Sixth and prior editions. It was moved to 323I in the 1973 Edition and to 323H in the 1987 Edition.)

These areas or enclosures are subject to collection of corrosive, poisonous, and explosive gases. Both safety and reliability are served by adequate ventilation.

323I. Mechanical Protection

(This rule was added in the 1973 Edition as 323J and was moved to 323I in the 1987 Edition.)

Rule 323I recognizes the practical problems of many installations. Open grates are often used to provide adequate ventilation. Even without any human mischief, such areas are subject to collection of debris that is blown or tracked onto the grates and falls inside. Facilities behind or under such grates should be so located or otherwise sheltered to limit damage by falling or blowing objects.

323J. Identification

(This rule was added as Rule 323K in the 1973 Edition and was moved to 323J in the 1987 Edition.)

To limit inadvertent entry into electric or communication manholes by personnel of other utilities, the manhole and handhole covers should identify either the type of utility, e.g., "ELECTRIC," or the ownership of the facilities.

Section 33. Supply Cable

(This section was added in the 1973 Edition. It includes Rule 295A of the Sixth and prior editions.)

330. General

(This rule was added in the 1973 Edition.)

This rule recognizes that the purpose of a cable installation will not be fulfilled, at least not without a decrease in safety, reliability, and economy, unless the cable system components are designed and installed to meet the mechanical, thermal, environmental, and electrical stresses expected during installation and operation.

However, Rule 330D recognizes that it is impractical to require such cable to withstand the direct effects of a fault, such as arcing, fire, exploding gases, etc., in the immediate area of the fault.

331. Sheaths and Jackets

(This rule was added in the 1973 Edition.)

Adverse environmental conditions will decrease the safety and reliability of cables unless appropriate measures are taken. The use of fully insulating jackets also can affect the grounding requirements in certain situations (see Rule 354).

332. Shielding

(This rule was added in the 1973 Edition.)

Rule 332 recognizes the variation in shielding techniques and materials that may be appropriate for use. The intent of the requirement for nonmagnetic metal is to reduce damage to the cable insulation or its shielding as a result of overheating by induction. Steel-clad copper concentric neutral cable meets this intention if it is designed and applied so as to avoid induction heating problems.

NOTE: The requirements for conductor shielding are a "should" rule, while the requirements for insulation shielding are a "shall" rule.

333. Cable Accessories and Joints

(This rule was developed in the 1973 Edition from Rule 295A of the Sixth and prior editions.)

Like cables themselves, the joints and accessories must withstand the mechanical, thermal, environmental, and electrical stresses expected during installation and operation.

Section 34. Cable in Underground Structures

(This section was developed in the 1973 Edition from Rules 291E and F, 292G, 293, and 295B of the Sixth and prior editions.)

340. General

(This rule was added in the 1973 Edition.)

The rules of Section 33 also apply to cables in underground structures. Above 2 kV to ground, an effectively grounded shield or sheath, or both, can effectively reduce catastrophic destruction of nonmetallic conduit and adjacent cables during a fault by speeding the operation of protective devices and reducing current flow to and in cables not involved initially in a fault.

While this rule is similar to Rule 350 for direct-buried cables, the sheath or shield is not *required* by this rule as it is in 350B. However, since Rule 332 requires shielding as specified by applicable cable specifications, the practical effect is that shielding, sheathing, or both are generally used above 2 kV.

341. Installation

(This rule was developed in the 1973 Edition from Rules 291E and F, 292G, and 293 of the Sixth and prior editions.)

Because of the limited amount of space in manholes, cables should be racked carefully and so spaced and located that they are readily accessible to workers. Experience has shown that, when cables are crowded together and have an inferior working space about them, the work performed will be inferior to that performed on cables that are readily accessible.

Because of the great cost of underground construction, it is appropriate to allow communication and supply cables to be located in jointly used conduit, if the proper precautions are taken in the manholes and the parties agree on installation and maintenance designs or procedures.

When a supply cable fails, the arc caused may damage other cables in the same manhole. If communication cables and supply cables are located close together in a manhole, trouble originating in the supply cable could damage the communication cable, possibly transferring electric potentials from the supply cable to the communication cable. When it is necessary for both classes of lines to use the same manhole, they naturally should be kept as far from each other as practical.

In a conduit system, the ducts in the center will dissipate heat less effectively than those on the sides or corners. Cable of heaviest capacity should, therefore, be placed in the corner positions and those of lowest capacity in the center positions. If ducts are grouped in large numbers, they cannot be relied upon to dissipate the heat properly unless detailed calculation is made. Therefore, the number of supply ducts in a bank may need to be limited.

The splicing of supply cables is a very important and particular operation. Joints usually give more trouble than any other part of an underground system. This usually results from failure of insulation, overheating due to poor contacts, and entrance of moisture through poor joints. Joints should be inspected to determine if they are heating excessively or if other defects are beginning to show. Joints should be located only in accessible places, not in ducts.

In order to reduce the possibility of damage to low-voltage cables by arcs due to failure of high-voltage cables, the two should be separated as much as practical.

Where practical in underground construction, supply and communication lines should, as with overhead construction, be given separate routes. The expense of providing separate routes for supply and communication lines, and the lack of room in a street, have often necessitated use of a multiduct, single conduit line by both utilities. With a single conduit line, a somewhat greater hazard exists than with separate conduit lines. However, that hazard is confined mainly to employees and can be largely negated through careful installation of facilities and adequate training of employees; the public rarely is endangered at any time by underground lines. Where the supply lines are of high voltage or of very large capacity, it is still more desirable to keep the two kinds of systems separate, if practical. However, when both systems are installed in a single conduit line, the requirements of these rules provide adequate safety for both workers and the public; see the additional discussion at Rule 320B.

CAUTION: The wording of Rule 341A6 prohibited most joint use of ducts by communication and supply conductors until, when revised in the 2002 Edition, it prohibited joint power/communication ducts unless all cables were operated and maintained by the same utility. Reference to control cable was removed. Multiple communication cables may be in the same duct, if all parties agree. This rule still applied if the two cable facilities are installed with random separation; Rule 354D2a(3) (Rule 354C2a(3) prior to 1993) and the EXCEPTION to Rule 354D2a(3) do not provide an exception to Rule 341A6. To limit potential confusion, the 2007 Edition added Rules 352E and 352F to directly specify such limits within Section 35.

This rule particularly stresses the care appropriate during installation and the support required to ensure that cable components will not be overstressed either during initial installation or while in service. Likewise, potential damage to one cable during installation or maintenance of another, and potential hazard to workers, can be limited by considering these requirements in rack placement and design.

Easily understandable identification of cables in manholes is necessary for both safety and maintenance efficiency. Identification may be made by use of metal tags, stenciling of the cable, or by charts showing the position of the cables, or other permanent means. When tags are employed for this purpose, a fire-resistant, noncorrodible material should be used and the markers should not be easily obliterated. A uniform method for installing cables in ducts should be followed throughout a system, as far as practical. For instance, it is customary to install the local power-distribution cables in the top ducts of a conduit line. This both facilitates their identification and permits the installation of an intermediate service hole, which requires access to these cables only, between manholes.

The 2002 Edition revised Rule 341B3b so that, like joint-use manholes, joint-use vaults now require marking of cables as to utility name and type of cable use.

342. Grounding and Bonding

(This rule was developed in the 1973 Edition from Rule 295B of the Sixth and prior editions.)

Equipotential systems aid both safety and reliability. The 2007 Edition was revised to emphasize the importance of grounding bare metallic shields, sheaths, or concentric neutrals that are exposed to personnel contact. As a practical matter, bonding of the items would be required; see the definition of *effectively grounded*.

343. Fireproofing

(This rule was added in the 1973 Edition.)

The need for fireproofing is a function of the environment around the underground installation.

344. Communication Cables Containing Special Supply Circuits

(This rule was added in the 1973 Edition.)

This rule is almost identical to Rule 224B2. It recognizes that many of the communication systems now in use require assistance from low-power amplifiers and other electrically powered equipment. If the power circuit is to be grounded, the grounded cable sheaths required by Rule 344D1 cannot be used as the grounded conductor.

Section 35. Direct-Buried Cable

(This section was developed in the 1973 Edition from Rules 290A and 294 of the Sixth and prior editions.)

350. General

(This rule was added in the 1973 Edition. In the 2007 Edition, Rule 350F)

The rules of Section 33 also apply to direct-buried cable. The rules of this section detail the arrangement and installation conditions required for safe installations. These rules essentially parallel those included in Section 32 for conduit systems and are expanded in order to recognize the special problems of direct-buried systems. For example, an effectively grounded, continuous metallic shield, sheath, or concentric neutral is required for direct-buried cables above 600 V. This speeds the operation of the system protection and limits the current flow to and on other cables, thus limiting the catastrophic effects of a high-voltage cable fault.

Rule 350C was added in the 1993 Edition to be consistent with Rule 354.

Rule 350E of the 1987 Edition (later Rule 350F and now Rule 384C) added the requirement to bond together all above-ground metallic pedestals, cases, etc., that are located within reach of each other, i.e., 1.8 m (6 ft). The intention is to limit the opportunity for a worker to touch two items that may be at significantly different voltage potentials. Where walls or barriers or grade level differences prevent simultaneous contact, the rule is not applicable.

Below 600 V, if an effectively grounded sheath or shield is not used around each cable, all the cables of the same circuit are required to be placed in the trench without intentional separation from each other; this does not prohibit more than one circuit of the type in random lay. The special requirements of Rules 344A1 through 344A5 also apply when communication cables containing special supply circuits are placed underground by direct burial.

Rule 350F of the 2007 Edition (previous Rule 350G) requires direct-buried supply cables operating above 600 V and all communication cables to be legibly marked. It was added in the 1993 Edition and initially carried a special effective date of 1 January 1994. The effective date was later extended to 1 January 1996. This allowed existing supplies of unmarked cable to be used and allowed production time for manufacturers to meet the new cable-marking requirements. The new marking requirements were developed over several years with the aid of individuals and groups from both the utility industry and manufacturers. Rule 350G recognizes that some cables may be too small or the jackets too thin to effectively mark. Communication service drops contain only a few pair of wires are generally considered as too small to mark, but larger communication cable, whether used in a line or service run, must be marked. The 1997 Edition clarified the ability to separate or combine sequential markings.

Rule 350G was added as Rule 350H in the 2002 Edition to clarify which rules apply to "cable-in-conduit" installations that are installed using direct-buried techniques. The requirements of Section 35 also apply to a single-duct conduit that is not part of a conduit system (i.e., that has no manholes or handholes). The term also was included because Section 33 applies. Neither Section 32 nor Section 34 applies to single duct conduit that is not a part of a conduit system.

351. Location and Routing

(This rule was developed from Rule 290A and those other rules of the Sixth Edition that were used to develop Section 32.)

The discussions of the rules in Section 32, especially Rule 320A, apply to the similar or identical requirements in Rule 351. Because direct-buried cables lack the protection of a conduit, they need additional care in installation in order to provide the same level of safety and reliability at an economical cost. In many ways, direct-buried cables are inherently safer, once their integrity has been violated, than systems in conduits. The direct-buried cable generally will short-out with the ground without creating a personnel hazard. Cables in conduits, while harder to breach accidentally, may not ground-out as easily once the conduit has been broken unless an effectively grounded sheath, shield, or neutral is directly available within the duct containing the faulted cable.

The 2002 Revision *requires* adherance to Rule 353 *or 354* when cables are installed parallel to and directly over or under another subsurface structure.

This rule includes separations for direct-buried cables from swimming pools; no such separations are required for conductors in conduits. In the 1984 Edition, it was recognized that such installations cannot always be located 1.5 m (5 ft) horizontally away from in-ground swimming pools. The revision makes this rule agree with NEC requirements. The rule recognizes that the desired separation may not be practical in subdivisions with small lots and allows supplemental mechanical protection to be used in lieu of separation to limit the opportunity for damage of the cable(s) from rock movement, above ground loads, etc. The Code intentionally did not specify the type of supplemental protection to be used in such circumstances (see Rule 012).

Because of questions about the applicability of Rule 351C1 to above-ground swimming pools, Rule 351C2 was revised in 2007 to limit installation of cables under the foundations of *other structures* unless the foundation is suitably supported to limit the likelihood of cable damage due to the structure load. Thus, above-ground pools and other structures are covered in the same manner previously specified for buildings and storage tank foundations. Rule 351C1 is now limited to in-ground pools.

Burial depths below railroad tracks were reduced in the 1981 Edition and the EXCEPTION and NOTE were added to Rule 351C3.

352. Separations From Other Underground Structures (sewers, water lines, fuel lines, building foundations, steam lines, other supply or communication conductors not in random separation, etc.)

(This rule was developed in the 1973 Edition from Rule 294 of the Sixth and prior editions. The title was changed from "Clearances" to "Separations" in the 1987 Edition. It was renumbered from 352 in the 2002 Edition.)

352. Installation

(This rule was added in the 1973 Edition. It was renumbered from 353 in the 2002 Edition.)

This is the direct-buried equivalent of Rule 321 and Rule 320B2; see the discussions of those rules. Rule 352 provides adequate protection for the installation and normal useful life of direct-buried cables. The depths of burial required by Rule 352D2 apply unless either of Rules 352D2a, 352D2b, or 352D2c apply. The table was revised in the 1984 Edition and the voltages were required to be phase to phase. The organization of the requirements was revised in the 1990 Edition and the burial depths were placed in Table 352-1 (then 353-1), thus causing the EXCEPTIONS to be renumbered from 353D2b, 353D2c, and 353D2d to 353D2a, 353Db, and 353Dc in 1990, now 352D2a, 352D2b, and 352D2c.

Rule 352D1 (Rule 353D1 in 1990–1997) indicates the requirement that must be met if Rule 352D2b (353D2b in 1990–1997) is used. The supplementary protection must be sufficient to protect the cable from damage imposed by expected surface usage. The extent of the supplementary protection required by Rule 352D2b (Rule 353D2b in 1990–1997) depends upon the depth that can be achieved and upon the surface usage. It is the responsibility of the design engineer to consider these factors in the analysis used to select the supplementary protection. In any such installation, the limiting requirements of Rules 94B5 and 354D2 (Rule 354C2 prior to 1993) continue in force. Transitions from the required burial depths of Rule 352D2, Table 352-1, (Rule 353D2, Table 353-1 in 1990–1997) to the lesser depths of Rule 352D2b (Rule 353D2b in 1990–1997) must also be supplementally protected. The possibility of electrolytic reactions between the concentric neutral and the supplemental protection are a design consideration.

The burial depths of Table 352-1 (Table 353-1 in 1990–1997) are intended for land application. For submarine cables, such depths may be insufficient to provide protection against damage from heavy anchors, erosion, etc. (see Rule 351C5). The 2002 Edition restricted installation of street and area lighting cables at an 18-in burial depth to only those places where conflicts with other underground facilities exists.

Rules 352E and F were added in 2007 to match Rules 341A6 and 7 to emphasize that direct-buried supply and communication cables are prohibited from being installed in the same duct (raceway) unless all are owned and maintained by the same utility. Communication cables of different owners may be installed in the same duct only if all utilities involved agree.

353. Deliberate Separations—Equal to or Greater than 300 mm (12 in) From Underground Structures or Cables

(This rule was developed in the 1973 Edition from Rule 294 of the Sixth and prior editions. The title was changed from "Clearances" to "Separations" in the 1987 Edition. It was renumbered from 353 in the 2002 Edition.)

Special care is required in locating direct-buried cables near other facilities. These rules are intended to provide (1) adequate room for maintenance of all facilities and (2) appropriate protection for each system from the effects of the other; see the discussion of Rule 320B, which is similar to this rule, and see Rule 354.

This rule was revised in the 2002 Edition to coordinate with changes to Rule 354—*Random Separation—Separation Less than 300 mm*

(12 in) from Underground Structures or Other Cables. The key to application of Rule 353 versus Rule 354 is whether a separation of not less than 300 mm (12 in) will be deliberately maintained. Both rules apply to separation of supply cables and communication cables from each other and from other underground structures.

There is no requirement that 300 mm (12 in) or more be maintained for a complete cable run, so long as Rule 354 is met for those circumstances where 300 mm (12 in) will not be maintained. However, if supply and communication cables are installed at less than 300 mm (12 in) at any point; it will require adherance to the restrictions of Rule 354d for extensive portions, if not all, of the cable.

The allowance of less than 300 mm (12 in) of separation on crossings by agreement (formerly contained in 352B4 before renumbering) was removed in 2002, since the entire rule now applies when 300 mm (12 in) separation will be maintained.

354. Random Separation—Separation Less Than 300 mm (12 in) From Underground Structures or Other Cables

(This rule was added in the 1973 Edition as Random Separation—Additional Requirements; the name changed in 2002.)

These special rules recognize the value to be gained in some cases by installing cables with less than 30 cm (12 in) of clearance between them. The rule specifies the conditions that are required to ensure safe and reliable service.

The rule was revised in the 1981 Edition to recognize the use of semiconducting jackets and to detail the requirements for such use.

The following EPRI Reports support the proposed changes: (1) EPRI EL-619 Project 671-1, (2) *Final Report* Volume 1—December 1977, (3) *Evaluation of Semiconducting Jacket for Concentric Neutral URD Cable*, (4) EPRI EL-619 Project 671-1, (5) *Final Report*

Volume 2—December 1977, and (6) *Cable Neutral Corrosion*. These reports demonstrate the following pertinent points:

(1) Bare and semiconducting jacketed concentric neutral cables exhibit approximately the same touch and step voltages when tested under equivalent conditions.

The equivalent performance of the bare and semiconducting jacketed cables is contingent on the radial resistivity of the semiconducting jacket being comparable to that of the earth and remaining essentially stable in service. Semiconducting jacket compounds are available commercially that satisfy these criteria.

(2) When tested under equivalent conditions, the touch voltage for a concentric neutral cable with insulating type jacket will be higher than for a bare or semiconducting jacketed cable.

The difference in touch voltages diminishes with increase in resistivity of the backfill soil in the area of the fault. However, the touch voltage will be lower for an insulating-type jacketed cable in a circuit with a high ratio of high side (impedance from the supply to point of fault) to total circuit impedance than for bare or semiconducting jacketed cable in a circuit with a low ratio of high side to total circuit impedance. For the same location with respect to the cable, the step voltages will be higher for the bare and semiconducting jacketed cables and the step voltages for the bare, semiconducting, and insulating-type jacketed cables tend to approach the same value with increase in backfill soil resistivity in the area of the fault.

To continue to meet the requirements of Rule 97 and to be able to interconnect the secondary neutral with the primary neutral and the primary arrester ground, primary neutrals must continue to be ***both*** *effectively grounded* and *multigrounded* throughout their length. This requirement also applies to concentric neutrals surrounding underground cables.

If the concentric neutral on an underground primary cable is exposed to the earth, or is covered by a semiconducting jacket with radial resistivity of 100 $\Omega \cdot$m or less, for a length of 30 m (100 ft) or greater, then the concentric neutral itself is considered to be an electrode and no additional ground connections are required. The radial resistivity value was changed from 20 Ω•m to 100 Ω•m radial resistivity in the 1984 Edition.

When the concentric neutral conductors are bare or covered with a semiconducting jacket, and one cable of a three-phase underground line faults to ground, the fault current can flow back to the source over all of the concentric neutral conductors. However, if a jacket of more than 100 Ω•m radial resistivity is used around the concentric neutral conductors, then

(1) the concentric neutral conductors of nonfaulted cables will be effectively insulated from the fault current, and

(2) the concentric neutral conductors will not be continuously grounded and, therefore, will have to be connected to ground four times in each 1.6 km (1 mi) of the entire line (Rules 96 and 97), just like the overhead neutrals and the neutrals separated from earth contact by a conduit.

Thus, using fully insulating jackets around underground supply cables results in additional limitations and grounding requirements, especially on three-phase supply circuits using a reduced neutral.

The traditional construction for exposed concentric neutral cables (and those covered with a semiconducting jacket of 100 Ω•m or less radial resistivity) has been to use a one-third-sized neutral around each phase conductor of a three-phase circuit, a one-half-sized neutral for a vee-phased circuit, and a full-sized neutral on a single-phase circuit. With such neutrals effectively in full contact with the earth, a full neutral return capability is available in the event of a fault in one cable.

In recent years, reduced neutral capacity has been used for underground circuits, as well as overhead circuits, where the neutral is sized to take the phase-load imbalance from unbalanced single-phase loads, as well as sized to be adequate for the magnitude and duration of expected fault-current flow.

In order to limit the potential flow of supply-circuit fault current over a communication cable sheath in random lay, the use of supply cables with a fully insulated jacket around the concentric neutral conductors in random lay with communication cables was limited in the 1987 and prior editions. Fully insulated jacketed supply cables were allowed in random lay with communication cables only if

(1)   each phase conductor was surrounded with a sheath, insulation shield, or concentric neutral that was sized for the magnitude and duration of the fault current and grounded four times in each 1.6 km (1 mi) to meet Rule 97; or

(2)   a separate conductor sized for the magnitude and duration of the expected fault current (either bare or covered with a semiconducting jacket of 100 Ω•m or less radial resistivity and interconnected at least four times in each 1.6 km [1 mi]) was also laid in random lay with the supply circuit to take the fault current. This option can also be used when direct-buried communication cables are intended to be placed in random lay with nonconducting conduits containing supply cables.

In the 1993 Edition, the previous paragraph under Rule 354 was expanded and renumbered as Rule 354A1, 354A2, and 354A3; the previous 354A, 354B, and 354C were renumbered 354B, 354C, and 354D. The 2002 Edition inserted a new Rule 354A2 to require separation of electric supply and communication cables and conductors from gas lines by at least 300 mm (12 in). Former Rules 354A2 and 354A3 were renumbered to A3 and A4.

In the 1990 Edition, Rule 354D (then Rule 354C) was revised to recognize another configuration that could be used for a three-phase circuit using fully insulated cables in random lay with communication cables. The additional configuration requires both (1) the use of at least a *half-sized neutral* around each phase conductor and (2) *eight* ground connections in each 1.6 km (1 mi) segment, instead of the usual four in each 1.6 km (1 mi) segment requirement. This provides the equivalent of a full neutral's current-carrying capacity away from the fault site, one half back toward the source and the other half away from the source. Because of the eight-grounds-in-each-mile requirement, the fault current only travels a few hundred feet in each direction before being connected back into the other half-neutrals on a two- or three-phase system. Single-phase runs require a full neutral for load purposes and, thus, meet the size requirements of this rule; however they still require the eight grounds in each 1.6 km (1 mi) segment.

> *NOTE:* The eight grounds in each 1.6 km (1 mi) segment requirement applies only to random lay systems meeting Rules 354D1 and 354D4. If the supply cables with insulating jackets around the grounded neutral conductors are located in a single-duct conduit, (1) this rule applies and (2) eight ground connections in each mile are required (see Rules 354D4 and 350H). If the supply cables are located in multiduct conduit meeting the requirements of Section 32, the conduit is required by Rule 322A2 to limit the effects of a supply cable fault, and the additional four grounding connections would not be required for this purpose.
>
> In areas of high soil resistivity, the effectiveness of each electrode is decreased by the high soil resistance. Thus, additional ground connections may be required for any circuit, underground or overhead, to meet the effective grounding requirements of Rules 96 and 97.

A single-phase circuit using a fully insulated jacket around a full neutral meets the previous requirements for random lay with communication cables. However, three-phase circuits using *fully insulated jacketed cable with a one-third sized neutral cannot be used in random lay with communication cables*, regardless of how many ground connections are made, unless the three cables are accompanied by a separate grounding conductor meeting the requirements of Rule 354D2a(3) (Rule 354C2a(3) prior to 1993).

In the 1993 Edition, Rule 354C was moved to 354D, reorganized, and augmented. Rule 354D1c is the previous 354C4. Rules 354D1d through 354D1g are new, as is Rule 354D4. Rule 354D4 allows supply cables with a concentric neutral covered by an insulating jacket to be installed within nonmetallic conduit and to be located in random lay with communications cables *if the supply cable meets Rule 354D3*, the one-half neutral/eight grounds in each 1.6 km (1 mi) segment rule. Old Rule 354C4 was deleted in this revision.

NOTE: These rules address fault conditions, not ordinary operation; a one-half sized neutral may not be appropriate for normal operation of many three-phase circuits, particularly where large loads on single-phase tap lines are present. A larger neutral may be required for operating purposes.

IR 517 issued 1 February 1999 made it clear that this rule intends that a common electrode be used for each subneutral of multiphase circuit cables at each of the required grounding points; separate, individual electrodes for each phase cable are not intended. However, where desired the common grounding electrode may be made up of multiple rods or other electrodes to achieve the desired ground conductance.

Rules regarding random separation were introduced when all buried communication cable was of shielded, metallic-conductor construction. Neither unshielded, metallic-conductor cable nor fiber-optic cable were available at that time. Joint communication supply tests and trial installations (early 1960s) were, in all cases, conducted with shielded cable bonded to the supply neutral. Acceptance of random lay as a safe construction practice was based on these tests and trial installations. There is no basis to allow extrapolation of the test and trial results to unshielded, metallic-conductor communication cable. Further, under fault conditions, the shield acts as a voltage divider to prevent full voltage from being impressed on communication equipment in the event that the supply neutral is badly corroded or open. This condition could occur with a supply cross on unshielded communication cable downstream of the communication pedestal, even though the pedestal is properly fused; see also NEC Rule 800-30, *Protective Devices,* for additional protection requirements on customers' premises.

IR 475 issued 24 February 1994 clarified that, due to lack of specificity in the code, Rule 012C (*"For all particulars not specified in these rules, construction and maintenance should be done in accordance with accepted good practice for the given local conditions."*) applies if there is no shield on a communication cable used in random lay with power cables. Until the 1997 addition of Rule 354D1e requiring a metallic shield under the jacket of communication cables having metallic components that were to be used in random lay with supply cables, it was the responsibility of the communication utility to determine if a shield is appropriate. As of 1997, such communication cables are prohibited in random lay with power cables.

As of the 2002 Revision, new exception to Rule 354D allows entirely dielectric fiber-optic communication cables to be installed with less than 300 mm (12 in) of separation, if all are in agreement and Rules 354D1a, 354D1b, 354D1c, and 354D1d are met; meeting Rule 354D2 or Rule 354D3 is not required for entirely dielectric fiber-optic communication as it is for other communication cables in random lay with supply.

Rule 354D requires that communication cables having cable sheaths or shields that are located *in random lay with supply cables* must bond the communication shields or sheaths to an effectively grounded supply conductor. There was no requirement in Rule 354D that the communication conductor in random lay have a sheath or shield, until Rule 354D1e was added in the 1997 Edition to require a continuous metallic shield around communication cables in random lay that contain metallic wires. However, this shield is only required when communication is in random lay with supply (see the EXCEPTION to Rule 354D1e).

Rule 354D2a(3) provides an EXCEPTION to allow short sections of cable to pass through a conduit (without being in continuous contact with the earth; see Rule 354D2a) and still retain the classification as direct buried, as in Figure H354D.

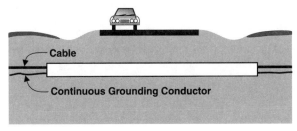

Figure H354D
Exception to Rule 354D2a(3)

Rule 354E was added in 2002 to specify clearances of direct-buried facilities from nonconductive water and sewer lines. Random lay is generally allowed.

Section 36. Risers

(This section was developed in the 1973 Edition from Rule 297 of the Sixth and prior editions.)

360. General

(This rule was developed in the 1973 Edition from Rules 297B and 297C of the Sixth and prior editions.)

Part 2 of the NESC covers the requirements for mechanical protection for supply conductors or cables in risers. This protection must continue at least 300 mm (12 in) below ground. Grounding of the riser guard must meet Rule 314.

361. Installation and
362. Pole Risers—Additional Requirements

(Rules 361 and 362 were developed in the 1973 Edition from Rules 297A and 297B of the Sixth and prior editions.)

Both Rules 361 and 362 are concerned with the effect on cable insulation and conductor continuity of above-ground activities and forces, including frost upheavals and pedestrian and vehicular traffic. The number and position of risers on a pole affect the likelihood of damage from area activity and interference with required climbing by line workers.

CAUTION: Placement of standoff brackets may enable unauthorized personnel to easily climb the structure. The above-ground portion of riser poles are covered by Part 2, as well as by these requirements. See the requirements of Rule 217A2 (217A1b prior to 1997) for readily climbable supporting structures.

---

363. Pad-Mounted Installations

(This rule was added in the 1973 Edition.)

Care should be taken with the design and placement of pads and pad-mounted equipment. Settling foundations under pads are a special cause of maintenance, if not safety, problems.

Similarly, cables should maintain appropriate burial depths, or have supplemental protection, until they are physically protected by the pad of pad-mounted equipment.

Section 37. Supply Cable Terminations

(This section was developed in the 1973 Edition from Rules 296B, 297D and 297E, and 298 of the Sixth and prior editions.)

370. General

(This rule was added in the 1973 Edition.)

Supply cable terminations are subject to the requirements of Rule 333 and may be subject to the clearance requirements of Parts 1 or 2.

This rule should not be interpreted to mean that *all* cable terminations must be installed with a mounting bracket secured directly to the body of the termination. Elastomeric terminations generally do not add enough weight to the cable to require a support bracket. For this reason, Rules 370B and 371A were revised in the 1981 Edition to use the term "installed" instead of "supported or secured."

The value(s) of the insulating medium or media used is critical to the maintenance of appropriate BIL (basic impulse insulation level) levels. Closely spaced terminations will require additional barriers or fully insulated terminations.

371. Support at Terminations

(This rule was developed in the 1973 Edition from Rule 292G of the Sixth and prior editions.)

One of the most common, and most avoidable, problems with cables is insufficient support at terminations (see the discussion of Rule 370).

372. Identification

(This rule was developed in the 1973 Edition from Rule 298 of the Sixth and prior editions.)

Easily understandable identification of cables is necessary for both safety and maintenance efficiency. Identification may be made by use of metal tags, stenciling of the cable, or by charts showing the position of the cables, or other permanent means. When tags are employed for this purpose, a fire-resistant, noncorrodible material should be used and the markers should not be easily obliterated.

NOTE: Although Rule 341B3b requires marking or tagging of all cables in joint-use manholes and vaults, this rule does not specifically require terminations to be tagged or marked in joint-use manholes or vaults. Such marking or tagging is usually not needed to distinguish supply terminations from communication, but it can be very useful if different circuits or owners are involved.

373. Clearances in Enclosures or Vaults

(This rule was developed in the 1973 Edition from Rules 296 and 297D of the Sixth and prior editions.)

Where metal sheathing is used on cables, it should be made continuous electrically and mechanically with the cases of equipment, such as switches and transformers. Where metal sheathing is not used, the conductors should enter cases of equipment through openings that have proper bushing or gaskets to ensure watertight joints.

Underground current-carrying parts exposed to contact in manholes and handholes are a source of great hazard and should not be allowed. Live parts of transformers, switches, fuses, lightning arresters, or other apparatus should be either enclosed completely or isolated or guarded as a protective measure. The horizontal and vertical clearances, and the guard zones and guarding requirements of Rules 124A, 124B, and 124C for electric supply stations may be used when considering underground vault installations, but they are not required.

Consideration should be given to the expected movement and activities of workers around exposed live parts in vaults, especially under emergency conditions. Note also the requirements for appropriate warning signs in Rule 411D and the approach distances to exposed live parts of Rule 441 of the NESC work rules.

374. Grounding

(This rule was added in the 1973 Edition.)

As for other equipment in underground installations, the exposed conducting surfaces of termination devices, etc., should be at equal potential with adjacent surfaces not intended to carry current. Grounding all of these conductive surfaces effectively is a key to a safe working environment.

Section 38. Equipment

(This section was developed in the 1973 Edition, partially from Rule 299 of the Sixth and prior editions.)

380. General

(This rule was added in the 1973 Edition.)

Types of equipment covered by the rules in Section 38 are defined in this rule. The characteristics of this equipment vary widely from an operation and maintenance point of view, as well as from a safety point of view. As a result, Rule 380B requires concurrence of all parties in a joint-use manhole before the installation of equipment therein.

Rule 380C requires supporting structures to be designed to take all expected loads and stresses; no overload factors are specified (see Rule 012C). Rule 380D was added in the 2002 revision to specify clearance from pad-mounted equipment and pedestals to fire hydrants. This is similar to the overhead requirement in Rule 231A.

381. Design

(This rule was added in the 1973 Edition.)

Like cables to which they are connected, underground equipment and mountings must be designed for the stresses expected to be imposed on them by the various environmental and operational conditions. These conditions include both the normal and emergency conditions, as well as applicable fault conditions. In the latter case, appropriate design of equipment and mountings will limit the effects of a fault to its site without concatenate catastrophic failures. Enclosers of fuses and interrupter contacts are expected to contain or otherwise limit the effect of arcs, gases, or other harmful resultants of normal, emergency, or fault conditions, without adversely affecting nearby equipment or personnel. When personnel are expected to use tools to connect or disconnect energized parts, appropriate clearances or barriers are required to limit the opportunity for a phase-to-ground or phase-to-phase fault.

Routine operation can be eased effectively by having uniform types of switch handles and directions of movement to eliminate confusion. Providing workers with clear indications of switch contact positions can be critical especially during emergency operation.

Rule 381D requires local provisions to make both remote and automatic controls inoperable when the operation of a switch or other piece of equipment needs to be blocked for work to be performed. The language was revised in the 1997 Edition to correlate with Rule 216E.

The intention of Rule 381G was clarified in the 1984 Edition. Rule 381G requires pad-mounted equipment to be locked or otherwise secured against entry. The 2002 Edition added other above ground equipment, including communication equipment to the title of the rule and Rule 381G1.

Since the 1973 Edition, Rule 381G2 has required two separate procedures to obtain access to live parts above 600 V in pad-mounted equipment. The first procedure was and is required by Rule 381G1 to be the opening of a door or barrier that is locked or otherwise secured against unauthorized entry. Opening of the first door or barrier is required to be completed *before* the second procedure was started.

Unfortunately, this language was *misinterpreted* to allow the use of a locked outer door that also had a penta-headed bolt on the door to fulfill the requirements of the two separate procedures. Such an installation does not meet the full requirements because there is not a second procedure *after* the outer door is opened. A second door or barrier is still required before exposing live parts above 600 V. A second door or barrier is not required if enclosed exposed parts do not exceed 600 V.

This rule is intended to recognize the fact that the accidents on record generally have not occurred to those who originally vandalize pad-mounted equipment. Usually the victim has been an inquisitive child who came along later.

Rule 381G requires that first an outer door or barrier must be opened. When that door or barrier is open, a conspicuous warning sign should make the entrant readily aware that this is "far enough." The rule required that a second *secured* door or barrier be removed before gaining access to live parts until the 1987 Edition, in which the requirement for security provisions on the second door or barrier was removed. This effectively recognized the excellent record of pad-mounted equipment with an interior barrier that pivots out of the way for interior work but is not otherwise secured. These are reasonable and practical safety measures that will, on the basis of known evidence, provide adequate safety for such equipment and the general public. It is not possible to foil a determined vandal, but it is practical to limit the adverse effects of the actions of such people and any who follow them. These requirements will do the latter and will limit the former.

Rule 381G does not apply to "dead-front" transformers where the connection is made with insulated elbow connectors. Note that so-called "dead-front" equipment that utilizes insulated separable elbows for the high-voltage connection is not considered to have exposed live parts above 600 V, but it meets the intention of the rule; in essence, the elbows provide the equivalent protection of the second barrier, since energized parts are not exposed.

IR 534 issued 23 May 2005 discussed application of Rule 381G2 to a below-ground fiber-reinforced *basement* with a fiber-reinforced cover mounted flush with the ground and secured with bolts above cable loops. If the enclosure is pad-mounted above ground, the removal or opening of two doors or barriers is required if the equipment is *live-front* with exposed parts energized to voltage above 600 V contained inside; the requirement does not apply to so-called *dead-front* equipment where the high-voltage parts are insulated. If cables in an underground basement installation are fully insulated with the grounded sheath, shield or concentric neutral remaining intact (whether continuous, spliced, or capped) and no termination is made to an exposed spade, the basement installation is similar to *dead-front* equipment. If terminations are made in the basement to exposed, energized spades, the installation is similar to *live-front* equipment. The language of Rule 381G2 does not mention basement-type installations; the rule was codified before such installations were common. Since particulars for entrance limitations for basement enclosures are not specified in the NESC, the requirements of Rule 012C for accepted good practice should be considered.

During the revision of the 1987 Edition, a proposal was made to require a "Mr. Ouch" sign on both the outside and inside of pad-mounted equipment. "Mr. Ouch" is essentially a caricature of a child figure reeling away from a malevolent-looking, face-like lightning bolt that was developed with careful review and trial with elementary school children. The basic difference between the proposed signs was that the inner sign had a red "danger" banner at its top, and the outer sign had an orange "warning" banner at the top. The two signs were so alike that even the NESC subcommittee members confused them and rejected the proposal. If the intended audience requires the use of a caricature-type pictorial, the use of sign motifs so similar as to be confused on both the inside and the outside is not appropriate. The recommendation is that there be an appropriate safety sign inside the equipment that would be prominently visible when the outer door is open. The reference was changed to the generic term *safety sign* from the previous generic term *warning sign* in the 1997 Edition to recognize that ANSI Z535 (which is an update and expansion of ANSI Z35 and Z53) has three level of signs involving notification of a hazard. DANGER implies a hazard that has a high probability of causing death or serious permanent injury, if not avoided. WARNING is also associated with a hazard that could cause death or serious permanent injury, but has a low probability of doing so. For this reason, a DANGER sign is appropriate inside pad-mounted equipment and, if used, a WARNING sign is appropriate on the exterior of the cabinet. CAUTION is reserved for association with a hazard that could cause minor or moderate injury or equipment damage; CAUTION is not appropriate for use on pad-mounted electrical equipment (see Handbook Appendix B).

With increasing frequency, utilities are placing NOTICE or GENERAL SAFETY signs on pad-mounted equipment in public areas to keep plantings far enough away to allow service of the unit and to limit the likelihood of dig-ins. However, the underground rules do not contain working space requirements like those in Rule 125, and such signs are not required by the NESC.

Note that to be consistent with Occupational Safety and Health Administration (OSHA) requirements and similar works, an inside sign should use the "DANGER" banner. Although OSHA regulations would suggest using a CAUTION sign on the exterior (because OSHA has not yet updated to ANSI Z535), OSHA recognizes the use of updated standards when they refer to previous ones, and WARNING is, therefore, appropriate.

382. Location in Underground Structures

(This rule was added in the 1973 Edition.)

This is a companion to Rule 323.

383. Installation

(This rule was added in the 1973 Edition.)

Rule 383 recognizes the practical constraints involved in installing and maintaining equipment safely and efficiently. These include control of movement of equipment during installation, isolation of live parts from workers and from exposure to harmful liquids, and the availability and control of operating mechanisms.

384. Grounding and Bonding

(This rule was added in the 1973 Edition. Former Rule 350F was moved to new Rule 384C in 2007.)

This is a companion to Rule 374.

Rule 350E of the 1987 Edition (later Rule 350F and now Rule 384C) added the requirement to bond together all above-ground metallic pedestals, cases, etc., that are located within reach of each other, i.e., 1.8 m (6 ft). The intention is to limit the opportunity for a worker to touch two items that may be at significantly different voltage potentials. Where walls or barriers or grade level differences prevent simultaneous contact, the rule is not applicable. Since the rule was located in Section 35 in the 2002 and previous editions, the specifics of the rule did not directly apply to pad-mounted enclosures of equipment served from conduit systems that included handholes and manholes; it only applied to enclosures supplied by direct-buried cables and, after 2002, to so-called *cable-in-duct*. This was questioned in IR 7 January 2002. As a result, the rule was moved to Section 38 where it would apply regardless of whether the cables were in conduit or direct buried.

Rule 384C requires bonding of a variety of conductive, grounded items if they are located within 1.8 m (6 ft) of one another. It does not require bonding between a communication pedestal and a vertical ground wire on an adjacent pole, even if the distance between the two is less than 1.8 m (6 ft). The parenthetical listing of pedestals, terminals, apparatus cases, transformer cases, etc., intentionally does not specifically include ground wires on poles. Neither does the rule prohibit such bonding (see Rule 012C). The purpose of the rule is to limit the potential hazard of a person bridging the distance between the two grounded items (see Figure 384C).

As a practical matter, it is rare that a hazardous level of voltage potential can exist between these items on systems with interconnected grounded conductors, especially where the pedestal of one system is located immediately next to the pole of the other system. However, if the grounded conductors of the systems are not interconnected, consideration should be given to the level of the potential voltage difference.

Particularly because the secondary side of pad-mounted power distribution transformers is on the right front side of the transformer, it is usual and customary for utilities in the same service area to agree to run a short piece of #6 AWG CU wire out from under the pad-mounted enclosures, coil it up and bury it at the right front corner of the enclosure. If another utility installs pad-mounted equipment within 6 ft, it can then easily find the bonding wire and connect to the existing grounded equipment enclosure. For earlier installations without such a bonding wire present, the second utility can run its own bonding wire over to the front right corner of the existing installation and inform the existing owner of the need to make the bonding connection. It is the responsibility of the successive installers of pad-mounted enclosures near existing equipment to cause the bonding to occur, if they install enclosures within 6 ft of the existing equipment.

Figure 384C
Power and communication pedestals requiring bonding

385. Identification

(This rule was developed in the 1973 Edition from Rule 299 of the Sixth and prior editions.)

The importance of indicating the multiple connection (network) of the apparatus covered by the rule is emphasized by the fact that, due to low-voltage feedback, the resulting excitation of the high-voltage side of individual transformer regulators or similar apparatus may be hazardous, even though such apparatus is disconnected from the high-voltage supply.

Section 39. Installation in Tunnels

(This section was added in the 1973 Edition.)

390. General

(This rule was added in the 1973 Edition.)

Section 39 supplements the requirements of Parts 2 and 3. Typically the facilities involved in underground construction, especially energized live parts, are not accessible to the public. They are in locked equipment or vaults, etc. However, in rare cases, such as some tunnel areas, personnel who are not qualified to be around this equipment may have access to the area. In these cases, the applicable requirements of Part 2, such as overhead clearances, also will apply. Just as in joint-use manholes, the agreement of all parties is required for the equipment to be installed therein, but this rule goes further to include both the design of the structure *and* the design of the installation within the tunnel.

391. Environment

(This rule was added in the 1973 Edition.)

Tunnels are generally special- and multiple-occupancy places that require special care and extension of existing requirements for related facilities. This rule details the special concerns for the utility system(s) components and workers relative to hazards created by other occupancies and vice versa.

Part 4. Rules for the Operation of Electric Supply and Communications Lines and Equipment

The first proposals for NESC requirements were for operating rules. It was recognized that, while safe work required equipment and lines are to be located in an appropriate place, there is no substitute for appropriate training, operating procedures, tools, and supervision.

There have been two major revisions to these rules since 1914: the complete rewrite of the 1973 Edition, and the reorganization of the 1990 Edition. In addition, Rules 441 and 446 were extensively revised and augmented in the 1993 Edition to reflect recent changes in, and desired changes for, the application of Occupational Safety and Health Administration (OSHA) regulations to *maintenance* of electric supply lines. The NESC Work Rules Subcommittee, various technical organizations, and OSHA personnel have been working toward unification of electrical worker safety requirements applicable to those working on electric supply and communication lines and equipment, whether for maintenance or construction purposes, into one document that could be used by all. That document is the NESC. The NESC now includes the basic technical information required to implement OSHA regulations. It is intended that these industry-consensus standard work rules will be adopted by OSHA and that OSHA experts and the NESC Work Rules Subcommittee will work together on future revisions. Thus, with the 1993 Edition of the NESC, the technical content of the NESC and OSHA regulations for both *construction* and *maintenance* have been harmonized. Due to OSHA administrative processing requirements, there may be a slight lag time between issuance of NESC revisions and OSHA adoption thereof. In 1997, this process continued with the extensive revision and additions to the fall protection requirements and clarification of the approach distance requirements and tagging requirements. See Appendix D for a comparison of major NESC and OSHA requirements.

Unlike the construction rules of Parts 1, 2, and 3, which can continue to be applied to existing facilities, the work rules of Part 4 of the current edition are required to be used in conjunction with *all* facilities, old or new. However, it is recognized that some states are slower in adopting Code changes and some discussion of recently superseded rules is appropriate. In addition, work rules are self-explanatory and obvious in their intent, generally, and may not benefit from lengthy discussion. As a result of all these considerations, less historical comment will be provided than can be found elsewhere in this work.

In the 1990 Edition, the NESC work rules were reorganized to place the work rules that are common to all employees within Section 42. The additional rules for communications workers are contained in Section 43, and the additional rules for supply workers are contained in a new Section 44. This organization eliminates duplication of text and makes both common requirements and different requirements immediately obvious and easy to modify in context.

The following list indicates rule movement that occurred in the reorganization of the 1990 Edition.

Rule Movement in the 1990 Code

1987 Rule #	1990 Rule #	1987 Rule #	1990 Rule #
420A	420A	421A2	442A2
420B1	420C1	421B1	421A1 & 421A2
420B2a	420C2a	421B2	421A3
420B2b	420C2b	421B3	421A4
420B2c	420C2c	421B4	421A5
420B3	420C3	421C	421C
420B4	420C4	421D1	442B
420B5	443G	421D2	442C
420C1	420B2	421D3	441D
420C2	420B1	421E	443E
420C3	420B3	421F	442D

Rule Movement in the 1990 Code *(Continued)*

1987 Rule #	1990 Rule #	1987 Rule #	1990 Rule #
420D	420D	421G	442E
420E	420E	421H1	442F1
420F	420F	421H2	442F2
420G1	420G	421H3	443C
420G2	420G	421I1a	421B1a
420G3	420G	421I1b	421B1b
420H	420H	421I2a	421B2b
420I1	420I1	421I2b	421B2b
420I2	420I2	421I3	421B3
420J1	420J1	422A1a	443A1a
420J2	420J2	442A1b	443A1b
420J3	420J3	422A2	422A2
420J4	420J4	422A3	422A3
420K1	420K1	422A4	422A4
420K2	420K2	422A5	422A5
420K3	420K3	422A6	422A6
420L	420L	422B	441A
420M	442G	422C	443B
420N	420M	422D	441
420O	420N	422E	443C
420P	420O	422F	443D
420Q	443J	422G	443F
421A1	442A1	422H	443H
422I	443I	425B4	422A2
Tables	Tables	425C1	423A
422-1-422-4	441-1-441-3	425C2	423B1

Rule Movement in the 1990 Code *(Continued)*

1987 Rule #	1990 Rule #	1987 Rule #	1990 Rule #
423A1	444A1	426A	423B2
423A2	444A2	426B1	423B3
423A3	444A3	426B2	423B4
423B	444B	426B3	442K
423C	444C	426B4	423C1
423D	444D	426C	445A2
423E1	444E1	426D1	445A4
423E2	444E2	426D2	423C2
423F1	444F1	426D3	423C3
423F2	444F2	426E1	423D1
423G1	444G1	426E2	423D2
423G2	444G2	426E3	423D3
423H	444H	426E4(a-d)	423D4(a-d)
424A1	445A1	426F1	423E1
424A2	445A3	426F2	423E2
424A3	445B	426F3	423E3
424A4	422B1	426F4	443A4a
424B	422B2	426F5	44344b
425A1	422C1	426G	423F
425A2	422C2	427A	446A
425B1	422C3	427B(1-3)	446B(1-3)
425B2	422C4	427C(1-4)	446C(1-4)
425B3	422A1		

Section 40. Purpose and Scope

Section 40 is similar to Sections 10, 20, and 30 and is, therefore, unremarkable except to remind the reader that these work rules, like the remainder of the Code, apply to the installation, operation, and maintenance of both public and private electric supply and communications utility systems. It is perhaps significant that in 1997 it was necessary to state in Rule 400 that all "reasonable" steps are required to be taken, but that it "is not the intent to require unreasonable steps" to comply with the work rules contained herein.

These rules are based upon the results of hundreds of millions of hours of both good and bad experience of those working on electric supply and communication facilities. It is, unfortunately, not possible for an employer to prevent every accident. It is, however, practical for employers to (1) train employees in safe methods of planning and executing work on and around these facilities, (2) provide employees with appropriate tools, equipment, and personal protective gear for the work to be done, and (3) provide appropriate supervision for required tasks.

The greatest responsibility is, *and must be*, on the workers themselves to use the training, equipment, tools, and protective gear provided. In the final analysis, it is the workers themselves who (1) have the control over their preparations, planning, control, and movements at any job site and (2) must identify potential hazards in each work site, plan their work sequences to de-energize or to cover up appropriate parts or conductors, plan their route on a structure to maintain safe clearances, and allow appropriate fall-protection measures to be employed.

These rules have been proven to be both practical and effective. The utility industry is one of the safest construction-related industries. Electrical workers, in particular, have proved to be among the safest of workers. This is due in large measure to high level of training and the care taken in planning and executing their work. When these rules are followed, workers are protected. Experience shows that there is so much redundancy built into these rules that, when a serious accident does occur, more than one rule was usually violated by the worker(s) involved. Often, the appropriate protective gear, tools, and equipment were already available on site to do the work properly and safely, but one or more of these was not used by the worker(s) involved. In other cases, special test equipment or protective equipment needed for the work was not preselected prior to the work, and the work was attempted without waiting for the required equipment. *There is no substitute for careful, thoughtful work that is well-planned and well-executed.* Identification of appropriate equipment and methods to be used, and clear, two-way discussion of these in a "tailgate" or other planning session, is critical *before starting work*, to assure that each member of the team understands his or her responsibilities.

Note that Rule 402 includes wording to require referral to Section 3—*References.* The wording is such that all the applicable reference documents *shall* be considered in applying the rules included in other portions of Part 4. This practice was adopted after considerable deliberation; many wanted each work rule spelled out in detail; others felt that only a mention should be considered sufficient. With the advent of the Occupational Safety and Health Administration (OSHA) regulations in 1972, and the new consensus ANSI Standards, it was decided that it was appropriate to use these ANSI Standards only as techniques in conjunction with the NESC rules.

The NOTE to Rule 402 included specific reference to the OSHA document, and the publishing of 29CFR1926, Subpart V, in 1972. This OSHA regulation only covered construction, however. The NESC recognized that its rules encompassed both construction and maintenance, and the NOTE was included to point out that there were differences in the two documents. The significant differences were in the "clearances," or minimum approach distances to be employed by trained workers around energized facilities at some voltage levels.

These differences were most pronounced in Table 441-1. The NESC included technical information upon which approach tables for worker safety could be developed for a series of voltage ranges. However, OSHA Subpart V only permitted this for voltages of 345 kV and above, where switching surges (transient overvoltages) are usually the controlling factor.

Because of this still unresolved situation, the 1990 Edition continued to refrain from giving instructions on how to construct new tables or how to control the transients to make the practice safe. At the time of final development of the 1990 NESC Edition, it was known that the OSHA proposed rule-making documents for maintenance activities were in final preparation, and hence no other changes were made to address this issue; such changes would be considered for the 1993 Edition. The 1993 Edition was revised to include the technical basis for the implementation of OSHA requirements for both *maintenance* and *construction*.

Section 41. Supply and Communications Systems—Rules for Employers

The NESC recognizes the responsibility of employers of persons engaged in the installation, operation, and maintenance of electric supply and communication lines to provide those employees with the training, equipment, and tools, and supervision that is appropriate for the intended work under the expected conditions. With few exceptions, these requirements have not changed substantially since the First Edition of the NESC.

The NESC also recognizes that, in order to meet its responsibilities under the Code, employers must designate appropriate personnel to be responsible for certain activities. The requirements to designate such personnel are contained in Section 41. The responsibilities of the designated employees are contained within Sections 42, 43, and 44, as applicable.

The Code is intentionally not specific as to all the types of protective gear the employer needs to supply to the workers. These matters are usually so site-specific as to be impractical to cite in specificity. Note that recognized standards for protective devices and equipment are referenced in Section 3 of the NESC.

410. General Requirements

Rule 410 requires employers to use positive procedures to secure compliance of employees with the rules. Generally this takes the form of training, supervision, and verification of employee knowledge. The 2002 Edition made Rule 410A2 more specific as to training and retraining. Similar to the more recent OSHA requirements, employers must ensure that employees working around exposed energized facilities demonstrate their knowledge of safe work practices or be retrained. This applies to all types of workers covered by these rules: supply, telephone, CATV, electric railway, and railroad signal workers.

It is recognized that the work covered by the Code involves the use of human beings, and while appropriate training, tools, and supervision can limit the opportunity for an accident to occur, they cannot *prevent* an inattentive employee from performing an unthoughtful and unexpected act that results in an accident. A requirement was added in the 1993 Edition for employers to train employees working in the vicinity of exposed energized facilities on the advantages and limitations of various types, combinations, and materials of wearing apparel. Because new materials are being developed and tested on an ongoing basis, the original requirement was general in nature. See also the discussion of Rule 420I.

After years of development and discussion, a new Rule 410A3 and a new Table 410-1—*Clothing and clothing systems—voltage, fault current, and maximum clearing time for voltages 1 to 46 kV* were added in the 2007 Edition to give specific guidance to code users. Rule 410A3 requires that, prior to 1 January 2009, the various exposures of employees to a potential arc, should an arc occur during work, be assessed as to the calorie/square centimeter that might be received by an employee from the electric arc. This determination must include the effect of available fault current, duration of the arc, and the distance of the employee from the arc, should an arc occur. If this level is greater than 2 cal/cm^2, then the employer is required to have employees wear clothing or a clothing system that has an effective arc rating not less than the anticipated exposure.

It is recognized that the level of exposure can be different when working on different equipment, even when the same fault current is available at the same voltage and with the same protective devices. If an arc occurs between points on a pole, the energy can radiate in all directions. The employee exposed to an arc on a pole would be exposed to less energy than an employee exposed to the same type of arc in an electrical panel, with a back that acts as a reflector of energy that would otherwise have radiated away from the worker.

A layered system of clothing may be used to protect employees from arc heating exposure; a layered system is often preferable to a single layer. This is particularly true when the basic clothing system is chosen for application to the majority of work to be performed by the employee and is augmented with additional layers whenever work with exposure to higher levels of heating is performed. For example, an employee wearing a shirt and pants suitable for exposure while working from a bucket or from a pole may wear additional protection when working in front of equipment that could act as a reflector, such as a pad-mounted transformer.

An EXCEPTION to Rule 410A3 allows the use of clothing or a clothing system with a 4 cal/cm^2 rating for work on secondary systems below 1000 V, without requiring an arc hazard analysis. NOTE 2 was added to the rule to recognize and emphasize the fact that arc energy levels can be high with faults on secondary systems, due to the relatively high available fault current. It is important that the work rules be followed and, where practical, engineering controls be utilized in addition to using the clothing or clothing system.

Table 410-1 provides the maximum clearing times in cycles that is allowed for four levels of voltages and four levels of available fault current for use with clothing or clothing systems at each of three different levels of calorie ratings. A variety of software programs are commercially available with which to calculate the expected arc energy exposure from different arc situations. Similarly, a variety of commercial testing facilities are available for testing of specific types or systems of clothing. The results of these efforts were used in the preparation of Table 410-1.

The values in Table 410-1 are set to reduce the amount or degree of injury that may result from exposure to an electrical arc but they may not prevent all burns from occurring. The prescribed system of assessment of expected exposures and use of appropriate clothing, when combined with prescribed work methods is expected to both (1) limit the opportunity for an arc to occur and (2) limit the extent of any injury to levels that are not life threatening. Unfortunately, it is not possible to prevent all injuries.

Rule 410B also recognizes that the injurious effect of an accident may be substantially mitigated in many cases through the timely use of appropriate rescue and first aid measures.

411. Protective Methods and Devices

In addition to its requirements relating to the devices and equipment to be used, Rule 411 recognizes that employees cannot perform intended work safely unless they understand which facilities are to be worked, the nature of the facilities, and the conditions of work and the methods to be used. Because of the good overall history of the use of a variety of methods of identifying facilities to be worked, no method is specified (see Rule 012).

Both Rule 411 and the employee rules recognize that employees on the work site are in the final, and sometimes best, position to identify site-specific conditions that could adversely impact the safe performance of the work with the intended methods and equipment. Notice also that, while this rule requires inspection or tests of protective devices and equipment, the employees are required by the employee work rules to inspect their personal protective gear prior to use.

Special note should be taken of Rule 411A2 concerning diagrams. It requires that line maps, switching information, and associated information be available to employees. Many errors and potential accidents would be avoided if more attention were paid to this rule and its implications. Knowing "on what" you are required to work does much to ensure safe activities.

Rule 411B provides a list of common protective devices and equipment used by line workers on or around energized lines and equipment. The list is a sample list of common items. Not all will be used on all jobs or even by all personnel. For some work, additional specialty gear will be used. Nothing in this rule requires the employer of the employees doing the work to own all of the protective devices and equipment provided for use by their employees. It can be rented or borrowed from a utility contractor, leasing firm, etc. However, under OSHA and the

NESC, it is the responsibility of the employer of the employees doing the work (see Rule 012B) to assure the appropriateness of the protective devices and equipment, regardless of ownership. Under other rules, it is (1) the responsibility of employers to train their employees on the use of these devices and equipment and (2) the responsibility of the employees to use them appropriately.

Whether furnished by the employee or employer, the various items used in the work must be appropriately tested or inspected. The employee rules of the later sections require employees to assure the integrity of the personal protective gear before use. For some items, the employer and the employee will each have testing or inspection functions. For example, rubber gloves are generally tested by the employer (or a contract laboratory) on a set schedule, but the employees must inspect them and test them for pinholes before each use.

The warning signs required in Rule 411D should contain such information as may be appropriate for employees entering the area; these signs are different from the public warning signs required by Rule 110A. The 1997 Edition *requires compliance* with ANSI Z535-1991 standards. The content is intentionally not specified. For example, where the utility has stations that differ in the included voltages, and the voltage is not apparent from the equipment bushings, etc., it may be appropriate to inform employees of the voltage to ensure the use of correct protective gear and work methods unless the work methods appropriate for the greater voltage are always used. Such signage may be especially important on vaults and other areas where the nature of the enclosed facilities is not apparent from the outside. The 2002 Edition modified the title and text to be clear that both safety signs and temporary safety tags, such as HOLD tags, etc., must be in the format of ANSI Z535.

In the 1993 Edition, the requirement for persons responsible for underground lines to be able to locate such facilities was added to Rule 411E for consistency with the requirements of Part 3.

Rule 411F—*Fall Protection* was added in the 1997 Edition. It should be noted that OSHA now requires annual inspection of personnel that climb. The retraining provisions of the NESC are key to effective implementation of a fall protection program.

Section 42. General Rules for Employees

(In the 1987 Edition, Section 42 contained both the general rules and the special rules for supply system employees, and Section 43 contained the rules for communications employees, most of which duplicated the general rules contained here. In the 1990 Edition, the general rules for all employees were combined here; the special additional rules for supply employees were moved to Section 44.)

Section 42 contains the general work rules to be followed by *all* employees while working on or near the lines and equipment of public or private electric supply or communications utility systems. They include specific responsibilities for those persons designated by the employer to perform a control or supervisory function, as well as the responsibilities of individual workers. The requirements for workers also apply to supervisory personnel when performing the covered functions.

The general rules for employees have withstood the test of time; they are essentially the same as those contained in the First Edition of the NESC.

420. Personal General Precautions

The 1987 versions of Rules 420 (supply) and 430 (communication) covered the same requirements for their respective types of workers. In the 1990 Edition, Rule 430 was merged into Rule 420, the inconsistencies were addressed, and the language was modified to be more specific as to locations and conditions covered.

420A. Rules and Emergency Methods

To achieve safe and proper completion of assigned work, employees must understand the safety rules that are to be followed when performing the various methods of work used to complete the required tasks. These include the identification methods, work sequences, cautions, inspections, tests, and other procedures to be employed, as well as proper application of appropriate tools, equipment, and protective gear.

Safety rules are designed to take into account expected working conditions and appropriate protection for the employees and equipment that might be expected by intended work. For example, NESC Rule 445 requires an employee who is making a protective ground connection to assume that, even though a test has indicated no voltage is present, the previously energized part may still be energized at some voltage potential, including its normal operating voltage or some different voltage. The part may inadvertently become energized through no fault of the employees. Thus the employee is required to use insulated handles or other suitable devices to maintain the appropriate distance or insulation level when making all the connections to the previously energized part.

Following the safety rules will limit the opportunity for damage to equipment or hazards to workers or the public arising from the work actions employed. However, it is not possible to eliminate all sources of potential problems and, in spite of the best efforts of employees involved, accidents may occur.

Experience has shown that prompt emergency assistance may limit the extent or lasting effect of injuries. It is incumbent upon individual employees to stay current on emergency procedures to be used under various conditions.

420B. Qualifications of Employees

Over the years a variety of methods, and combinations of methods, for training utility employees have proven effective. No matter how much, or how little, classroom training is given, there is no substitute for the on-the-job training and progression of work assignments that occur in the field. This includes both the opportunity for employees to learn by assisting qualified workers and the opportunity for supervisors to assess the knowledge, capability, and work habits of the employee. Line workers are expected to progress through an orderly advancement in the complexity of assigned tasks before being allowed to perform those tasks without direct supervision. However, even the best supervision cannot replace the good, common sense of the individual employee—the common sense to be sure to understand the work to be performed, the tools and procedures to be used, and the protection to be employed, as well as the common sense to stay out of the vicinity of supply facilities when not performing an assigned task.

Rule 420B requires supply workers to *perform only directed tasks*. All too often, workers are injured because they "jump the gun" and start working on are near energized parts, or start preparatory work near energized parts, before the protection systems are in place and they have been directed to begin work. Likewise, employees are required to request further instructions if they do not understand the proposed work, including parts or lines to be worked, location of energized parts versus de-energized and grounded parts, tools and equipment to be used, and the sequences and procedures to be followed. Careful adherence to, and full completion of, safety procedures is necessary to assure safe completion of work.

It is recognized that it is essential to include refresher training on existing and new work practices for all work operations and activities. This training ensures the continued qualification of employees in their trade.

The OSHA rules also implement the requirement that specific attention be given to first aid activities and cardiopulmonary resuscitation (CPR) training.

Rule 446A again emphasizes training in the specialized area of live working activities. It directs certain training requirements.

The 2007 Edition added a specific requirement for operators of mechanized equipment to be qualified to perform those tasks. Many of the ANSI standards for operating mechanized equipment, such as ANSI B30.5 for operation of cranes, require that operators must pass tests in the specific type of equipment being operated. The NESC definition of qualified is consistent with these American National Standards by requiring qualified workers to both (1) be *trained in* and (2) have *demonstrated adequate knowledge of the installation, construction, or operation of lines and equipment and the hazards involved, including identification of and exposure to electric supply and communication lines and equipment in or near the workplace.*

420C. Safeguarding Oneself and Others

The requirements of Rule 420C are, in effect, application of good common sense by electric supply and communication workers. One of the most effective accident prevention measures is for workers to continually determine the state of the working environment and adjust work methods and personal movements accordingly. It is incumbent upon employees on a work site to maintain a "heads up" attitude, i.e., to look for job-site conditions that could adversely affect the safety of the employees, the public, or the equipment (whether at that site or on another affected part of the system) during the performance of the work; to consider the effect of their own actions on themselves and others; and to avoid placing themselves in a position to be adversely affected by the actions of others. When another person is seen to be located near a potential hazard or performing work in a potentially hazardous manner, Rule 420C1 requires workers to warn the person(s) in danger. This rule applies to utility workers who, in the normal course of their duties, happen to see another person in danger near utility lines or equipment. There is no duty to search out persons who might (or might not) someday violate OSHA regulations and place themselves in danger, as has been alleged by some.

Rule 420C2 applies only to covered defects not corrected at the time by the workers. This rule matches Rules 214A4 and 313A4. Records are required only for items not corrected, and then only until corrected.

To ensure that employees understand the prohibition against taking conductive objects closer to energized equipment than the distances specified in Rule 441, Rule 420C5 was added in the 1990 Edition.

420D. Energized or Unknown Conditions

It is imperative that employees who are undertaking to work on or near electric supply equipment and lines determine *before they begin work* which are energized and which are de-energized. Any line or equipment that is not positively determined by inspections or tests to be de-energized must be considered as energized, and the work must proceed accordingly using appropriate protective measures. When around supply facilities that do not meet the requirements of Rule 444 and, thus, cannot be considered to be de-energized, it is particularly important that employees plan their work movements to avoid bringing a portion of their body or a conductive object within the approach distances of Rule 431 (communication) or Rule 441 (supply), as applicable.

420E. Ungrounded Metal Parts

Ungrounded metal cases and parts of equipment and devices can be energized by loose jumpers, falling wires, internal insulation failure around conductors and windings, etc. In many cases, they can remain energized indefinitely and, therefore, must be treated as energized to the highest voltage to which they are subject.

420F. Arcing Conditions

Generally, the approach distance requirements and clear insulation distance requirements of other rules are large enough to keep portions of the body safely back from portions of a switch or other device that might create an arc during operation. However, special care should be taken to position oneself as far as is practical away from such devices to limit potential injury if catastrophic failure of the device occurs, a fuse blows during reenergization, etc.

420G. Liquid-Cell Batteries

Employees working in the vicinity of liquid-cell batteries are required to recognize in their work habits the three potential hazards presented by these systems: (1) fire resulting from ignition of hydrogen generated during charging, (2) skin or eye damage from electrolyte contact, and (3) damage to personnel or the batteries resulting from an accidental short. In addition, the OSHA has now published detailed information concerning hazardous materials, including batteries.

420H. Tools and Protective Equipment

This is the corollary to Rules 411 B and C; employees are required to use and inspect the protective gear supplied by their employer. It is incumbent upon employees to utilize available methods and systems to protect themselves when performing assigned work. The employee who will use personal protective gear is in the final, best position to ensure that this gear is in good condition, and to ensure that it is both properly placed and properly used. A large number of accidents involving both overhead and underground facilities could be avoided if the employee simply uses the appropriate tools and protective equipment in the manner intended. This includes wearing the required rubber gloves and/or sleeves when in close proximity to energized parts; installing the ground or structure end of a grounding set first, before proceeding with the connection to the previously energized part; inspecting and verifying that both ends of a safety strap are secure in their "D" rings before leaning into it; using care in checking the strength of supports to be used during work operations; and verifying that the work position selected is appropriate for the entire work operation.

Obviously, all provided protective gear will not be needed or be practical for every job, but careful, thoughtful choice of tools and protective equipment for each work segment will reduce accidents. For example, insulation line worker's overshoes don't help much for a single person in an insulation bucket, but they can help if two people are in a two-person bucket. Insulation line worker's overshoes have been shown to be a practical way to limit step and touch potentials for ground workers around trucks and equipment that can become energized by falling conductors, etc.

420I. Clothing

Wearing appropriate clothing has been shown to limit the lasting effects of being caught in a flashover. The level of injury sustained by a worker involved in a flashover can be substantially reduced by careful selection of the clothing materials to be worn (to avoid fusing melted plastic to the skin, skin damage due to burning clothing, and unnecessary exposure of the skin to heat during a flashover). Loose clothing can catch on exposed hardware or extend into energized areas. A requirement was added in Rule 420A2 of the 1993 Edition for employers to provide training on the advantages and limitations of various wearing apparel. This is the companion rule to require the employees to use appropriate clothing.

Metal articles exposed to the heat of a flashover tend to retain the heat next to the skin and act like a branding iron; they also may increase electrical flow through the body due to better skin contact. Metal articles worn under rubber gloves may provide a pressure point or chafe the glove, thus increasing the opportunity for failure at that point.

It is interesting to note that the First Edition of the NESC recommended that workers keep their sleeves down, avoid wearing unnecessary metal articles, and avoid wearing loose clothing.

The 2007 Edition specifically required workers exposed to electric arcs to wear clothing or a clothing system in accordance with Rule 410A3. Thus, in 410A3 the employer is required to consider the calorie exposure from electric arcs from various assigned work and train the employees to use appropriate clothing for protection; this rule requires the employees to wear that clothing. Notice that Section 42 applies to both supply and communication workers. Communication workers on joint-use poles, or working from aerial buckets on joint-use poles with supply lines or equipment should dress accordingly, in case of inadvertent contact with energized facilities.

420J. Ladders and Supports

Many accidents could be avoided by using care in checking the strength of supports to be used and in selecting work positions. Checking to see that the right support is used in an appropriate place and has the required strength is necessary before elevating oneself above ground.

420K. Fall Protection

(This rule was completely revised in the 1997 Edition as a complement to recent and proposed OSHA changes.)

The 1997 revision of the fall protection requirements followed an extensive review of fall protection problems by OSHA and the NESC Work Rules Subcommittee. The rule is practical and flexible. It recognizes both the good and bad history. Several key words were introduced. New definitions were provided for *fall arrest system, fall prevention system, fall protection program, fall protection system (hardware), harness, lanyard, positioning device system, positioning strap, qualified climber, transferring, transitioning,* and *worksite.*

All climbers above 3 m (10 ft) must now be attached to equipment or a structure at work sites, rest sites, and in aerial devices, helicopters, cable carts, and boatswain's chairs. The 1997 Edition recognizes that (1) it is not always practical (or the safest procedure) to remain attached to equipment or structures when performing some climbing operations, *transferring* to or from a structure, or *transitioning* across obstacles and (2) climbing is a learned art or skill. A *qualified climber* is permitted to climb, transfer, or transition unattached. However, personnel learning to climb are required to remain attached, thus limiting what they can do, until they are "qualified." Rule 420K includes extensive information and warning about the hazards of incomplete engagement, or subsequent disengagement, of snaphooks. The rule prohibits connecting snaphooks to each other or using 100% leather positioning straps.

420L. Fire Extinguishers

Using the wrong fire extinguisher on an energized part may be more of a safety problem than letting the fire burn until the part can be de-energized.

420M. Machines or Moving Parts

It is the responsibility of employees working on or in the vicinity of remotely controlled or automatically operated equipment to *satisfy themselves* that (1) appropriate measures have been taken to limit the opportunity for automatic or remote operation to occur while they are performing their work, and (2) if practical, their work position will not place them in an unsafe area if such operation occurs.

420N. Fuses

Since fuses, by their very nature, involve the creation and extinguishment of an electric arc, employees installing or removing such fuses should only do so when appropriately protected by location and protective tools and gear.

Particular caution should be exercised when closing a fused device when it is known or suspected that there may still be a fault present on the electrical equipment. Such closure stresses fuses to the fullest, and precautions should be taken to recognize that even the fuse holder may fail. Such precautions include shielding the face (over and above wearing protective glasses), and positioning oneself out of possible explosive gases expelled by the blowing of the fuse. Protective glasses should provide both mechanical protection and ultraviolet light spectrum filtering (see ANSI Z87.1-1989 *Practice for Occupational and Educational Eye and Face Protection*).

420O. Cable Reels

This rule was added in the 1990 Edition to recognize that accidental rotation of a reel is also a hazard to be considered when selecting blocking methods.

420P. Street and Area Lighting

(In the 1990 Edition, former Rules 286G2 and 286G4 were moved here to form new Rule 420P. The remainder of 286G moved to 232B4.)

Nonmetallic ropes are often used for lowering equipment of certain types of street lamps. Most often, they are positioned inside the supporting pole, and out of sight. The deterioration of these ropes is not due so much to wear as to the action of the elements. Because their strength may be reduced materially, due to decay of the interior fibers, even though they may still appear to be sound, the necessity for systematic inspection is evident. Nonmetallic ropes may be particularly susceptible to the effect of climatic changes and in some localities may have a life as short as two years. They are, therefore, a possible hazard, not only to passersby but also to workers. It is not safe to use material that would deteriorate rapidly under ordinary bad weather conditions, high humidity for extended periods of time, or even from reasonable amounts of smoke, dust, etc., to lower or suspend luminaires. At locations where large amounts of deleterious gases or dust are present, as near chemical works, blast furnaces, cement mills, etc., special materials should be used, and they should be inspected more frequently than in other locations.

Because metallic lowering equipment is conductive, and the breaking down of the luminaire cut-out may cause the lowering equipment to become charged with a voltage as high as the potential of the line, an insulator should be installed in the metallic chain or cable out of reach of the ground. Because of this problem, it may be advisable to employ nonconductive suspension ropes, especially where the luminaires are fed from HVDC circuits.

Even if series lamps are always handled from insulating stools or platforms, it is advisable (although not required) to use disconnectors, where practical, to ensure that the part being lowered or worked is dead. An opening in a series street light transformer circuit may have a voltage present equal to the total voltage of the supplying or high side source. The voltage across only one lamp is relatively low, but with no flow of current to limit this voltage, there is a high probably of a large open-circuit voltage.

420Q. Communication antennas

This rule was added in 2007 to specify limits on radiation levels received by employees working on or near communication antennas operating in the range of 3 kHz to 300 gHz. Sources for information on regulatory limits are given in a NOTE. In some situations, antennas on communication or supply line structures may have to be turned off to allow workers to enter an adjacent work area. If the antennas will be needed during storm or other emergencies and it is not desirable to turn them off for required restorations of the utility structures or supported facilities, the locations of such antennas should be carefully chosen to limit adverse impacts on utility restoration or emergency communications. See Rule 235I2 for the physical clearances of antennas from lines. Such clearances are not designed to provide the worker clearances required by Rule 420Q.

421. General Operating Routines

The general operating routines provide for control of access to the work site and control of the work to be performed to ensure that the safety rules are observed. It is recognized that the person in charge of the local work site is in the best position to ensure that access to the work site is limited to authorized persons, that the activities of authorized persons remain within the scope of their qualifications, and that the proper tools, devices, and work methods are used.

This rule is a companion to Rule 420. It is recognized that both the person in charge and the employees under his or her supervision must do their part to ensure that the actual conditions of the work site are considered and that appropriate precautions are taken for the work to be performed. Proper planning and supervision can promote safe and well-executed work, but adherence of the employees to all of the safety rules is vital to the successful completion of those goals.

422. Overhead Line Operating Procedures

Key elements of safety that run throughout the NESC are (1) the use of personal protective equipment, including gloves, sleeves, coverup, etc., as required; (2) maintenance of the required minimum approach distances (clearances) to parts of different voltage potentials; and (3) full consideration of the strength requirements associated with the work operations. This rule addresses the forces that can be expected to be applied during line work and the exposure of workers to high voltages when poles, structured components, wires, or cables are being handled. Common sense requirements are specified.

The act of grounding vehicles and equipment is often overlooked, or not followed to the letter. For example, when conductive booms on trucks are placed near energized equipment, there is a very great potential for accidental energization of the trucks. Grounding *may not* in itself provide enough protection, since the resistance of the earth achieved through ground rods is usually high. Use of a system neutral for grounding is discussed in IEEE Std 1048™ *IEEE Guide for Protective Grounding of Power Lines* (ANSI) along with other engineering practices; this document is excellent reading on this subject.

423. Underground Line Operation Procedures

This rule addresses the safety requirements that are particular to underground lines. Special emphasis is placed on ensuring that the quantity and quality of air present in underground personnel areas is suitable for safe work during the time personnel are present.

There was no change in the 1990 Edition from the 1987 Edition concerning Rule 423B—*Ventilation*. It seems that the intent of the rule is now understood; it requires ensured air supply of proper quality. However, there is a concern that the "source" could become contaminated if it is from the surface air. With forced ventilation on the street, it is very easy to force carbon monoxide fumes from vehicles into the manhole.

The question of smoking in manholes has been raised many times. While it is recognized that some cable sheath or joint soldering operations may require open flames, it is also recognized that significant damage has been done to cables by cigarettes being laid on them, ashes being dropped into splices, etc. Therefore, smoking is not allowed in manholes. Rule 423C4 was added in the 2002 Edition. Visibly exposed gas or other fuel lines shall be protected by clearance or barriers from torches or open flames used in underground splicing work.

When power-driven rods are used to obtain an opening or thread a line through a duct, they are often highly stressed. When they break, the equipment can be tossed around in a manhole; it is not appropriate for personnel to be where they can be endangered by such equipment breakage.

Rule 423D2 was added in the 2007 Edition to require exposure of existing utilities before using guided boring or directional drilling methods where the bore path will cross the other facilities. In recent years, several incidents have occurred in which existing conduits or pipes have been damaged by using these methods blindly; the results have been millions of dollars of damage to facilities, including damaged pipelines, power transmission cables, power distribution cables, communication cables, water lines, and other facilities, as well as lengthy service interruptions.

Rule 423D6 was added as 423D5 in the 2002 Edition. Shoring or other methods are required where a cave-in hazard exists or the trench or excavation is in excess of 1.5 m (5 ft) in depth.

Section 43. Additional Rules for Communications Employees

Most of the safety requirements for communications workers are included in Section 42. These additional requirements for communications workers address potential problems arising from working near supply facilities. The requirements of Section 43 are in addition to those of Section 42 and apply only to employees working upon communications facilities located in the communications space. See Section 44 for the additional requirements for supply workers working in the supply space. Note that *Rule 224A1 prohibits communication workers from working in the supply space unless they (1) are qualified to do so, (2) use the supply employee work rules, and (3) have permission of the supply utility to do so.*

The approach distances to energized parts required of fully trained communications workers are similar to, but generally slightly greater than, those required of supply workers; they were updated in the 1990 Edition. The voltage ranges in Table 431-1 are phase-to-phase, unless stated otherwise. These ranges were further harmonized with OSHA in the 1997 Edition.

Rule 431 was revised in the 2002 Edition to create 431A and 431B. Rule 431A requires communication workers that are repairing communication lines damaged in storms to treat those communication lines that are joint use with supply lines as if energized, unless the supply lines are appropriately de-energized and grounded.

Rule 431B requires communication workers to use the same altitude correction factors as supply workers (see Table 441-5). In 2007, the values in Table 431-1 were revised to include minimum approach distances for installations at elevations up to 3600 m (12 000 ft); as a result, the altitude correction factors now apply above 3600 m (12 000 ft) elevations, instead of the previous 900 m (3000 ft).

Before climbing joint-use poles, communications workers were required prior to the 1993 Edition to check the structure for conflicts with supply facilities to limit the opportunity for inadvertent contact with supply facilities or accidentally energized facilities during work. In the 1993 Edition, Rule 432 was revised to eliminate confusion; the language now requires careful positioning to maintain working clearances during work.

The requirements of the NESC do not intend for communications workers to climb or work around the level of the lowest electric supply conductor, unless special provisions are taken to limit the possibility for inadvertent electrical contact (such as the installation of guards or protective barriers).

The EXCEPTION to Rule 432 was limited in 2007 to work near voltages 140 kV and below where a rigid, fixed barrier is installed between the supply and communication facilities.

When supply cables are within joint-use manholes, an employee is required by Rule 433 to remain on the surface to render aid in the event of a supply-cable fault involving personnel in the manhole.

Because of the importance of cable sheath continuity to the safety and satisfactory operation of the cable, some means of ensuring sheath continuity is required by Rule 434 whenever underground cable is being worked. All necessary grounds should be checked to ensure they are intact and connected, since they provide protection should an electrical fault happen anywhere in the vicinity during the course of the communications worker's activities.

Section 44. Additional Rules for Supply Employees

(This section was created from portions of Section 42 in the general reorganization of the 1990 Edition to specify the additional rules applicable to construction, operation, and maintenance of supply systems. Section 42 was extensively revised in 1993 to coordinate with additional OSHA requirements for maintenance of supply systems.)

The general rules for supply and communication workers are contained in Section 42. These additional rules address the special requirements for working on or in the vicinity of energized supply facilities. Particularly emphasized are approach distances to be maintained from surfaces of differing potential when personnel are not insulated from such voltage difference, sequences to be followed for the control of energization, and the work with lines and equipment. Supply employees include both utility and contractor employees performing the covered work on or near supply facilities or in the supply space. See Section 43 for the additional rules applying to communication employees working only in the communication space. Section 44 also applies to communication workers working on communication facilities located in the supply space if permitted by Rule 224A1. To do so, the communication workers must be qualified to work in the supply space, adhere to all supply work rules, and have the permission of the supply utility to work in its space.

441. Energized Conductors or Parts

(These requirements were contained in Rule 422B in the 1987 Edition. Rule 441 was extensively revised in the 1993 Edition to coordinate with additional OSHA requirements for maintenance. Of particular interest are additions of calculation procedures required to determine allowable reduced approach distances where the maximum expected transient overvoltage is known and controlled.)

The intention of this rule is to ensure that the employee will not simultaneously come into contact with two surfaces of significantly different potentials, e.g., 50 V. This requires either (1) that the required approach distances of the body (and conductive parts in contact with the body) from parts at different potentials be maintained, or (2) that insulation be used between the employee and the item at a different voltage potential.

Other areas of Section 44 contain the requirements to be met if the normally energized part is to considered as de-energized.

The approach distances of this rule were revised in the 1990 Edition. They were extensively revised in the 1993 Edition based upon new flashover information. For the first time, the approach distances were based on consensus standards. IEEE Std 516™ *IEEE Guide for Maintenance Methods on Energized Power Lines* (ANSI) and IEEE Std 4™ *IEEE Standard Techniques for High-Voltage Testing* (ANSI) (used for voltages below 72.5 kV) provided the technical information to establish the electrical component of the approach distance. As the NOTES to IEEE Std 516 indicated, the NESC Tables 441-2 and 441-3 also did *not* include any allowances for inadvertent movement; these had to be added, although no values were specified.

The 2007 Edition extensively revised the minimum approach distances in the tables to include an extended range of maximum anticipated per-unit overvoltage factors for specified voltage ranges. Certain types of equipment can experience overvoltages greater than the 3.0 per-unit overvoltage factor limit previously specified in the tables. In addition, appropriate altitude correction factors were also added in the 2007 tables.

Another change in the 1990 Edition was the consolidation of the clearance for both insulated-tool (hot-stick) work and bare-handed work techniques into one series of tables. It was thus recognized in the NESC that the distance from the worker at ground potential to the energized part should equal that required from a grounded surface to the worker when the worker is at the line voltage, e.g., doing bare-hand work.

In the 1993 Edition, the rule was extensively revised and reorganized to coordinate with revisions in OSHA requirements. The technical basis for calculating reduced approach distances for work on circuits where the maximum expected transient overvoltage (from switching or any other source) is *both* known and controlled was added for the first time. This work was accomplished jointly with OSHA and provided the methodology for calculating approach distances to be used by the NESC and OSHA. Thus, in the 1993 Edition, the NESC and OSHA approach distances coincide.

441A. Approach Distance to Live Parts

(Rule 441A1 of the 1990 Edition is contained within Rule 441A1 in the 1993 Edition; Rule 441A2 of the 1990 Edition was moved to 441A4 in the 1993 Edition. The remainder of Rule 441A is new with the 1993 Edition.)

Rule 441A1 of the 1993 Edition is comprised of Rules 441 and 441A1 of the 1990 Edition; it was also reorganized and expanded to more clearly delineate the requirements. The voltages of Table 441-1 are phase-to-phase unless otherwise noted.

IR 540 issued 16 December 2005 clarified the use of Footnote 2 of Table 441-1 and the columns in the table. Footnote 2 applies only to Column 1 (i.e., to the column containing the voltage category *row* headings); Footnote 2 requires using the phase-to-phase voltage to find the correct row, even if the line being worked is a single-phase line off of a three-phase line and no other phases are present. Neither Footnote 1 nor Footnote 2 apply to the columns containing the minimum approach distances (MADs) for employees to energized conductors or parts.

The choice of the *distance-to-employee* part of Table 441-1 depends upon the physical orientation of the workers to the energized conductor or conductors. For the single-phase line, the MAD in the phase-to-ground column may be used because there is no possibility of contact with another phase conductor. However, if the line to be worked is multiphase, the phase-to-phase column must be used if the worker will be positioned either (1) between phase conductors or (2) near enough to another phase conductor such that there is any possibility of contact with that conductor. The application of both the voltage ranges and the MADs shown in Table 441-1 have been harmonized with OSHA requirements. See also the following discussion of Rule 441A3d below for ratings of cover-up equipment to be used during the work.

Rule 441A1 in the 1993 Edition specifically prohibits contact with exposed parts operating at 50–300 V without complying with either the de-energization or insulation requirements. This is the first time that employees have effectively been prohibited from touching 120/240V secondary while working on a wood pole, without also using rubber gloves or equivalent protection. Previous editions were silent at voltages below 1000V. The minimum approach distances (MAD) of the related tables are calculated using recognized methodologies for the maximum voltages expected to be encountered in the work area. That methodology is explained in Rule 441A7.

The 2002 Edition revised Rule 441A1a to require a line to be *grounded*, as well as *de-energized*. This requirement also appears in other Rules in Part 4 and coordinates with the new 2002 definition of *de-energized*, which is *disconnected*. Both disconnection and grounding are required for the employee not to have to use insulation methods. This change was made to stress the need for grounding to protect against voltages which may be or become present due to induction, improper switching, or contact of the worked facility with or by energized facilities. Mere disconnection will not protect against such hazards.

The 2007 Edition of Rule 441A1 added 441A1d to clarify that using bare-hand live work procedures of Rule 446 is also one of the options to allow work within the MAD distances.

The new EXCEPTION to Rule 441A2b(1) of the 1997 Edition clarified the intended use of rubber sleeves with rubber gloves. When working on meters and other equipment of 0–750V, and the only live parts exposed are those being worked for such cases, the insulating gloves provide all the protection that is necessary for safe work. However, if other parts of any voltage are exposed nearby, rubber sleeves or insulating coverup would be required. A similar EXCEPTION to Rule 441A2b(2) relaxed the requirement to cover all exposed grounded lines, conductors, or parts in the work area if (1) insulated tools or gloves are used and (2) the voltage is 750V or less.

In the 1997 Edition, the requirements for guarding and gloving above 300 V were placed in a new Rule 441A3 and the remaining rules were renumbered.

The prohibition against contact with ungrounded parts of 51–300 V was moved from 441A1 to 441A2 in 1997. Rule 441A3 in the 1993 Edition added specific guarding and rubber-gloving requirements, including the requirement to *either* wear rubber sleeves with the gloves or rubber up all parts within reach when working on or near voltages from 301 V to 72.5 kV. Note that coverup insulation is required to extend throughout the expected work area and slightly beyond so that the opportunity for inadvertent contact is limited. For the first time in 1993, this coverup requirement includes grounded objects, wires, conductors, equipment cases, etc. It is clear from Code documents that a wood pole is not required to be covered when working between the pole and an energized part. However, it is not clear that a metal pole is exempt from the coverup requirement. There is no practical difference between a steel pole and a grounding conductor. No provision allows laying an energized conductor temporarily on a wood arm or against a wood pole during work procedures without an intervening insulating materials.

Whether or not the worker is gaffed into a pole, it is important that conductive items in the work area be guarded or insulated so that uninsulated parts of the worker do not contact ground. A worker is not usually at ground potential even when gaffed into a pole due to secondary insulation from the wood and/or the worker's footwear. Insulated pole platforms and aerial lifts provide further insulation from ground. This protection would be short-circuited if the worker's body were to contact a grounded part.

Rule 441A3b of the 1997 and 2002 Editions (441A2b of the 1993 Edition) required employees to wear appropriate rubber gloves when "in the vicinity of" energized conductors or parts. "In the vicinity" was intended to mean within reach of the employee with arms in the extended position. Because of confusion as to the intention of the term *in the vicinity of*, the 2007 Edition changed the language to "within reach or extended reach of." Further, the previous rules referred to being "in the vicinity of *energized conductors or parts*;" the new rule refers to being "within reach or extended reach of the *minimum approach distances to live parts*" for the voltages involved.

In essence, it is not enough to position oneself so that the energized parts themselves cannot be touched—the position must not allow reaching into the MAD unless the employee is wearing insulated personal protective gear.

In addition to wearing rubber gloves when using the rubber glove working method, either the employee must wear rubber sleeves, *or* all exposed energized conductors or parts must be covered with insulating protective equipment, except conductors or parts temporarily exposed for work and maintained under positive control (not left free to move). The protective equipment must extend beyond the employee's maximum reach in the anticipated work position.

In essence, Rule 441A3b (441A2b in 1993) requires gloves to be worn at all times when in the vicinity of energized conductors or parts. In addition, sleeves must be worn until conductors or parts are covered with protective equipment, unless protective equipment can be installed (or removed) without violating minimum approach distances. Sleeves are not required while protective equipment is in place, provided that energized lines or parts temporarily exposed to perform work are maintained under positive control. Conversely, the rules do not prohibit wearing sleeves while protective equipment is in place.

The 2002 Edition modified Rule 441A3b to specify maximum use voltages for rubber insulating equipment. The new Table 441-6 specifies the class of equipment to be used with each maximum use voltage. The voltage used to choose from the table is based upon the exposure of the employee, not the maximum voltage of the circuit. Footnote 1 of Table 441-6 and its two EXCEPTIONS clearly specify use of phase-to-ground voltage, if the employee exposure is limited to that level. When rubber-gloving systems that are above 15 kV phase-to-phase, additional protection (such as an insulated bucket or platform) must also be used. An EXCEPTION to Rule 441A3b was added in 1997 to allow work performed on electric equipment (such as meters, etc.) of 0–750 V to be performed while wearing gloves without sleeves if only the live parts being worked are exposed.

The 2007 Edition added Rule 441A3d to specify the ratings for cover-up equipment to be used for the rubber glove working method. This rule is intended to further clarify the use of Table 441-6 in determining the maximum use voltage required for personal protective insulating gear. In particular, Rule 441A3d helps the user determine when cover-up equipment must be rated for the phase-to-phase voltage and when it can be rated for the phase-to-ground voltage. Insulating cover-up to be *used for the purpose of limiting exposure to phase-to-phase voltages* must be rated for the full phase-to-phase voltage required. Other cover-up equipment may be rated for the phase-to-ground voltage.

In some work areas on dual-circuit line structures, where workers may be potentially exposed during work to a circuit of a higher voltage, such as if a worker would have to be positioned between conductors of the two different circuit voltages during the work, the insulating cover-up to be *used* on conductors or parts other than the one being worked *for the purpose of limiting exposure to phase-to-phase voltages* must be rated for the full phase-to-phase voltage of the *higher* voltage circuit(s) in the work area. However, where a circuit of higher voltage is present on the structures but is *outside* of the work area and, thus, is not exposed to workers in the work area, such circuit is not required to be covered. As a result, the insulating cover-up equipment to be used on the circuit conductors or parts *in* the work area or the approach route to the work area are not required to be rated for the voltage of any circuits outside of the work area.

Rule 441A4 of the 2002 Edition (Rule 441A3 in 1993–97) was originally added in the 1993 Edition to specify how to address transient voltages above 72.5 kV. The approach distances in the tables are based upon the expected voltages, including transient overvoltages. Where transients are known or can be controlled to a known level, approach distances may be able to be reduced from the normal values shown in the applicable tables. In the 2007 Edition, these requirements were completely revised and expanded into Rules 441A4 and 441A5.

Rule 441A4 specifies two methods of determining the MAD for live work, depending upon whether the maximum anticipated per-unit overvoltage factor T has been determined by engineering analysis (see Rule 441A4b). If the requirements of Rule 441A4b cannot be met, then the general methods of Rule 441A4a must be used: i.e., using conservative values for T in Tables 441-2, 441-3, and 441-4. One of the methods of reducing the maximum anticipated per-unit overvoltage is the use of temporary transient overvoltage control devices (TTOCD) meeting the analysis and testing requirements of new 2007 Rule 441A5.

Altitude correction is addressed in Rule 441A6 (Rule 441A2 prior to 1993; Rule 441A4 in 1993; Rule 441A5 in 1997 and 2002). In the 1993 Edition, the former Rule 441A2 was revised, substituting tabled values for the previous formulas, and it was moved to Rule 441A4 (now 441A6). The altitude correction factor applies only to the electrical component of the required approach distance, i.e., the table value less the 0.6 m (2 ft) of mechanical clearance. At high altitudes, the total required approach distance (TAD) in feet is given by the following formula, where TV = table value and ACF = altitude correction factor from Table 441-5.

$$TAD = 2 + ACF(TV - 2)$$

Rule 441A7 (Rule 441A6 in 1997 and 2002; Rule 441A5 in 1993) explains the derivation of the values in revised Tables 441-1 through 441-4 and identifies the technical basis and calculations required to determine required approach distances for voltages not included in the tables. Results of calculations are rounded up. Interpolation between table values is not allowed. The MAD in Tables 441-2, 441-3, and 441-4 were taken from IEEE Std 516.

Rule 441B of the 1993 Edition added requirements for approach distances when working on insulators (see Figure H441B). The rule prohibits inserting conductive objects in the air-gap distance designed into a switch if one end is energized. Normal approach distances must be maintained to energize parts of switches when working on a grounded portion of a switch. It is specifically to short-out the first insulator on the grounded end of an insulator string when working on the insulator string assembly. If the work is performed bare-handed, the first insulator on the energized end of the string may also be shorted out.

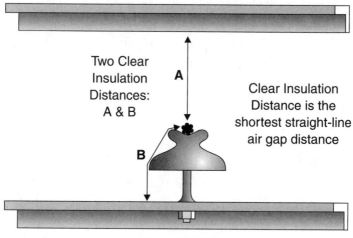

Figure H441B
Clear insulation distance

The 2002 Edition allows employees working on insulators on lines above 230 kV to temporarily short out up to three insulator units as part of work procedures, so long as approach distances of Rule 441 are met. The 2007 Edition requires testing of each insulator before shorting out the appropriate end insulator in the string to assure that adequate insulating capacity will remain in the string during the work.

In the 1993 Edition, Rule 441C specified for the first time the distances along insulated tool handles that employees must leave between themselves and energized work (see Figure H441C).

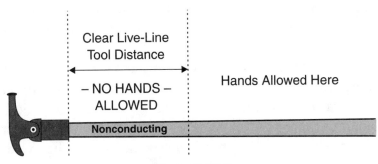

Figure H441C
Clear live-line tool distance

442. Switching Control Procedures

When switching could endanger line workers or system operation (such as switching of station equipment, transmission lines, or interconnected feeder circuits) intentional, specific control of switching operations is required.

For the safety of the system and the employees working thereon, suitable reporting and control systems are required to ensure that lines and equipment are neither mistakenly nor inadvertently energized or de-energized.

Central to the control of the system is the authority of the designated persons and effective communication to and from that person. The importance of clear communication and record keeping during switching operations cannot be overemphasized. Neither can the importance of following the exact sequences required for requesting switching action for de-energization, proper tagging of all points of control, and restoration of service.

These procedures are designed to provide a logical system for ensuring that employees are clear before lines, circuits, or equipment are re-energized. It is recognized that the vast majority of the operations of system-protection devices are caused by conditions that are transient in nature. Rules 442D and 442F prohibit reclosing circuits that are tagged until all workers have reported clear; local rules that recognize expected causes of such outages will determine whether one or more reclosings will be performed to see if the initial problem has gone away.

Rule 442G requires each employee receiving an oral message concerning the switching of lines and equipment to (1) immediately repeat the message back to the sender and (2) obtain the identity of the sender. Likewise, each employee sending such an oral message must (1) require the message to be repeated back by the receiver and (2) secure the receiver's identity. Many accidents can be avoided by requiring messages to be repeated back to ensure the sender that the proper message was received.

EXAMPLE: Switching circuit 324A is not the same as switching circuit 342.

IR 522 issued 28 February 2001 clarified application of Rule 442G. The basics apply even between members of the same crew working at the same location. Obviously the identity of each person is known to the other when talking directly to one another. If the crew chief has received an order and repeated it to the dispatcher in accordance with Rule 442G, and both line workers on the crew have discussed the operation, such as in a tailgate meeting at the job site, and both line workers understand their part in the assignment, the crew chief can instruct the other line worker to perform the switching at the appropriate time. It is not necessary at that point to repeat the messages back and forth between the two workers, but the crew chief is responsible for determining that the crew is at the proper location, has identified the proper switches, understands the operation to be performed, and has opened the switches. The other line worker is responsible for knowing what is to be done (open switches) when instructed to do so by the crew chief.

The 1997 Edition added EXCEPTION 2 to Rule 442B to allow suspension of the normal control and coordination by the *designated person* when catastrophic service interruptions occur due to earthquake, hurricane, etc. Strict observance of specified additional requirements, including observance of Rules 442A, 442D, 443, and 444, is required by this EXCEPTION to the reporting and coordination requirements.

Guidance regarding procedures for personnel protection lock-out/tag-out practices are found in ANSI Z244.1 *Safety Requirements for the Lock Out/Tag Out of Energy Sources.* Rule 442E was revised in the 1993 Edition to clearly emphasize that SCADA tagging alone does not satisfy the rules. There must be a physical tag at all points of possible control. Also note that a requirement was added in Rule 216E of the 1993 Edition to provide on each overhead line switch a means of rendering remote or automatic operation capabilities inoperable. These rules clearly intend that all sources of control be limited in capability in order to reduce the likelihood that lines that are being worked as de-energized or with reclosing features deactivated can be inappropriately switched. Rule 442E was again revised in the 1997 Edition to take advantage of recently available control and signal systems. As long as (1) the specified display and control is maintained at SCADA sites, (2) appropriate control deactivation and display is maintained at reclosing device locations, and (3) communication between the sites confirms operations at the reclosing site, automatic reclosing features of a reclosing device may be changed entirely by SCADA, without the previous requirement for someone to physically go to the reclosure site to place a tag.

443. Work on Energized Lines and Equipment

This rule covers special considerations required when working with or around several specific items of equipment and line components. Chief among these is ensuring that the employee remains insulated from surfaces of different voltage potentials. Since many of these items are so obvious or self-explanatory, only limited comments follow.

Rule 443A2 is consistent with the treatment required of covered conductors under Rule 230D. Coverings on conductors are not intended to be considered as "insulating" unless so marked and their integrity verified. Although such coverings may appear to be perfect, all too frequently that is *not* the case.

Rule 443A4 assumes that a supply cable that cannot be positively determined to be de-energized *may actually be energized* by positive connection to the supply source or by induction, either due to electric or magnetic field coupling. This rule requires employees to verify de-energization and to employ personal protective methods and devices suitable for the maximum voltage that could be present at the work location. Until the part is grounded, it cannot be treated as de-energized and, therefore, it cannot be approached within the required approach distances, touched, or handled without insulating tools and equipment.

There are many tasks required for maintenance and repair of lines and equipment that can safely be performed by one person during daylight hours; however, some of these tasks cannot be safely performed at night or in inclement weather. Rule 443B recognizes that the vision of an employee may be limited when working under conditions of reduced lighting. While a troubleman may remain safely isolated and safely perform many of the simpler tasks with insulated tools, etc., it is generally not appropriate for him or her to perform work requiring the body to be located in the vicinity of conductors that are or may be energized, unless the additional visual observation and assistance of a fellow worker is available. It is incumbent upon a single worker, or even a small crew for that matter, to call for additional help when the safety of any worker(s) might be jeopardized by working alone.

Proper operation of energized switches requires a smooth, continuous motion. Rule 443C requires a continuous motion to limit the opportunity for the switching to stick in the half-open position or for prolonged arc heating of the switch components.

Rule 443D does not always require employees to work from below the level of energized parts because it recognizes that

(1) upper conductors of multicircuit or vertically aligned circuit structures need work at times, and

(2) work on higher-voltage facilities may necessitate a position above lower-voltage facilities.

In most cases, however, it is practical to insulate such facilities with line hose, blankets, etc., to ensure that a fall will not bring the worker in contact with such energized facilities.

Because it is not always possible to verify that enclosed switches (like oil circuit breakers) are indeed open, blade disconnect switches are usually used on either side of enclosed switches. Rule 443E requires a load-break switch or "load-buster" device to be opened before disconnect switches that are designed to interrupt charging current, but not load current or fault current. It is often necessary to use a tong-type ammeter to measure the current present at the time to ensure that its value is within the rating of the device being operated.

Rule 443F recognizes problems that could occur if a jumper were placed on a theoretically de-energized part before being grounded; if the part is actually energized, the not yet attached end of the jumper would also become energized, causing potential conflicts.

Rule 443G works in concert with Rule 441 to ensure that employees will have limited opportunity to contact energized parts. Since barriers are often used in switchgear to separate areas with energized parts, rather than using air-insulation distance in their protective design, appropriate personal protection is required when the barriers are removed to service parts normally protected by the barrier. When the secondary of a current transformer is opened and the primary side is energized, very high voltages may be present on the secondary under these open-circuit conditions. Rule 443H prohibits opening the secondary circuit unless the entire apparatus is de-energized.

Rule 443I requires positive procedures to drain stored energy from capacitors before other work is performed on these installations. The capacitor(s) must first be disconnected from the power source, and then grounded and short-circuited. Next, *each* individual capacitor must be shorted to ground to remove its charge.

Special maintenance procedures are required by Rule 443J for gas-insulated equipment (see also IEEE C37.122, *IEEE Standard for Gas-Insulated Substations* [ANSI]). The toxic byproducts of arcing can cause personal or environmental problems if not properly handled.

When employees are working in a manhole that contains supply facilities, the possible effects on those employees if a supply cable fails can be catastrophic. Since prompt assistance of injured employees may be able to limit the extent or permanency of injury, Rule 443K requires an attendant to remain on the surface out of harm's way while work is being performed. The rule acknowledges that it may be appropriate for the attendant to provide short-term assistance, such as help in moving tools or equipment into the manhole in preparation for the work.

Rule 443L requires prompt removal of unintentional grounds on delta circuits. If another ground connection occurs on a different phase before this one is removed, the resulting ground fault will experience a phase-to-phase level of voltage and, in some cases, a large fault current.

444. De-energizing Equipment or Lines to Protect Employees

When one worker is required to depend upon another for switching, and when equipment or lines to be worked could be accidentally energized through switching, equipment failure, conductor contact, etc., it is imperative that all points of re-energization be controlled and that appropriate grounding be installed to limit the effects on workers should accidental re-energization occur. This requires getting permission for switching when others might be affected, having a knowledge of possible sources of energization (such as co-generation facilities), physically tagging all control points and switches that could be used to re-energize the lines or equipment, and rendering inoperable all applicable switches and disconnectors.

Tagging records must include the name of the requester (or that person who took responsibility under Rule 444F2) so that this can be checked against the name of the person who requests re-energization. Although some companies duplicate all required information on the tag itself, this information does not necessarily have to be on the tag as long as the record system maintains control linkages between the tag and the information, such as tag numbers, etc. "Tag" means a physical tag; a mere indicator on a SCADA system screen is not enough. A tag must be applied at every location where a control could be operated. There are specific requirements for obtaining, recording, and checking the switching information at each point in the required sequence of events.

Rule 444D allows two options for placement of grounds to protect workers. The first is the older method of placing a ground connection at each side of the work location as close as practical to the work location. The second, newer method is placing one ground at the work site. These methods are often referred to as *double-point grounding* and *single-point grounding*, respectively. The NESC has never used the term *double point* grounding, but it used the term *single point* grounding from its introduction to the Code in the 1984 Edition until replacement with the term *worksite* ground in the 2007 Edition. In many circumstances worksite grounding may be able to limit the level of voltage to which a worker might be exposed in the event of an unauthorized re-energization or other accidental energization of the line, lightning impulse transferred to the line, etc., better than double point grounding. However, in some circumstances, worksite grounding may be impractical. The choice between worksite grounding and double point grounding should be carefully made.

445. Protective Grounds

Central to the protection of those working on lines and equipment is that such lines and equipment remain de-energized and free of hazardous induced voltages. Other rules specify the requirements for control of the switching; here the concentration is on the grounding methods employed to limit the adverse effects on workers if re-energization accidentally occurs. Critical to employee protection in case of accidental re-energization is routing the fault current around the workers and holding the voltage difference between or across parts in contact with employees to the lowest practical level.

The size of the grounding conductor or device and its connection to an effective ground must be matched with the available fault current and the circuit-protection system to limit the amount and duration of current and heat affecting nearby workers.

The test for voltage is required *before the grounding device is connected to the supposedly de-energized part*. The time between the voltage test and the connection of the grounded grounding device to the "de-energized" conductor or part should be kept to a minimum to lessen the opportunity for accidental re-energization between the voltage test and the connection. Where the room is available to bring the grounding device into the work area and connect the "grounded end" or "structure end" to ground without creating a hazard in the process, it is good practice to do so and connect the "ground end" of the grounding device to the available ground before making the voltage test. This decreases the time interval between the voltage test and the attachment of the free end of the grounded grounding device to the part to be grounded. However, the Code recognizes that it is more often appropriate to make the voltage test first, in order to limit the opportunity for a hazard to be created while moving the grounding device into position.

When making the test for voltage, the minimum approach distances specified in Rule 441 must be followed. Also, since the line or circuit *may* still be energized when the grounding device is connected to it, the employee should also maintain at least this minimum approach distance during this work operation.

The worker must remember that, if the line or circuit is energized at supply source voltage, then there will be a large arc resulting from the establishment of a "fault" during the act of grounding. To preclude injury, sufficient distance between the worker and the arc is essential, as is the wearing of eye protection that provides both mechanical protection and ultraviolet filtering.

Additional guidance is provided in IEEE Std 1048 *IEEE Guide for Protective Grounding of Power Lines* (ANSI).

The 2007 Edition revised Rule 445B to address the same concerns about inadvertent contact of the ground cables with energized parts during removal of the cables that is expressed in Rule 445A for installing the ground cables. In order to limit the opportunity for a worker to be presented with induced voltages by removing the grounded end while the other end is still connected to a conductor, the conductor end of the ground cable is required to be removed first.

446. Live-Line Work

These rules formerly referred to *bare-hand* work, to contrast it with insulated hand work. These special rules apply where it is desired to have the employee be at the same voltage potential as the *energized* conductor or part to be worked and to insulate, guard, or isolate the employee from surfaces at other voltage potentials. "Bare-hand" work is typically only used at voltage so high that either (1) rubber insulating gloves and sleeves are not available or (2) the gloves and sleeves are so thick as to be difficult to manage.

While an insulated aerial device is still used to insulate the worker from ground, a conductive bucket liner may be used within the bucket to ensure bonding of the employee to the energized conductor or part, as long as the liner is part of the electrical circuit. When this practice is followed, conductive boots or overshoes must be worn. Note that a variety of methods may be used to limit exposure to other voltages through contact or induction. See also IEEE Std 516 *IEEE Guide for Maintenance Methods on Energized Power Lines* (ANSI).

In the 1993 Edition, Rule 446 was revised to specifically refer to applicable standards for testing insulation of aerial devices, specify insulation distances, and require protective clothing for electrostatic shielding purposes.

447. Protection Against Arcing and Other Damage While Installing and Maintaining Insulators and Conductors.

(This rule was renumbered to 276 from Rule 277 in the 1997 Edition and moved here in the 2002 Edition.)

The basic requirement for insulators is contained in this rule, which is a general statement of the engineering principles involved. it is not specific and thus permits sufficient latitude in designing a supply line to meet all the various conditions that must be considered.

The requirements were expanded in the 1977 Edition to prohibit damage to parts of the supporting structures, insulators, or conductors. The original rule only prohibited such damage as would allow the conductors to fall.

Appendix A—Reviewers and Policy

Reviewers of the NESC Handbook

The following NESC Subcommittee Chairs and/or Secretaries reviewed the sections of the NESC Handbook for which their respective subcommittees were responsible. Their assistance is greatly appreciated.

Reviewers:	Handbook Edition(s):
O. C. "Chuck" Amrhyn	2 and 3
Charles J. Blattner	1
Charles C. Bleakley	4, 5, and 6
Archie W. Cain	2 and 3
D. J. Christofersen, P.E.	5 and 6
Johnny B. Dagenhart, P. E.	4, 5, and 6
Frank A. Denbrock, P. E.	1, 2, 3, 4, 5, and 6
Eric K. Engdahl, P.E.	6
Gary R. Engmann, P. E.	2, 3, and 4
David G. Komassa, P.E.	5
Wayne B. Roelle, P. E.	1, 2, 3, and 4
Paul S. Shelton	1
John T. Shincovich	1
William A. Thue, P. E.	1, 2, and 3
James R. Tomaseski	4, 5, and 6
J. M. Van Name, P. E.	2 and 3

Review Policy

The information contained in IEEE Standards Information Network/IEEE Press publications is reviewed and evaluated by peer reviewers of relevant IEEE Societies and/or Standards Coordinating Committees. To assure the accuracy of The National Electrical Safety Code Handbook, the chair or secretary (or both) of the relevant NESC technical subcommittees has reviewed and contributed to the discussion of each part of the NESC. The editor addressed all of the reviewers' comments to the satisfaction of both the IEEE Standards Information Network and those who served as peer reviewers for this document.

The quality of the presentation of information contained in this publication reflects not only the obvious efforts of the authors/editors, but also the work of these knowledgeable peer reviewers. The IEEE Standards Information Network/IEEE Press acknowledges with appreciation their dedication and contribution of time and effort on behalf of the IEEE.

Appendix B — Safety Signs

(This Appendix is adapted by permission of Clapp Research, Inc. from Meeting the ANSI Z535 Safety Sign Standards *by Allen L. Clapp, IEEE Representative on the ANSI Z535 Committee and Chair of the Z535.2 Subcommittee.)*

Introduction

The National Electrical Safety Code® (NESC®) includes various requirements for safety signs to inform line workers or members of the public of specific hazards or conditions to be avoided. Utilities routinely use general safety signs and specific hazard signs to inform appropriate personnel of their responsibilities. The NESC does not include specific requirements for sign language or format. The following American National Standards contain requirements applicable to *environmental* and *facility* safety signs and accident prevention tags.

ANSI Z535.1—*Safety Color Code*

ANSI Z535.2—*Environmental and Facility Safety Signs*

ANSI Z535.3—*Criteria for Safety Symbols*

ANSI Z535.5—*Safety Tags and Barricade Tapes (for Temporary Hazards)*

Another related standard, ANSI Z535.4—*Product Safety Signs and Labels*, applies to products, not facilities. A sixth related standard, ANSI Z535.6 *Product Safety Information in Product Manuals, Instructions, and Other Collateral Materials* applies to the use of safety messages in collateral materials, such as product or training manuals. The first five of the new standards were formed in 1991 and extensively revised from the previous Z35.1-1972—*Specifications for Accident Prevention Signs*, Z35.2-1968—*Specifications for Accident Prevention Tags*, and Z53.1-1979—*Safety Color Code for Marking Physical Hazards*. They were further refined in 1998, 2002, and 2006 to recognize recent research on safety sign efficiency and promote harmony with international standards.

These coordinated standard criteria apply to every temporary or permanent sign or tag on a utility system, regardless of whether it is on a fence, tower, piece of equipment or cabinet, or whether it is inside a station or out in an area accessible to the public. Some of these signs and tags are appropriate for product manufacturers to put on or in equipment; others are appropriate for utilities to place within or at the periphery of their facilities.

The 1997 NESC referred directly to the ANSI Z535.1–1991 through Z535.3–1991 safety sign standards. The 1991 Editions of the ANSI Z535 standards were the latest approved editions at the time of voting on the 1997 NESC. Similarly, the 2002 NESC refers to the 1998 Z535 standards and the 2007 NESC refers to the 2006 ANSI Z535 standards. The 1998 revisions to the five ANSI Z535 standards were approved for balloting after the 1997 NESC vote and the 2001 Z535 standards were balloted after the 2002 NESC.

This Appendix is a modified version of the Appendix B that had been proposed for the 1997 NESC. Through coordination with the ANSI Z535 Committee, many of the concerns of the NESC Committee were addressed in the 1998 changes to ANSI Z535. Thus, the proposed NESC Appendix B was not added to the Code book, but has been updated and presented in this Handbook. Since each of the 1998 changes was deemed by the ANSI Z535 Committee to have been allowed as primary or alternate designs by the 1991 ANSI Z535 standards, meeting Z535–1998 standards (1) meets 2002 NESC requirements both (2) meets the 1991 ANSI Z535 standards (and, thus, the 1997 NESC), and (3) prepares the signs for staying in compliance with future ANSI Z535 changes under consideration. The older format of ANSI Z535.2-1991 was still allowed by Z535.2-1998 as an alternate. However, the old format was deleted in the 2001 Z535 standards and should not be used for new safety signs and labels.

ANSI Z535.2-1991 included specific requirements for the format of environmental and facility safety signs of the type used for utility facilities. These included distinctive shapes and colors for the backgrounds of the *signal words* on DANGER (the so-called "raccoon mask" of the red oval in a black rectangular background) and WARNING (the truncated diamond) signs. ANSI Z535.2-1991 *allowed* the use of the alternate format for product safety signs and labels given in ANSI Z535.4-1991. The 1998 revisions show a *preference* for the cleaner, simpler, signal word panel first introduced for products in Z535.4-1991 (see Figure B1). This is supported by research that shows greater readability and hazard association value (HAV) associated with more referential color and larger letters. Z535.2-2002 discontinued use of the older, alternate style of signal word header.

Figure B1
Signal Word Changes

Safety Colors

Color schemes have been developed to promote universal identification of different types of potential hazards. In some cases, the use of color alone may serve as the identifier, such as safety red for fire extinguisher locations and safety yellow for traffic aisles, stumbling and tripping hazards, etc. In other cases, color needs to be combined with a signal word, message panel, and/or pictorial to convey the seriousness associated with the hazard identified.

The Safety Colors Standard (ANSI Z535.1) fully coordinates color schemes with other requirements contained in ANSI Z535.2–.5, as well as other standards, including radiation warning, hazardous materials transportation, and ambulance colors. Table B1—*Intended Uses of Safety Colors* shows common applications of the individual safety colors. The Color Safety Standard specifies color mixes for each safety color.

Comparison of Requirements of Facility Signs v. Product Signs

The standards for environmental and facility safety signs (Z535.2) and the standards for product safety signs and labels (Z535.4) are similar; the latter were developed from the former. However, there are significant differences.

Environmental and facility signs are usually larger and contain less information, in order to be clearly understood at longer distances. Product safety signs are generally used at closer distances and are likely to contain more information of more specific nature.

An *environmental safety sign* is a sign or placard in a work or public area that provides safety information about the immediate environment. A *product safety sign* is a sign, label, or decal permanently affixed to a product that provides safety information about that product.

Environmental and facility safety signs are intended to communicate the presence of environmental hazards to the observer in the area. Product safety signs and labels are directed at "persons using, operating, servicing, or in proximity to, a wide variety of products." Permanent safety signs or labels affixed to a product are intended to "warn against potential exposure to hazards inherent in the normal use of or associated with the product, or which might be created during reasonably anticipated product use."

Temporary safety signs or tags are used to alert observers in the area to temporary, but potentially hazardous, changes in the environment or to a facility, such as switches in a circuit opened to protect workers downstream. Temporary safety signs or tags *affixed to a product or its container* are intended to warn against temporary hazards created by situations such as shipment, setup, service, or repair. Both types of temporary tags are to be removed when the potential hazard no longer exists.

Table B1
Intended Uses of Safety Colors

Color	To Identify:	Common Application
Safety Red	DANGER; STOP	Containers of flammable liquids; bars, buttons, or electrical switches used for emergency stopping of machines; fire protection equipment or apparatus
Safety Orange	Hazardous parts of machines; Intermediate level of WARNING	Emphasizing parts of machinery that can cut, crush, etc., when guards are removed and hazard is exposed; Marking exposed parts of pulleys, gears, etc.
Safety Yellow	CAUTION (can use solid yellow, yellow/black stripes, or yellow/black checkers for maximum contrast with background.	Physical hazards of tripping, falling, stumbling, being caught between; cabinets for storing flammable materials; containers for corrosives or unstable materials
Safety Green	SAFETY, emergency, egress, location of FIRST AID and SAFETY EQUIPMENT	Gas masks, first aid kits or dispensary; stretchers; safety deluge showers; safety signs and bulletin boards; emergency egress routes
Safety Blue	SAFETY INFORMATION on informational signs and bulletin boards; special means for railroads	Mandatory action signs for wearing personal protective gear, such as a hard hat.
Safety Black	Traffic, housekeeping markings	Clear traffic lanes (often yellow outlined); materials storage areas (often white outlined); combinations of black and yellow are preferred for traffic markings; combinations of black and white are preferred for informational markings
Safety Purple, White, Gray, Black and Brown	not assigned	not assigned

Environmental and facility safety signs also differ slightly in the information *required* to be presented. The information requirements of each are presented in Table B2.

Table B2
Information Required on Signs

Required Alert Information	Required by ANSI Z535.2 for **Environmental and Facility Safety Signs**	Required by ANSI Z535.4 for **Product Safety Signs and Warning Labels**
A Specific Hazard	yes	yes
The degree or level of seriousness of the hazard	yes	yes
The probable consequence of involvement with the hazard	Allowed, but not required in 1991; required in 1998, unless the probable consequence is obvious from other messages or the context of use	yes
How to avoid the hazard	yes	yes

This information is communicated with signal words and message panels. The use of symbol/pictorial panels is optional; they may be used to supplement *or substitute for* the message panel. Symbol/pictorial panels are often used to convey the consequence(s) of not avoiding the accident.

There are several key differences in the revised 1991 and 1998 Z535 requirements for environmental and facility signs, but the main one is that the older ANSI Z35.1-1972 (upon which OSHA signs were based) used a CAUTION sign to cover both situations that are now covered by the separate WARNING and CAUTION signs of the newer ANSI Z535. Table B3—*Classification of Signal Words* shows the five classifications used for Environmental and Facility Safety Signs. Product Safety signs and labels generally use only DANGER, WARNING, and CAUTION classifications.

One of the main intentions of ANSI Z535 is to limit the proliferation of DANGER signs and promote use of the DANGER signal word only in places where it truly applies. The addition of the WARNING sign category provides a useful separation of information. With a DANGER sign, there is an *imminent hazard,* i.e., you are in the area where the hazard is located. If you do not avoid the hazard, the result is death or serious injury. With the WARNING sign there is a *potential hazard,* i.e., you are safe where you are, but if you go further, you will be in a DANGER area or situation. This contrasts with the CAUTION sign, which is associated with potential minor or moderate injuries or, until 2002, with equipment damage. As of the 2006 Edition, the preferred signal word for equipment damage is *NOTICE;* CAUTION without the safety alert symbol is still allowed by the 2006 Edition as an alternate to *NOTICE* by the 2006 Edition.

Table B3
Classification of Signal Words

Sign Type	Classification
DANGER	Indicates an imminently hazardous situation that, if not avoided, will result in serious injury or death.
WARNING	Indicates a potentially hazardous situation that, if not avoided, could result in serious injury or death.
CAUTION	Indicates a potentially hazardous situation that, if not avoided, may result in minor or moderate injury. Can also be used as an alternate to *NOTICE* to alert against unsafe practices that can result in equipment damage.
NOTICE	Preferred to alert against unsafe practices that can result in equipment damage.
SAFETY INSTRUCTIONS OR SAFETY EQUIP-MENT LOCATION	Indicates general instructions relative to safe work practices or procedures, or the location of safety equipment.

Both DANGER and WARNING are only used when the probable consequence is death or serious injury, if the hazard is not avoided. See Table B4—*Attributes of Environmental and Facility Safety Signs.* The difference is that the DANGER signal word *is only to be used* where the hazard is imminent—where there is a high probability associated with the hazard. WARNING is to be used when the probability associated with the hazard is low. Thus, WARNING is appropriate outside of area where the potentially serious or deadly hazard is located, such as on a substation fence, padmount transformer door, etc. DANGER would be appropriate if a safety sign is appropriate *inside* the area or casing containing the hazard.

In essence, the CAUTION sign is analogous to the WARNING sign, except that the CAUTION sign is associated with lesser injuries. Both refer to potential hazards, not immediate hazards; both are to be located outside of the hazardous area and are to be used to warn against a hazardous movement or practice.

All signal words associated with personal injury hazards (DANGER, WARNING, and CAUTION) must be preceded by the *safety alert symbol* (exclamation point in a triangle)—the international symbol for a personal safety hazard. The safety alert symbol is not allowed on CAUTION signs used as an alternate to *NOTICE* to address equipment damage.

Because of the relative difference in the hazard, a CAUTION sign could be located on a container of material that could cause minor injury. In contrast, if contact with the material might be in the DANGER category, then a DANGER sign will be appropriate on the container, with a WARNING sign outside the area in which the container resides.

Although OSHA still refers to the superseded ANSI Z35, OSHA allows and encourages the use of revised standards. Eventually OSHA requirements can be expected to change to use the DANGER, WARNING, and CAUTION differentiations, rather than the present DANGER and CAUTION.

In general, a DANGER, WARNING, or CAUTION category of hazard requires a signal word panel. Thus, these signs are generally in the format of a 3-panel sign or a 2-panel sign (see Figure B2). Where used, the signal panel is at the top of the sign. Message panels and symbol panel backgrounds should be white (for contrast).

Almost all signs require a signal word panel. See ANSI Z535.2 for exceptions.

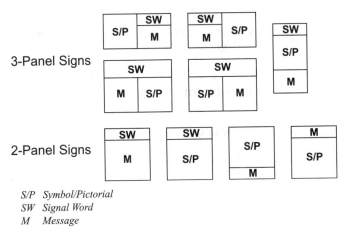

S/P   *Symbol/Pictorial*
SW   *Signal Word*
M   *Message*

Figure B2
Permitted Arrangements

The attributes for Product Safety Signs and Labels are essentially the same as, and may be substituted for (and vice versa), the attributes of Environmental and Facility Safety Signs.

The various editions of ANSI Z535 standards reflect recent research. The 1991 signal word format previously preferred for product signs and labels became the preferred format for environmental and facility safety signs in the 1998 Edition, although the older formats of ANSI Z535.2-1991 could still be used. The 2002 Edition of the ANSI Z535 standards discontinued the use of the old formats. The 2007 Edition was the first to state a preference for using the *NOTICE* signal word to alert users to potential equipment damage; this change was made to further differentiate personal injury signs from equipment damage signs.

The preferred background of the signal word is a rectangular solid, without a distinctive background shape.

If a symbol or pictorial is used, it is black on a white background or vice versa. Other colors may be used for pictorial emphasis, such as red for fire.

Message panels can have white letters on a black background, if required for contrast with the background upon which the sign is placed.

Table B4
Attributes of Environmental and Facility Safety Signs

Type of Sign	DANGER	WARNING	CAUTION	*NOTICE*	SAFETY INSTRUCTIONS; SAFETY EQUIPMENT LOCATION
Hazardous situation	Imminent	Potential	Potential	Potential	—
Result, if not avoided	Death or serious injury	Death or serious injury	Minor or moderate injury	Equipment damage	—
Purpose	Hazard Identification	Hazard Identification	Hazard Identification	Hazard Identification	General Instructions; Proper Procedures; Location of Equipment
Signal Panel					
Signal Words:	DANGER	WARNING	CAUTION	*NOTICE*	SHUTDOWN PROCEDURES; EYEWASH; etc.
Other words allowed	No	No	No	No	Yes
Letter style	Sans-serif, Erect, Upper Case	Sans-serif, Erect, Upper Case	Sans-serif, Erect, Upper Case	Sans-serif, Italics, Upper Case	Sans-serif, Erect, Upper Case
Letter Color	White	Black	Black	White	White
Background Color	Red	Orange	Yellow	Blue	Green
Message Panel					
Letter Style	Sans-serif, Erect, Upper Case or mixed	Sans-serif, Erect, Upper Case or mixed	Sans-serif, Erect, Upper Case or mixed	Sans-serif, Erect, Upper Case or mixed	Sans-serif, Erect, Upper Case or mixed
Letter Color	Black or Red	Black	Black	Blue or Black	Green or Black
Symbol Panel					
Panel shape	Square	Square	Square	Square	Square
Symbol Color	Black or Red	Black	Black	Blue or Black	Green or Black

Lettering

Lettering on signs is required to be of a size that will be readable by a person with normal vision at a safe viewing distance from the hazard. Safe viewing distance for the signal word considers a reasonable hazard avoidance reaction time. The safe viewing distance for the message panel does not include reaction time and is obviously less than that for the signal word, since the signal word letters may be twice the height of the message panel's letters.

Sign Placement

Signs are required to be so placed as to alert the viewer in sufficient time to take evasive action to avoid the hazard. The sign should not be hazardous in itself (i.e., keep them out of the climbing area near poles). The placement should be such as to allow viewing even when nearby doors, panels, etc., are moved or removed.

Neither ANSI Z535 nor any other national standard provides guidance to determine appropriate sign spacing on perimeter fences. Such guidance is available in the 1985 edition of the Westinghouse Product Safety Label Handbook and other publications. In general, the readability of a sign deteriorates significantly at viewing angles more than 60 degrees from a perpendicular line to the sign, or 30 degrees from the plane of the sign; see Figure B3 and Figure B4 on sign readability for message words and signal words. This information, combined with the letter height requirements of ANSI Z535.2 can be used to calculate appropriate sign spacing.

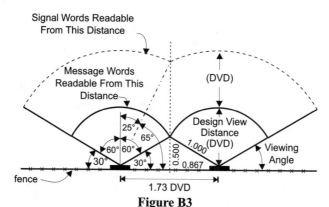

Figure B3
Sign Readability—Signs Spaced for Message Word Readability

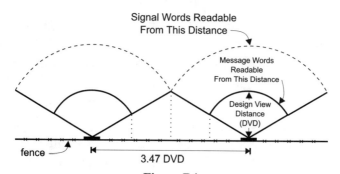

Figure B4
Sign Readability—Signs Spaced for Signal Word Readability

If the signal word letters are twice the height of the message letters, as required by ANSI Z535.2, the observer can read the signal word twice as far away as the message. If the signs were so close together that the *message* letters can be easily read all along the fence (as in Figure B3), the signs would be spaced at 1.7 times the message viewing distance. The signal word would then be effectively spaced at a viewing angle of only 25 degrees or less as the observer approaches a fence. If the signs are spaced that close together, the area can look cluttered, depending upon overall sign size.

On the other hand, if the signs are spaced to yield a 60-degree viewing angle for the signal word, the signs would be spaced 3.5 times the message viewing distance apart (as in Figure B4); ***this would leave a significant gap between the areas from which the message could be easily read.***

ANSI Z535.2 requires that the message height be used to determine the design viewing distance for the safety signs. As a result, message letter height should generally be chosen to assure that the message can be read from any angle of approach

In general practice, signs have typically been spaced at intervals of 2 to 3 times the message viewing distance with good results. For example, if the message letters are designed to be readable from 30 m (100 ft) away, the signal word will be readable from 60 m (200 ft) or more away, and the sign spacing of 60–90 m (200–300 ft) is reasonable. However, such spacing makes the signs difficult to see in low light.

NESC Rule 110 requires warning signs at all entrances and on all sides of supply stations. In practice, warning signs placed at corners of a station so as to be seen from likely avenues of approach may be all that are appropriate, even for large fenced areas. However, the size of the signs may be so large as to present too much of a wind area (i.e., potential structural damage to the fence in a storm). As a result of these issues, and the fact that smaller signs are usually less expensive, it is usually appropriate to have multiple signs on the sides of medium and larger station fences.

If signs are placed behind fence mesh, the mesh will restrict the realistic viewing angle, requiring larger or more signs to achieve the same coverage provided by smaller and fewer signs located on the exterior of the fence.

Sometimes fences will have slight *jogs* in the fence line to include or exclude some ground facility, provide an area for a truck to pull off the road while someone opens the gate, etc. Such jogs in the fence are usually only a few panels wide and are not considered to be a separate *side* requiring an additional sign.

Safety signs relating to a potential personal safety hazard that are placed on the outside of equipment, such as a pad-mounted transformer, need only be placed on the door (entrance) side of the equipment. However, for worker convenience and safety, it is often appropriate to provide additional signs on other sides of the equipment to limit landowners from planting shrubs close enough to interfere with workers or digging over the cables.

Criteria for Safety Symbols

Symbols or pictorials that have been validated by research may be used with other ANSI Z535.2, Z535.4, or Z535.5 requirements to replace or augment text messages. ANSI Z535.3—*Criteria for Safety Symbols* includes both approved symbols and standard criteria for evaluating symbols. To meet the test criteria, responses to the symbol must be at least 85% correct, with no more than 5% critical confusions. Symbols that have not passed these criteria may be used to supplement text, but may not be used to replace text messages.

Symbols of principal interest to designers of signs for utility facilities include (a) a wire with a hand (or other body part) shown disjointed by a lightning bolt type of jagged area, (b) a circle with a slanted bar through it overlaying a person, to symbolize "no entrance", (c) a falling person, and (d) a stumbling person as shown in Figure B4. Combinations, such as shown in (e) are frequently used, especially where multiple hazards exist.

a. Electrical   b. No admittance   c. Falling Person

d. Stumbling Person   e. Electrical/Falling

Figure B5
Symbols of Principal Interest to Utilities

Figure B6 illustrates (1) the differences between the types of signs covered by the ANSI Z535 standards and (2) typical sign language and configuration for utility use. These examples are *not* intended to be all-inclusive, nor are they mandatory.

Figure B6
Typical Safety Signs for Specific Hazards

Appendix C—Metric Conversions

(Metric Conversions used in the NESC *and* NESC Handbook)

Table C1—Length

ft	m	mm	ft	m
0.1		30	16.5	5.0
1.0		300.0	17.0	5.2
1 ft. 4 in.		410.0	17.5	5.3
1 ft. 11 in.		580.0	18.0	5.5
2.0	0.6	600.0	18.5	5.6
2.5		750.0	19.0	5.8
3.0		900.0	20.0	6.1
3 ft. 1 in.		940.0	21.0	6.4
3.5	1.07		22.0	6.7
3 ft. 8 in.	1.12		22.5	6.8
4.0	1.2		23.0	7.0
4 ft. 4 in.	1.32		23.5	7.2
4.5	1.4		24.0	7.3
5.0	1.5		24.5	7.5
6.0	1.8		25.0	7.6
6 ft. 4 in.	1.9		25.5	7.8
6.5	2.0		26.5	8.1
7.0	2.13		27.0	8.2
7.5	2.3		28.5	8.7
8.0	2.45		30.0	9.0
8 ft. 6 in.	2.6		30.5	9.3
9.0	2.7		31.0	9.4
9.25	2.8		31.5	9.6
9.5	2.9		32.0	9.8
10.0	3.0		34.5	10.5
10.5	3.2		36.0	11.0
11.0	3.4		36.5	10.9
11.5	3.5		37.5	11.4
12.0	3.6		38.0	11.6
12.5	3.8		40.5	12.3
13.0	4.0		42.5	12.7
13.5	4.1		50.0	15.2
14.0	4.3		60.0	18.3
14.5	4.4		100.0	30.0
15.0	4.6		150	45
15.5	4.8		300	90.0
16.0	4.9		800	250.0
			1000.0	300.0
			1500.0	450.0
			3300.0	1000.0

Table C1—Length continued

in	mm		cm	m		miles	km
0.2	5.0					1.0	1.6
0.4	10.0			0.01			
0.5	13.0						
1.0	25.0						
1.25	31.0						
1.5	38.0						
2.0	50.0						
3.0	75.0						
4.0	100.0						
5.0	125.0						
6.0	150.0						
7.0	175.0						
8.0	200.0						
9.0	230.0						
10.0	250.0						
11.0	280.0						
12.0	300.0						
16.0				0.41			
18.0	450.0						
24.0	600.0						
26	660.0						
30.0	750.0			1.0			
40.0			107				
42.0	1070.0			1.27			
50.0							

Table C2—Area

in^2	cm^2		ft^2	m^2
4	25.8		2.0	0.185
36.0	230.0		0.5	0.046
72.0	465.0		5.0	0.47
151.0	975.0			
288.0	1860.0			
611.0	3940.0			
720.0	4650.0			

Table C3—Force

lb	kg	kN	N
60			267
80			356
90			400
120			534
125			556
230			1023
700		3.1	
2000.0		9.0	

Table C4—Pressure

lb/in^2	kPa
150.0	1030.0

lb/ft^2	Pa
4.0	190.0
6.0	290.0

Table C5—Temperature

^{0}F	^{0}C
0	−20
15	−10
30	−1
32	0
60	15
120	50
125	52
200	93
300	150

Appendix D—Cross References of Major OSHA and NESC Requirements

Topic	NESC	OSHA-Operations & Maintenance	OSHA-Construction
Inspection & Testing-Prot. Devices	411C	1910.137(b)	1926.951(a),(b)
Medical Training & First Aid	410B	1910.269(a),(2)	1926.950(e), 1926.50(c)
Emergency Procedures Training	410B	1910.269(b)(1)	1926.950(e), 1926.50(c)
De-energizing, Switching	442	1910.269(d)	1926.950(d)
Guarding Open Manholes & Opening	423A	1910.269(e)(5)	1926.956(a)
Attendant on Surface	443K	1910.269(e)(7), 1910.269(t)(3)	1926.956(b)
Testing for Gas	423B	1910.269(e)(6)(11)(12)(13)	1926.956(a)(3)
Flames in Manholes/Vaults	423C	1910.269(e)(12)	1926.956(b)(2),(3)
Fall Protection	420K	1910.269(g)(2)	1926.951(b), 1926.959
Ladders and Supports	420J	1910.269(h)	1926.951(c)
Power Driven Equipment-UG	423F	1910.269(t)(4)	no clear match
Live-Line Equipment	446B	1910.269(j)	1926.955(e)
Approach Distances	441	1910.269(l)(2)	1926.950(c)
Working Position	443D	1910.269(l)(4)	no clear match
Making Connections	443F	1910.269(l)(5)	no clear match
Capacitors	420I, 443I	1910.269(w)(1)	no clear match
Fuses	420N	1910.269(l)(7)	no clear match
Ungrounded Metal Parts	420E	1910.269(l)(9)	1926.950(b)(2), 1926.954(a)
Deenergizing lines & Equipment	444	1910.269(m)	1926.950(d)
Protective Grounds	444D	1910.269(n)(3)	1926.954(e)
Installing/Removing Protective Grounds	445A, 445B	1910.269(n)(4,(5),(6),(7),(8),(9))	
Switching Procedures	442	no clear match	no clear match
Installing/Removing Wires & Cables	422C	1910.269(q)(2)	1926.955(c),(d)
Substation Fences	110A	1910.269(u)(3)	1926.957(c),(g)
Current Transformers	443H	1910.269(w)(2)	no clear match
Area Protection-Traffic	421B	1910.269(w)(6)	1926.200(g), 1926.201

Appendix E—Application of the National Electrical Safety Code Grandfather Clause

1. Discussion

The early editions of the NESC each required that all existing facilities either be brought into compliance with the new edition or guarded, unless the administrative authority said it wasn't worth the cost of doing so. This was contained in Rule 101 for Electric Supply Stations and Rule 201 for Overhead (including the limited rules for underground risers).

The last edition of the NESC containing these requirements for electric supply stations was the 5th Edition of 1941. For overhead and underground it was the 6th Edition of 1961. See the code language for these editions shown below.

The first editions to add the so-called *grandfather clause* to allow existing facilities to remain in compliance with the edition that previously applied (instead of the new edition) were the 1971 revision of Rule 101 of Part 1 (Stations), the 1973 codification of new Rule 302 in new Part 3 (Underground) and the 1977 revision of Rule 202 in Part 2 (Overhead). See the applicable language of these editions shown below

As a result of these changes, an existing installation can now stay in compliance with the previously applicable edition after a new edition is issued. However, ***even if an overhead or underground line was originally constructed under the 5th Edition, the oldest edition that could be applicable is the 6th Edition of Part 2, because that edition required existing facilities to be brought into compliance with it or guarded. Similarly, for power plants and substations installed before 1941, the 5th Edition of Part 1 is the earliest that could apply.***

The 1981 Edition was the first modern edition in which all Parts of the NESC were revised at the same time. As a part of that revision, the scope, application and definition rules were moved from the individual Parts of the code to a new Section 1 that applied to the entire code.

The application rules were contained in new Rule 013, as shown below. Rule 013 was clarified in the 1990 and 2002 editions. Rules 013B2 and 013B3 form the limitations and application of the grandfather clause.

Under Rule 013B2, an existing facility (including maintenance replacements) is not required to change when a code requirement changes, but *it is allowed to be changed* under Rule 013B1 to meet the new edition.

The basic mechanism for applying code editions falls under the application of Rule 013B3. If an existing installation has either (1) its structure replaced for maintenance purposes, (2) an item on the structure replaced, (3) an item added to the structure, or (4) items on the structure altered (such as relocating items to accommodate required clearance to a new item), the resulting installation must meet either (1) the present edition of the NESC or (2) the edition that was previously applicable. The previously applicable edition may be either (1) the edition applicable at the time of original construction (for electric supply stations constructed after 1941 and overhead or underground lines constructed after 1961) or (2) a subsequent edition with which it is in compliance.

As a practical matter, utilities seldom have older construction standards available while working on structures or inspecting structures. Construction standards are typically updated as each new code edition becomes effective. When they inspect after working on an installation, the inspection is usually done with the current utility standards. Further, most of the older installations meet the requirements of modern editions of the code. As a result, most installations are generally in compliance with current standards as they change from time to time, unless there are major difficulties in meeting new code requirements. If existing facilities meeting one code edition are brought into compliance with a later edition, it is not intended that an earlier edition be reapplied at a later time.

Significant problems can occur with some older installations because of new additions to the installation. Later codes allow many things to occur that were not allowed or specified by earlier editions. If someone installs something in or on an older facility in a manner that is routinely done by today's standards, but not allowed by the earlier standard, the installation no longer complies with the earlier standard. Thus, *a new addition to an existing installation could result in taking the existing installation out of compliance with the grandfathered edition—if care is not taken to assure that new additions are only made in compliance with the restrictions of the grandfathered edition.* In such a case, the edition that is current at the time of the addition would be required of the resulting installation.

Example 1: Consider that, prior to the 1990 Edition, the NESC required a supply cable conduit installed down a pole to be covered by a nonmetallic covering where it runs through the communication space. The 1987 and prior editions did not allow the later option of bonding the grounded cable messenger to the grounded conduit, instead of covering the conduit with a nonmetallic covering.

Scenario 1A: *Consider an overhead line power line that currently meets the 6th Edition and needs to remain in compliance with that edition to meet building clearance requirements. One of the poles has a primary underground riser running up the pole. The cables are protected by a galvanized steel conduit. No communication cables are attached.* If a communication utility with a joint-use agreement were to attach a cable to the pole and bond the grounded cable messenger to the metal riser conduit, as is allowed by current code, the installation would no longer meet the 6th Edition.

Scenario 1B: *Consider an overhead joint-use line with existing power and communication cables.* If the power utility added a metallic conduit down the pole and bonded it to the cable messengers, that would take the installation out of compliance with the 6th Edition.

Unless older facilities that are intended to be kept in compliance with an earlier edition are identified in some fashion, and unless the employees know how to keep the facility in compliance with the earlier edition, the result may be noncompliance with the code. Further, when an older edition does not specify a particular requirement, but a new edition does specify a requirement, the requirement specified in the current edition must be met if changes occur, in order to comply with the previous edition. This is because Rule 012C requires accepted good practice for given local conditions for all particulars not specified in the code. The current edition specifies good practice.

Example 2: A lake is constructed under an existing power line that was installed and is being maintained under the 6th Edition. The first accidents involving the relatively new high-masted catamaran sailboats occurred in 1969 and 1970. The 6th Edition did not contain specifications for clearances above water. The first edition to do so was the 1977 Edition.

If sailboat clearances had been addressed by the utility for lines above water before the 1977 Edition was published, and (1) the line cleared the water by enough to clear the masts of the expected sailboats, but (2) not by the full amount of the new code requirement, the line would still be in compliance with the 6th Edition, since the line would clear the expected sailboat and, thus, meet good practice at the time.

The NESC specified accepted good practice for sailboat clearances over water for the first time in 1977. Thus, if (1) the line was built before the 1977 Edition and application of the 6th Edition of the NESC was desired to be maintained, and (2) the lake was built after the 1977 Edition specified clearances above water, the NESC water clearances in effect at the time of the post-1977 lake construction under the line (1977 or later, as applicable) would have to be met, in order to meet the good practice rule of the 6th Edition.

Notwithstanding the above, if the US Army Corps of Engineers administers an over-water clearance requirement above the lake, that requirement may exceed the NESC clearances and would be specified in the permit. The Corps first adopted the NESC in 1977.

The following excerpts show how the NESC rules have changed. Emphasis has been added.

2. Language of NESC Requirements

NESC 5th Edition; NBS Handbook No. 31, issued 8 May 1940, Approved ASA 8 May 1941
Part 1. Electric Supply Stations
RULE 101. APPLICATION OF THE RULES AND EXEMPTIONS
A. Application and Waiving of Rules
   The rules are intended to apply to all installations, except as modified or waived by the proper administrative authority or its authorized agents. They are intended to be so modified or waived in particular cases wherever any rules are shown for any reason to be impractica-

ble such as by involving expense no justified by the protection secured; provided equivalent or safer construction is secured in other ways, including special working methods.

Other methods of construction and installation than those specified in the rules may also be made as experiments to obtain information, if done where supervision can be given by the proper administrative authority.

B. Intent of Rules

The intent of these rules which constitute a minimum standard will be realized:

1. By applying the rules in full to all new installations, reconstructions, and extensions.
2. By altering existing installations as needed in a manner approved by administrative authority.
3. The time allowed for bringing existing installations into compliance with the rules will be determined by the administrative authority

NESC 6th Edition, ASA C2.2-1960, NBS Handbook 80, issued 1 November 1961

Part 2. Overhead Electric Supply and Communication Lines

Editors Note: *Part 2 included requirements for underground risers in Section 29 in this and previous editions.*

RULE 201. APPLICATION OF THE RULES AND EXEMPTIONS.

A. Intent, Modification

The rules shall apply to all installations except as modified or waived by the proper administrative authority. They are intended to be so modified or waived whenever they involve expense not justified by the protection secured or for any other reasons are impracticable; or whenever equivalent or safer construction can be more readily provided in other ways.

B. Realization of Intent

The intent of the rules will be realized:

1. By applying the rules in full to all new installations, reconstructions, and extensions, except where for special reasons any rule is shown to be impracticable or where the advantage of uniformity with existing construction is greater than the advantage of construction in conformity with the rules.

2. By placing guards on existing installations or otherwise bringing them into compliance with the rules, except where the expense involved is not justifiable.

NOTE: The time allowed for bringing existing installations into compliance with the rules as specified in 2 will be determined by the proper administrative authority.

NESC 6th Edition, ANSI C2.1-1971, NBS Handbook 110-1, issued June 1972

Part 1. Electric Supply Stations

RULE 102. APPLICATION OF THE RULES AND EXEMPTIONS APPROVED BY ANSI 14 JULY 1971

A. Application

The rules shall apply to all installations except that they may be modified or waived by the proper administrative authority when shown to the impracticable. In such cases, equivalent or safer construction shall be secured in other ways, including special working methods. Methods of construction and installation other than those specified in the rules may also be made as experiments to obtain information, if done where proper supervision can be administered.

B. Intent of Rules

The intent of these rules, which constitute a minimum standard, will be realized by applying the rules in full to all new installations, alterations, reconstructions, and extensions. Rules in this Code which are to be regarded as mandatory are characterized by the use of the word "shall." Where a rule is of an advisory nature, it is indicated by the use of the word "should." Other practices which are considered desirable but not intended to be mandatory are stated as recommendations. It is realized that conditions may exist which necessitate departures from such recommendations. Notes contained herein are for information purposes only and are not to be considered as mandatory or as part of the Code requirements.

Editor's Note: *By removing the requirement to guard existing facilities meeting the previous edition that do not also meet the new edition, the grandfather clause was effectively created.*

1973 NESC ANSI C2.3-1973 issued 20 July 1973
Part 3. Underground Electric Supply and Communication Lines
302 INTENT AND APPLICATION OF RULES

B. New Installations, Reconstruction, and Extensions

These rules shall apply to all new installations, reconstructions, and extensions except that they may be waived or modified by the proper administrative authority when shown to be impractical. In such cases, equivalent or greater safety shall be secured in other ways including special working methods. Methods of construction and installation other than those

C. Existing Installations

These rules do not apply to existing installations except as may be required for safety reasons by the proper administrative authority.

1977 NESC ANSI C2.2-1977 issued 28 February 1977
202 INTENT AND APPLICATION OF RULES

B. Application of Rules

1. New Installations, Reconstructions, and Extensions

These rules shall apply to all new installations, reconstructions, and extensions except where they may be waived or modified by the proper administrative authority when shown to be impractical or when equivalent of safer construction can be more readily provided in other ways. Methods of construction and installation other than those specified in the rules may be used experimentally to obtain information if done where proper supervision is provided.

2. Existing Installations

Existing installations, including maintenance replacements, which comply with prior editions of this code need not be modified to comply with these rules except as may be required for safety reasons by the proper administrative authority. A replacement for a supporting structure, however, shall conform to the current edition of Rule 238C.

Editor's Note: *This was the first edition to state the intended requirement that any existing installation not meeting the new edition must meet the previously applicable edition. Starting with the 1981 Edition, an effective date of 180 days after publication was specified. A new edition may be used upon issuance.*

1981 NESC, ANSI C2-1981 issued 5 September 1980
RULE 013. APPLICATION

Editors Note: *This rule is unchanged in 1984 and 1987 Editions and applies to all Parts of the NESC*

A. New Installations and Extensions

1. These rules shall apply to all new installations and extensions, except that they may be waived or modified by the administrative authority. When so waived or modified, equivalent safety shall be provided in other ways, including special work methods.

2. Types of construction and methods of installation other than those specified in the rules may be used experimentally to obtain information, if done where qualified supervision is provided.

B. Existing Installations

1. Existing installations including maintenance replacements, which comply with prior editions of the code, need not be modified to comply with these rules except as may be required for safety reasons by the administrative authority.

2. Where conductors or equipment are added, altered, or replaced on an existing structure, the structure or the facilities on the structure need not be modified or replaced if the resulting installation will be in compliance with the rules which were in effect at the time of the original installation.

1990 NESC, ANSI C2-1990 issued 1 August 1989
Rule 013. APPLICATION

Editor's Note: *This rule is unchanged in 1993 and 1997 Editions.*

A. New Installations and Extensions

1. These rules shall apply to all new installations and extensions, except that they may be waived or modified by the administrative authority. When so waived or modified, equivalent safety shall be provided in other ways, including special work methods.

2. Types of construction and methods of installation other than those specified in the rules may be used experimentally to obtain information, if done where qualified supervision is provided.

B. Existing Installations

1. Where an existing installation meets, or is altered to meet, these rules, such installation is considered to be in compliance with this edition and is not required to comply with any previous edition.

2. Existing installations including maintenance replacements, which comply with prior editions of the code, need not be modified to comply with these rules except as may be required for safety reasons by the administrative authority.

3. Where conductors or equipment are added, altered, or replaced on an existing structure, the structure or the facilities on the structure need not be modified or replaced if the resulting installation will be in compliance with either (1) the rules which were in effect at the time of the original installation, or (2) the rules in effect in a subsequent edition to which the installation has been previously brought into compliance, or (3) the rules of this edition in accordance with Rule 013B1.

2002 NESC, ANSI C2-2002 issued 1 August 2001
RULE 013. APPLICATION

A. New Installations and Extensions

1. These rules shall apply to all new installations and extensions, except that they may be waived or modified by the administrative authority. When so waived or modified, equivalent safety shall be provided in other ways.

EXAMPLE: Alternative working methods, such as the use of barricades, guards, or other electrical protective equipment, may be implemented along with appropriate alternative working clearances as a means of providing safety when working near energized conductors.

2. Types of construction and methods of installation other than those specified in the rules may be used experimentally to obtain information, if done where
   a. Qualified supervision is provided
   b. Equivalent safety is provided, and
   c. On joint-use facilities, all affected parties agree.
B. Existing Installations
   1. Where an existing installation meets, or is altered to meet, these rules, such installation is considered to be in compliance with this edition and is not required to comply with any previous edition.
   2. Existing installations including maintenance replacements, which comply with prior editions of the code, need not be modified to comply with these rules except as may be required for safety reasons by the administrative authority.
   3. Where conductors or equipment are added, altered, or replaced on an existing structure, the structure or the facilities on the structure need not be modified or replaced if the resulting installation will be in compliance with either (1) the rules which were in effect at the time of the original installation, or (2) the rules in effect in a subsequent edition to which the installation has been previously brought into compliance, or (3) the rules of this edition in accordance with Rule 013B1.